Lecture Notes in Artificial Intelligence 9835

Subseries of Lecture Notes in Computer Science

More information about this series at http://www.springer.com/series/1244

Naoyuki Kubota · Kazuo Kiguchi
Honghai Liu · Takenori Obo (Eds.)

Intelligent Robotics and Applications

9th International Conference, ICIRA 2016
Tokyo, Japan, August 22–24, 2016
Proceedings, Part II

 Springer

Editors
Naoyuki Kubota
Tokyo Metropolitan University
Tokyo
Japan

Honghai Liu
University of Portsmouth
Portsmouth
UK

Kazuo Kiguchi
Kyushu University
Fukuoka
Japan

Takenori Obo
Tokyo Metropolitan University
Tokyo
Japan

ISSN 0302-9743 ISSN 1611-3349 (electronic)
Lecture Notes in Artificial Intelligence
ISBN 978-3-319-43517-6 ISBN 978-3-319-43518-3 (eBook)
DOI 10.1007/978-3-319-43518-3

Library of Congress Control Number: 2016946926

LNCS Sublibrary: SL7 – Artificial Intelligence

Printed on acid-free paper

This Springer imprint is published by Springer Nature
The registered company is Springer International Publishing AG Switzerland

Preface

The Organizing Committee of the 9th International Conference on Intelligent Robotics and Applications aimed to facilitate interactions among active participants in the field of intelligent robotics and mechatronics and their applications. Through this conference, the committee intended to enhance the sharing of individual experiences and expertise in intelligent robotics with particular emphasis on technical challenges associated with varied applications such as biomedical application, industrial automations, surveillance, and sustainable mobility.

The 9th International Conference on Intelligent Robotics and Applications was most successful in attracting 148 submissions addressing the state-of-the-art developments in robotics, automation, and mechatronics. Owing to the large number of valuable submissions, the committee was faced with the difficult challenge of selecting the most deserving papers for inclusion in these lecture notes and presentation at the conference. For this purpose, the committee undertook a rigorous review process. Despite the high quality of most of the submissions, a total of 114 papers were selected for publication in two volumes of Springer's *Lecture Notes in Artificial Intelligence* as subseries of *Lecture Notes in Computer Science*, with an acceptance rate of 77 %. The selected papers were presented at the 9th International Conference on Intelligent Robotics and Applications held during August 22–24, 2016, in Hachioji, Tokyo, Japan.

The selected articles were submitted by scientists from 22 different countries. The contribution of the Technical Program Committee and the reviewers is deeply appreciated. Most of all, we would like to express our sincere thanks to the authors for submitting their most recent work and to the Organizing Committee for their enormous efforts to turn this event into a smooth-running meeting. Special thanks go to the Tokyo Metropolitan University for their generosity and direct support. Our particular thanks are due to Alfred Hofmann and Anna Kramer of Springer for enthusiastically supporting the project.

We sincerely hope that these volumes will prove to be an important resource for the scientific community.

June 2016

Naoyuki Kubota
Kazuo Kiguchi
Honghai Liu
Takenori Obo

Organization

Advisory Committee

Jorge Angeles — MgGill University, Canada
Zhongqin Lin — Shanghai Jiao Tong University, China
Hegao Cai — Harbin Institute of Technology, China
Imre Rudas — Obuda University, Hungary
Tianyou Chai — Northeastern University, China
Shigeki Sugano — Waseda University, Japan
Jiansheng Dai — King's College London, UK
Guobiao Wang — National Natural Science Foundation of China, China
Toshio Fukuda — Meijo University, Japan
Kevin Warwick — University of Reading, UK
Fumio Harashima — Tokyo Metropolitan University, Japan
Bogdan M. Wilamowski — Auburn University, USA
Huosheng Hu — University of Essex, UK
Ming Xie — Nanyang Technological University, Singapore
Han Ding — Huazhong University of Science and Technology, China
Youlun Xiong — Huazhong University of Science and Technology, China
Oussama Khatib — Stanford University, USA
Huayong Yang — Zhejiang University, China

General Chair

Naoyuki Kubota — Tokyo Metropolitan University, Japan

General Co-chairs

Xiangyang Zhu — Shanghai Jiao Tong University, China
Jangmyung Lee — Pusan National University, Korea
Kok-Meng Lee — Georgia Institute of Technology, USA

Program Chair

Kazuo Kiguchi — Kyushu University, Japan

Program Co-chairs

Honghai Liu — University of Portsmouth, UK
Janos Botzheim — Tokyo Metropolitan University, Japan
Chun-Yi Su — Concordia University, Canada

Special Sessions Chair

Kazuyoshi Wada   Tokyo Metropolitan University, Japan

Special Sessions Co-chairs

Alexander Ferrein   University of Applied Sciences, Germany
Chu Kiong Loo   University of Malaya, Malaysia
Jason Gu   Dalhousie University, Canada

Award Committee Chair

Toru Yamaguchi   Tokyo Metropolitan University, Japan

Award Committee Co-chairs

Kentaro Kurashige   Muroran Institute of Technology, Japan
Kok Wai (Kevin) Wong   Murdoch University, Australia

Workshop Chair

Yasufumi Takama   Tokyo Metropolitan University, Japan

Workshop Co-chairs

Simon Egerton   Monash University, Malaysia
Lieu-Hen Chen   National Chi Nan University, Taiwan

Publication Chair

Takenori Obo   Tokyo Metropolitan University, Japan

Publication Co-chairs

Taro Nakamura   Chuo University, Japan
Chee Seng Chan   University of Malaya, Malaysia

Publicity Chair

Hiroyuki Masuta   Toyama Prefectural University, Japan

Publicity Co-chairs

Jiangtao Cao   Liaoning Shihua University, China
Simon X. Yang   University of Guelph, Canada
Mattias Wahde   Chalmers University of Technology, Sweden

Financial Chairs

Takahiro Takeda	Daiichi Institute of Technology, Japan
Zhaojie Ju	University of Portsmouth, UK

Track Chairs

Lundy Lewis	Southern New Hampshire University, USA
Narita Masahiko	Advanced Institute of Industrial Technology, Japan
Georgy Sofronov	Macquarie University, Australia
Xinjun Sheng	Shanghai Jiao Tong University, China
Kosuke Sekiyama	Nagoya University, Japan
Futoshi Kobayashi	Kobe University, Japan
Tetsuya Ogata	Waseda University, Japan
Ryas Chellali	Nanjing Tech University, China
Min Jiang	Xiamen University, China
Shinji Fukuda	Tokyo University of Agriculture and Technology, Japan
Boris Tudjarov	Technical University of Sofia, Bulgaria
Hongbin Ma	Beijing Institute of Technology, China
Sung-Bae Cho	Yonsei University, Korea
Nobuto Matsuhira	Shibaura Institute of Technology, Japan
Xiaohui Xiao	Wuhan University, China
Zhaojie Zu	University of Portsmouth, UK

Local Arrangements Chairs

Takenori Obo	Tokyo Metropolitan University, Japan
Takahiro Takeda	Daiichi Institute of Technology, Japan

Web Masters

Naoyuki Takesue	Tokyo Metropolitan University, Japan
Yihsin Ho	Tokyo Metropolitan University, Japan

Local Arrangements Committee

Shinji Fukuda	Tokyo University of Agriculture and Technology, Japan
Kazunori Hase	Tokyo Metropolitan University, Japan
Takuya Hashimoto	The University of Electro-Communications, Japan
Mime Hashimoto	Tokyo Metropolitan University, Japan
Narita Masahiko	Advanced Institute of Industrial Technology, Japan
Koji Kimita	Tokyo Metropolitan University, Japan
Taro Ichiko	Tokyo Metropolitan University, Japan
Hitoshi Kiya	Tokyo Metropolitan University, Japan

Lieu-Hen Chen   Tokyo Metropolitan University, Japan
Daigo Kosaka   Polytechnic University, Japan
Yu Sheng Chen   Tokyo Metropolitan University, Japan
Osamu Nitta   Tokyo Metropolitan University, Japan

Contents – Part II

Assistive Robotics

Intelligent Space

Sensing and Monitoring in Environment and Agricultural Sciences

Human Data Analysis

Robot Hand

Contents – Part I

Robot Vision and Sensing

Planning, Localization, and Mapping

Interactive Intelligence

Cognitive Robotics

Bio-inspired Robotics

Smart Material Based Systems

Mechatronics Systems for Nondestructive Testing

Social Robotics

Location Monitoring Support Application in Smart Phones for Elderly People, Using Suitable Interface Design

Julia Szeles$^{(\boxtimes)}$ and Naoyuki Kubota$^{(\boxtimes)}$

Graduate School of System Design,
Tokyo Metropolitan University, Tokyo, Japan
{perecka,kubota}@tmu.ac.jp
http://www.tmu.ac.jp/

Abstract. This paper proposes a location monitoring application for elderly care reminding them to do things they do not wish to forget related to their locations. The paper also introduces the importance of the user interface and emphasizes the difficulties in the interface design for various generations. In the first section, this paper shows the importance of smartphone application design for elderly care, and presents the various types of disease where the proposed application can be used. The second section introduces the difference between regular application design and application design for elderly people in smartphone applications. In the following sections, we are going to propose the application in detail, representing a suitable interface design for the elderly users. The introduced application can be applied into robot partners, to extend the ability of caregiving The experiment part discusses the result of program and we will discuss the outcome of interface research among elderly users, using the above introduced interface units and the functions of the application.

Keywords: User interface · Monitoring application · Dementia · Alzheimer's disease · Robot partner

1 Introduction

Recently the rate of elderly has risen in the super-ageing societies such in Japan causing a serious society problem. According to the Japanese government annual report about the ageing population of the country, the 26 % of the Japanese people were 65 years old or over in the year of 2015. It is a fact that, Japan is the world's fastest ageing society [1]. In the year of 2013, there were approximately 2 million households in Japan and in these households at least one person was 65 years old or older. About 31 % of the elderly, lived with their family members and 26 % of them lived alone [2,3].

Living alone as an elderly person without any medical or social support could lead to serious disasters such as serious injury or even death. The lack of social

© Springer International Publishing Switzerland 2016
N. Kubota et al. (Eds.): ICIRA 2016, Part II, LNAI 9835, pp. 3–14, 2016.
DOI: 10.1007/978-3-319-43518-3_1

interaction can lead to complete isolation. Some of the alone living elderly people are unable to fully take care of themselves. Some of them have hard time to pay their bills alone, wash the dishes or take out the trash by themselves. These problem could be solved easily, if Japan had enough human resources for elderly support. Unfortunately, like the United States, Japan is also facing a serious aging problem and losing the manpower and money to pay for this monumental demographic shift. But there is one solution to both the human and financial caregivers [4]. The developed countries are becoming more focused on replacing the human caregivers with robot caregivers. Over the past couple of years, robot caregivers have been developed, it has grown more conventional and cheaper, producing products with great functionality.

However, it is a well-known fact that many people in the older generation have been struggling with new technologies. Nevertheless, some seniors often reject to be involved with the digital world [5]. Some of them feel nervous around digital devices and they have persuaded themselves that these modern technologies were not made for them. According to previous researches [6] elderly people feel more confident if the smartphone applications are funnier or fun to deal with.

This paper proposes an application for elderly people who have difficulties remembering events and things. The application monitors the location of the user and if the user gets close to a position which has been stored before it will alert the user and remind him/her to a specific event or thing the user has to do in that area which was saved before. There are more and more applications for smartphones to support elderly care [7–9] especially to help patient who has dementia. Those applications are focusing on giving the most optimized route to the user or to help them find their way to somewhere. The recent location monitoring applications are mostly made for the caregivers or the family members where they can monitor the movement of the patient [10,11]. Which also means, that the interface design in those applications are not fully optimized to the elderly. There are reminder applications for users with dementia however the above mentioned applications are functioning like a calender application depending on time and not depending on the position of the user [12]. The proposed application can be convenient for elderly and to people who struggle with dementia, Alzheimer's disease or even for users who have trouble keeping things in mind such as shopping list or paying bills. It can also be easily implemented to robot partners which run on an iOS system. The proposed application connects the reminder part and the location monitoring part by using a suitable interface design for seniors which can be useful for elderly and people who are struggling with dementia because the application is alerting them to do things related to their position. Moreover the recently released monitoring applications for dementia patients are functioning only on a smartphone device but the introduced application can be connected to collaborate with robot partners too.

This paper is organized as follows. The first section proposes the symptoms and the problem of dementia and Alzheimer's disease using recent surveys, introducing the number of individuals who can find this application helpful. It also emphasizes the importance and the difficulties of the interface design for seniors and explains a way of a suitable interface design. The second section introduces the application itself with the used methods and design interface patterns and it also shows the application implementation into robotpartner. The experiment section proposes the result of the application and the feedbacks of the interviewed elderly people.

1.1 Dementia and Alzheimer's Disease

Many people use the words "dementia" and "Alzheimer's disease" regularly to describe memory lapses. However they are not the same things. People can have a form of dementia which is completely unrelated to Alzheimer's. Younger people can develop dementia or Alzheimer's disease, the risk increases depending on the ageing. Nevertheless neither of them are considered to be normal as the part of ageing. People with dementia has often experience mild forgetfulness in their daily life. They regularly have trouble with keeping track of time and they can occasionally lose their way and intention. As dementia develops, forgetfulness and confusion grow, and it becomes harder to recall names, faces and events. Personal care becomes a problem, such as taking daily medicines, personal hygiene, or taking out the trash. People who struggle with dementia have repetitive questioning habits, inadequate hygiene, poor decision-making, and they often lose their way in familiar places such as their own neighbourhood or house [13].

According to the World Alzheimer Report 2015 [14] nearly 46.8 million people are living with dementia worldwide in 2015. There are 10.5 million people in Europe and 22.9 million in Asia who struggle with dementia. There can be expected, that these numbers will be doubled in the near future. Alzheimer is the most common form of dementia. Dementia is basically a group of symptoms and Alzheimer is a disease. By definition, Alzheimer is a progressive disease of the brain that slowly impairs memory and cognitive function.

Different cases of dementia can be overwhelming for the families of affected people and for their caregivers too. Dementia can cause emotional, physical, economic and social pressure to families and caregivers. Not to mention, that supporting a dementia patient in the family requires a lot of time and energy. Some families just cannot handle the stress which comes with it or they cannot spare enough time for them because they also have to do their daily obligation [15]. Dementia is a crucial problem in the 21th century world. Because of the lack of human resources it cannot be easily solved, but because of the technological advances, human caregivers can be replaced by electric devices such as robots or smartphones which can provide a simpler and cheaper way of daily life support for elderly people.

The introduced application can be helpful to users who have dementia and would like to be reminded of things regarding of their location and the introduced application can also be implemented into robot partners to support elderly in a wider range.

1.2 Difficulties in Interface Design for Elderly People

Some elderly do not wish to use the recent smart phone applications because they say, it is very difficult to handle and they would rather not want to get involved with the new technology. Although, according to previous [16] studies elderly people would like to use the now existing smartphone applications if it would have a suitable graphical interface in order to support them. It is also beneficial if the application is fun to deal with and the interface is interesting not just easy to use.

In order to make more comfortable application, the designers should understand the dynamic natural changes associated with aging. Those effects can be diminished vision, varying degrees of hearing loss or hand eye coordination. A lot of people have loss in color perception or degree of color blindness that accompanies their dimmed vision [17]. Problems with eye vision of elderly start to become obvious when they reach the age around 60 years old. Besides, elder people have less sensitivity to color contrast than younger people. Because of the color contrast, they have difficulties to distinguish the differences between the blue and the green range. This leads us to the conclusion that two colors that may look very different to an individual with normal color vision can be less distinguishable to someone whit partial sight impairment.

Touchscreens and multi touch are commonly used in many real life situations. From tablets to ticket automata or ATMs. Elderly people, act more comfortably with touch interface because they can interact with the actual program immediately and more naturally [18]. Moreover, simple text characteristics are preferred in elder readers. The reason behind this, is that elderly people are used to newspapers, magazines or books which usually use normal text presentation. Bigger font size can mean easier readability, but too big one requires more space and in smartphone applications, developers have limited space to display information.

2 Methods and Design

In this section, we would like to discuss the methods and the design pattern of the application which is called the *SeniorCar* application. This application is designed on the iOS software stack produced by Apple. The application was written in iOS and uses *CoreData* to store persistent data. *CoreData* functions like a regular *SQL* database but unlike an *SQL* database, which queries the data by the rows, the *CoreData* handles the items as objects. We choose the iOS platform opposed to others because of the ability to easily thread background running process, the polished *MapKit* Apple map, the easily usable *CLLocationManager*, and the compatibility with other iOS devices. We also chose to create our application in iOS because the priority platform in Japan is the iOS platform [19].

Recent GPS monitoring applications are more focused to help out the caregivers or the other family members than the actual patient who is struggling with dementia [20]. Those applications send a short message and notification with sound or vibration to the patient and to the nearby caregivers or family members, if the patient has wandered off too far or has left the "safe zone" [21]. On the other hand the *SeniorCar* application was specifically created in order to support the alone living elderly people and the alone living elderlies who suffer from constant memory leaps. The *SeniorCar* application has already included some features which we are going to purpose in the following section.

2.1   Software

The *SeniorCar* application is for elderly people who cannot walk long distances properly and have some trouble finding the right direction and they often getting lost. The application mostly emphasizes the importance of social interactions. It offers various place choices such as parks, nearby shopping malls or cultural places marked by pins in order to encourage the elderly user to go out and interact with other people. The main purpose of this application is to avoid social isolation and to have more interaction with other individuals to reduce the factor of loneliness. When the user chooses which place does he/she want to go, the application designs the fastest and simplest route to the destination. The recent application can be plugged into a car like device which can take the user to various places and by following the created track, the user can arrive to the given destination. In this paper, we developed the *SeniorCar* application in order to support elderly users, including dementia patients to reduce the trouble of memory lapses.

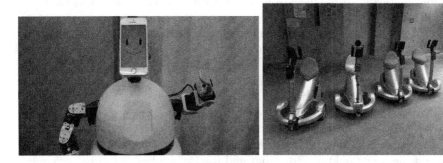

Fig. 1. The iPhonoid robot partner can be seen on the left side and the senior car device can be seen on the right side of the picture

The purposed application can be implemented into robot partners too which are running on iOS. In our previous research, we developed a robot partner called the iPhonoid [22]. The iPhonoid is robot partner which was created to support the elderly in their daily life. The iPhonoid and the iPhonoid body can be seen in Fig. 1. The iPhonoid is basically an iPhone attached to a robot body. When the user removes the iPhonoid from the robot body, the iPhonoid switches off itself into pocket mode, which enables the user to carry it anywhere and detect the location of the device. If the user connects the iPhonoid to the Senior car device, the *SeniorCar* application is automatically turned on and the above introduced feature can be used while the user is driving the senior car device, which can be seen in Fig. 1. It spares the time and the energy from the elderly to change from one application to an other because the system can automatically detect which of the above introduced devices have plugged in, the iPhonoid body or the senior car. The proposed application can be switched into the iPhonoid application form the SeniorCar application and backwards too using *URLScheme* by pressing a button.

2.2 Monitoring Application

The people who suffer from constant memory lapses have hard time remembering general and also important daily things or routines. Such as paying, phone/electricity/water or gas bills or even remembering what daily food products they should buy in the local grocery store or what medicines they should get from the drug stores. They might forget that they should check their post box or submit some important documents at the post office or in the city hall before they expire. There can be also a specific person they should call, like family members or important events to check on.

In the *SeniorCar* application users can make notes in specific coordinates on the map. The user can choose the location of the marker by touching the map on the interface. They can add title and subtitle to the created marker by choosing the right elements from the a table list. When they create a marker, it is going to be automatically stored in the *CoreData* database. With the help of the GPS location monitoring, the application can detect the position and movement of the user in the background thread and when the user gets close to a marker the application notifies the user by sending a message to the phone with sound and vibration. The markers can be easily deleted by the user by pressing the *Delete* button on the marker.

It is important to realize that the smartphone devices can run out of power pretty soon. Seniors who might not be familiar how to check the current state of the battery life may have problem noticing if the device is nearly out of power. Sure, the iOS system itself notifies the user if the power is too low but it might be too late if the user would like to go out and use the application. If the proposed application is running it checks the battery log in every 10 s and if the battery life is less than 20 % it notifies the user and suggest them not to go out before charging their device.

This application can be most useful for people who have Alzheimer's disease or constant dementia or they simply just have the habit to forgot things occasionally. The application basically reminds the users what to do in details, at specific places, and it also has the ability to guide them to given places. The proposed development is also can be used while the user is driving the senior car itself. The application is monitoring the user's location and send the notification reminding the user.

2.3   User-Friendly Interface

As people age they might have more trouble seeing tiny things, and as it has been discussed before, that elderly sometimes have problem distinguishing color contrasts. Which leads us to the conclusion, that designers need to pay more attention when they design user interface for elderly care. According to our previous research [23] we can realise that the combination of black and blue button was not suitable for elderly. It is complicated for them to recognize the difference between black and blue because of age related yellowed view. The safe approach is to keep colors bright and bold. A work from *Tiondall-Ford, 1997* reported that combination between graphics and words gives benefits to different cognitive abilities. Although, younger users prefer interface with less text and more graphical icons, elderly people gain better performance with word-based interface than graphic based interface. In conclusion, it is better to use buttons with text as well and not only icons as it is shown in Figs. 2 and 3.

According to our previous research the elderly users prefer bigger size characters on the screen. Based on our research, we can say that, bigger size font is better and more readable. On smartphone applications there are cases where the user needs to put input data into the device by typing it with the keyboard. Elderly users also have trouble typing characters on smartphones because these

Fig. 2. The buttons are big enough for elderly

Fig. 3. The buttons have images and text written on them too

devices have a small sized screen and the touch interface makes it even more difficult for them [23].

The best selection for that is to make various selections available for the users so instead of setting the input by typing, they can choose from elements from a table or list. In the purposed application to add a new marker to the map, the user has to set the title and the subtitle of the marker. The function is shown in Figs. 4 and 5, in Fig. 4 the user can choose from five title selections by touching them in the table: *shopping, pharmacy, paying checks, meeting with someone or post office.*

According to the selected item in the list, the application is going to offer another list, as it can be seen on Fig. 5, where the user can add elements as the subtitle of the marker on the map by pressing the "+" button. When the user is finished, adding the elements to the list, they can interact with the interface with buttons and save the created list into the database by pressing the "Create" button. The elderly users need to have some kind of feedback of their actions which were taken on the screen therefore a message box will pop up every-time to offer a choice to check if, the user really wants to store the chosen elements to the database. It can be seen in Fig. 8. If the user presses the "Create" button, the application stores the items and displays it on the map as it can be seen

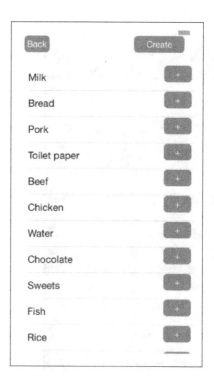

Fig. 4. The user can choose from five title choices

Fig. 5. Subtitle elements can be stored, as many as the user prefers

in Fig. 6. When the user gets close to a marker on the map, the application is going to notify the user by sending a message on the device with a sound and a vibration. The result can be seen in Fig. 9.

On the application, the user can see the time and the date almost every time and by the "Back" button, the user can always get away from the recent *ViewController* easily.

We made the user interface more user-friendly by putting a button on the screen which automatically zooms to the place where the user is. Elderly users have hard time interacting smoothly with the touchscreen if they have to use multi-touch gestures such as dragging, pointing, steering or crossing on the screen. The user needs to use two fingers to zoom in and to zoom out in the map. One finger interaction is faster than two fingers interaction therefore users do not have to zoom to their place because the application does it for them by itself when it loads the map. The map can be moved by one hand dragging gesture easily. When we create an user interface for elderly, designers should avoid complex gesture patterns.

Fig. 6. When the device gets close to any of this markers, the application is going to notify the user (Color figure online)

Fig. 7. GO button is hard to see because of the blue contrast and not border or background color (Color figure online)

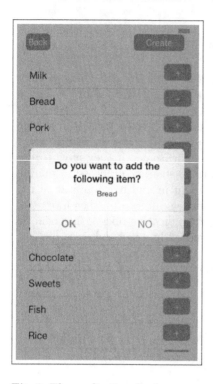

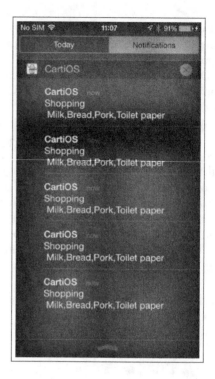

Fig. 8. The application displays a message box if the user has added a new item to the list

Fig. 9. The application sends a notification every time a user gets close to a marker on the map

The user location is zoomed immediately after starting the application but if the user dragged away too much, he/she can always return by pressing the "My location" button. The button was also created with the suitable color and suitable size to satisfy the needs of the elderly as it is shown in Fig. 6. In Fig. 6 we can see the stored marker displayed on the map with the right title and subtitle. By touching them, the user can see the information of it and can delete the selected marker by touching the "Delete" button. In the previous version, we set the color of the button to be blue and we did not set any background color at all. It is shown in Fig. 7 that the previous version of the interface design made the button less recognizable according to our previous research [23].

3 Experiments

In order to test the effectiveness of our application, we asked five elderly persons to test our application and share their opinion with us about the program itself and about the interface too. Each of the asked elderly persons were over 60 years old and they do not have much of an experience with smartphone applications. Every one of them said, that the interface in the Fig. 6 is much easier to see and handle than the interface on the Fig. 7. The participants said, they had hard time seeing the "GO" button on Fig. 7 and also the "Back" and the "AR view" button because of the blue contrast. They also said that the "My location" button which can be seen in Fig. 6 is very helpful because they do not have to scroll every time they would like to find their location. In Fig. 5 the "Delete" button is easy to see because of the color and easy to understand the purpose of that. The table view layout style to select the different titles and subtitles was also a great success because the asked elderly participants said, it is much better for them than typing. Only one participant could not figure out the purpose of the "+" button in Fig. 4 alone. Three participants also complained that the size of the buttons and the list elements in Figs. 4 and 5 could be bigger. All of the participants found the pop-up window very helpful, which popped up every time they added an extra subtitle to the marker Fig. 8.

4 Conclusion and Future Work

We developed a monitoring feature to an existing application which supports the elderly users who have hard time remembering daily things. This application can be useful not just for elderly or the dementia patients but also to people who have a constantly hard time remembering things. In our research, we focused on creating a suitable interface for elderly patients, but we also paid attention not to make it uncomfortable for younger users. We checked the existing papers about Healthcare applications in order to make our own, which contains something new and useful for elderly users. Furthermore, we could say, that our experiment was successful with the interface design and with the monitoring feature too based on the user feedbacks. The above introduced application can be used in iPhonoid robot partner and in the Senior car device too.

Acknowledgement. This paper and the application was also supported by Zoltan Lippai.

References

1. Projection of the Japanese population by major age group. Technical report. http://www.ipss.go.jp/p-info/e/S_D_I/Indip.asp#t_3. Accessed 18 May 2016
2. The Wall Street Journal. http://blogs.wsj.com/japanrealtime/2015/06/16/fear-of-dying-alone-the-state-of-japans-aging-population/. Accessed 18 May 2015
3. Boulos, M.N.K., Wheeler, S., Travers, C., Jones, R.: How smartphones are changing the face of mobile participatory healthcare an overview with example from eCAALYX. Biomed. Eng. **10**(1), 1 (2011)
4. Why Robots Are the Future of Elder Care. https://www.good.is/articles/robots-elder-care-pepper-exoskeletons-japan. Accessed 18 May 2016
5. Older Adults and Technology Use. http://www.pewinternet.org/2014/04/03/older-adults-and-technology-use/. Accessed 18 May 2016
6. Smith, S.T.: Electornic games for aged care and rehabilitation (2009). (Journal, Accessed 30 May 2016)

7. Luxton, D.D., McCann, R.A., Bush, N.E., Mishkind, M.C., Reger, G.M.: Integrating smartphone technology in behavioral healthcare (2011). (Journal, Accessed 30 May 2016)
8. Asghar, M.Z., Jamsa, T., Pulii, P.: A mobile system to remotely monitor travelling status of the elderly with dementia (2013). (Journal, Accessed 30 May 2016)
9. Zuehlke, P.: A functional specification for mobile eHealth (mHealth) Systems (2014). (Journal, Accessed 30 May)
10. Talaei-Kohei, A., Solvoll, T., Ray, P., Parameshwaran, N.: Policy-based Awareness Management (PAM): case study of a wireless communication system at a hospital (2011). (Journal, Accessed 30 May 2016)
11. Lv, Z., Xia, F., Guowei, W., Yao, L., Chen, Z.: iCare: a mobile health monitoring system for the elderly (2010). (Journal, Accessed 30 May 2016)
12. Hartin, P.J., Nugent, C.D., McClean, S.I., Cleland, I., Norton, M.C., Sanders, C., Tshanz, J.T.: A smartphone application to evaluate technology adaptation and usage in person with dementia (conference proceedings) (2014)
13. Japan's age-old problem. http://www.theguardian.com/world/2007/apr/17/japan.justinmccurry. Accessed 18 May 2016
14. Rees, G., Felcher, S.: World Alzheimer Report 2015. Technical report, The Global Impact of Dementia (2015)
15. The Dementia/Alzheimer's Connection. http://www.healthline.com/health/alzheimers-disease/difference-dementia-alzheimers#Connection1. Accessed 18 May
16. Boontaring, W., Chutimaskul, W., Chongsuphajaisiddhi, V., Papasratom, B.: Factors influencing the Thai elderly intention to use smartphone for e-Health services (2012). (Journal)
17. Is a big button interface enough for elderly users? http://www.idt.mdh.se/utbildning/exjobb/files/TR1091.pdf. Accessed 18 May 2016
18. Card, S.K., Moran, T.P.: The psychology of human-computer interaction (1983). (Journal)
19. iOS8 vs Google Android Users are the real differentiators. http://alchetron.com/iOS-8-vs-Google-Android-L-Users-Are-The-Real-Differentiators-1475-W. Accessed 18 May
20. Talaei-Khoei, A.: An awareness framework for agent-based mobile health monitoring (2009). (Journal, Accessed May 30)
21. Sposaro, F., Danielson, J., Tyson, G.: iWander: an android application for dementia patients (2010). (Journal)
22. Woo, J., Botzheim, J., Kubota, N.: Conversation system for natural comminucation with robot partner (2014). (Journal)
23. Komatsu, R., Tang, D., Obo, T., Kubota, N.: Multi-modal communication interface for elderly people in informationally structured space (2011). (Journal)

Evaluating Human-Robot Interaction Using a Robot Exercise Instructor at a Senior Living Community

Lundy Lewis[1]([✉]), Ted Metzler[2], and Linda Cook[2]

[1] Computer Technology, Southern New Hampshire University,
Manchester, NH, USA
l.lewis@snhu.edu
[2] School of Nursing, Oklahoma City University, Oklahoma City, OK, USA
{tmetzler,lcook}@okcu.edu

Abstract. A NAO humanoid robot and a Paro animaloid robot are taken to a senior living community for a study in human-robot interaction. The humanoid robot was programmed to perform autonomously (i) a warm-up routine in which the robot directs the participants to ask it to perform various tasks and (ii) an exercise routine in which the robot invites the participants to participate in various physical exercises. The Paro robot is then passed around among the participants. The participants included six elderly residents, three nurses/caregivers, and two administrators. The elderly residents are categorized with respect to cognitive awareness and physical capability. We tabulated video data to measure several dimensions of human-robot interaction among these diverse participant groups. Our findings suggest that while senior residents moderately accept the robots and nurses and administrators are enthusiastic about them, more work needs be done on the auditory capability of the robots.

Keywords: Elder care · Quality of life · Socially assistive robot · Exercise · Intelligent architecture · Human-Robot interaction

1 Introduction

It is well-known that the number of elders will increase dramatically over the next few decades while the number of care-givers will decrease, thus resulting in a social problem to which researchers, entrepreneurs, and governments are currently seeking possible solutions. Reasons for the impending problem include (i) a better understanding of healthy habits such as food intake and physical/mental exercise, (ii) advances in medicine and treatments of physical ailments such as hip and knee replacements, (iii) a better understanding of supportive environments such as mobility enhancement and transportation, (iv) fewer offspring from the current generation of potential parents, and (v) better social security for the elderly, among other reasons. Current trends among elders are aging-in-place in the elderly's familiar home or alternatively joining a senior living community. Fewer seniors elect to live with their offspring.

© Springer International Publishing Switzerland 2016
N. Kubota et al. (Eds.): ICIRA 2016, Part II, LNAI 9835, pp. 15–25, 2016.
DOI: 10.1007/978-3-319-43518-3_2

Nonetheless, elderly people will have difficulty performing routine tasks due to natural mental and physical decline. Robots show promise to assist the elderly with these tasks and thus help improve their quality of life. As an example, a tele-presence robot may allow a person to visit friends, relatives, and sights of interest remotely, or conversely allow friends and relatives to visit the elderly person. Simple virtual robots may remind the elderly of events or tasks such as medicine-taking. Humanoid and animaloid robots may provide companionship and conversation, or perhaps advice and therapy.

Some researchers are experimenting with humanoid robots as motivators for physical exercise [1]. Similarly, the pilot study described here involves a session with the humanoid robot NAO serving as an exercise coach on the premises of the Golden Oaks Senior Living Community in Oklahoma, USA [2] on October 23rd 2015. The goals of the study were to assess the acceptability of the robot among senior residents, caregivers, and administrators. The robot was programmed to be interactive and autonomous. The interaction is both verbal and physical, in which the robot reacts to the subjects' verbal input and the subjects react to the robot's verbal directives and physical demonstrations. It first provided an introductory, warm-up routine and then proceeded to lead the group through exercises focusing on leg, arm, feet, hands, neck, eyes, and full-body.

The authors described the details of the pilot study in a prior workshop paper [3] wherein we discussed the importance of context in studies on the acceptance of robots as social assistants for the elderly [4], our iterative methodology (Soft Systems Methodology [5, 6]), design principles, implementation of behaviors into the NAO robot, and a preliminary analysis of the results. In this paper we provide a further in-depth analysis of video data with respect to human-robot interaction (HRI). In Sect. 2 we discuss the parameters of the study, including the environment, the participants, and design concepts. In Sect. 3 we describe the architecture and implementation of the routines in the NAO Choregraphe development environment. In Sect. 4 we review a preliminary analysis of the study [3], discuss our subsequent analysis on human and robot responsiveness during the study, and our analysis of focus sessions with all participants. The paper concludes with comparisons with other work in the field and plans for further study.

2 The Parameters of the Study

2.1 The Environment and Participants

The subjects under study included six elderly residents, three nurses, and two administrators. The elderly residents were selected by the nurses and administrators with the goal of having seniors with varying cognitive and physical deficits. All eleven subjects participated in the exercises. Figure 1 shows the environment for the study. Table 1 characterizes each of the six elderly residents Rn.

Fig. 1.   Robot coach-to-senior interaction during an arm exercise

Table 1.   Characterization of the senior residents

	Cognitive awareness	Physical capability	Physical dependencies	Stamina & determination
R1	Low to moderate	Low	Walker	Low
R2	High	High	Wheelchair	High
R3	High	High	Wheelchair	High
R4	Very high	Very high	None	Very high
R5	Very low	Very low	Wheelchair	Very low
R6	Very high	Moderate	None	Very high

2.2   Method of Assessment

To assess the participants' engagement with the robot and the acceptability of the robot, we established several high-level design concepts as listed below, based on work reported in [1]. Next, we posited observable visual characteristics of the participants and mapped these characteristics into the design concepts. By video analysis, we simply counted and tabulated the observable characteristics.

Motivating. The degree to which the participants enjoy the engagement with the robot, indicated by observables *Attentive, Smiling/Laughter, Head-Nodding, Leaning Forward*, and *Intensity*.

Fluid and Highly Interactive. The degree to which the participants and the robot react to each other including verbal and physical reactions, indicated by observables *Cooperation, Imitation, Initiation, Intensity*, and *Participation*.

Intelligent. The degree to which participants perceive the robot as an entity that can be trusted as a competent human being rather than a toy, indicated by observables *Cooperation, Imitation, Initiation, Intensity,* and *Participation.*

Task-Driven. The degree to which participants perceive the robot as working towards a goal via various relevant tasks, indicated by observables *Cooperation, Imitation, Initiation,* and *Participation.*

Since our study will be carried out over multiple sessions where robot behaviors and routines will evolve over the sessions, the robot itself needs to be portable and programmable in a reasonable amount of time. Thus, we add two additional design concepts that are technology-driven:

Programmability. The degree to which the robot can be re-programmed in light of new suggestions and ideas offered by the participants, to be determined in future iterations of exercise therapy as the study unfolds.

Portability. The degree to which the robot and accompanying apparatus can be transported to different locations and set up in a reasonable amount of time without technical difficulties, to be determined in future iterations of exercise therapy as the study continues.

3   Robot Platform, Software, and Behavior Programming

The robot platform is the humanoid NAO robot. See Fig. 1. The technical specifications of the robot can be viewed at [7]. The robot is portable and offers several methods of behavior programming: (i) visual programming, (ii) Python and C++ programming, (iii) posture and gesture programming, and (iv) conversational programming [3]. These methods support the technology-driven design concepts of Programmability. The robot is capable also of random fluid gestures when listening or speaking to users, thus contributing to the design concepts Motivating, Fluid and Highly Interactive, and Intelligent.

The robot behaviors were implemented as intelligent agents and the overall design is based on the subsumption architecture [8]. The agents reside on levels such that higher levels subsume lower levels. If higher-level agents cannot function, the lower-level agents may continue to function.

For the exercise behavior, we developed (i) an introduction agent whereby the robot describes explains what he plans to do, (ii) a conversational agent to interact with the participants, (iii) a multiplicity of agents representing specific forms of exercise, and (iv) a transition agent that completes a specific exercise and initiates conversation with the participants to determine the next form of exercise. The agents communicate via message-passing between the input and outnodes nodes of the agents. The agents execute in linear fashion rather than distributive, collaborative fashion. Future behaviors may require collaborative decision-making. Figure 2 illustrates the exercise routine in accordance with the agent-based, subsumption architecture. Portions of the figure a circled in red for purposes of explanation.

The introduction agent on level 2 is passive. It speaks to the participants about the exercise agenda: "OK, let's do some exercises. We will exercise our head, arms, legs, hands, eyes, and feet. When you are ready we will try Tai Chi. First, let's get in the sitting position like I am now. Ready? OK, what would you like to do first? Head, arms, legs, hands, eyes, or feet?" After the introduction, control is passed to the conversational agent.

The conversational agent on level 1 has a reactive component and a deliberative component. The reactive component listens for audio input from participants and tries to detect a word it understands. The deliberative component is a simple rule-based system whereby the detection of a word routes control to a particular exercise. If the conversational agent doesn't detect an audio input within 5 s, it selects a random exercise and passes control to the selected exercise agent on level 0. If that doesn't work, a robot operator may start an exercise agent manually by clicking on the first box comprising the agent. In this way, the operator may take control of the robot in tele-operated fashion rather than abort the session. In our study we did not have to resort to tele-operation the robot, although it is conceivable that future instances might require tele-operation. It is comforting to know that this feature is available although one would hope not to have to use it.

The eight rows in level 0 comprise the exercise agents for specific forms of exercise, save row 7 row which is a "stop" agent. As an example, we have drawn a red circle around the behaviors comprising the leg exercise agent in row 3. The seven boxes in this row instruct the robot to perform the following verbal and physical communication with the participants: "Let's stand and then squat down, but be careful". Try it only if you feel comfortable. Here we go. Watch me. Stand. (The robot slowly moves from the sitting position to the standing position). Squat. (Robot slowly moves to a squatting position.) Back up. (Robot moves to the standing position, and so on).

The order of the exercises from top to bottom in Fig. 2 are head, arms, legs, Tai Chi, hands, eyes, stop, feet. The order in which the exercises are programmed is insignificant since exercises will be determined by the conversational agent on level 1. One can see that other forms of exercise are rather easily programmed and embedded into the existing structure, thus contributing to the design concept Programmability. A harder challenge would be to program the robot to gauge the success, progress, and/or motivation of individual participants and to give such individuals special treatment and advice. This feature is a subject for future research. It is yet to be determined whether such a feature is well-advised.

The transition agent on level 0 concludes each exercise with a compliment for the participants by randomly saying some synonym for "good," e.g. 'very good', 'excellent', 'beautiful', 'cool', 'very cool', 'most cool', 'splendid', or 'wonderful', then saying "Let's sit back and relax (at which point the robot goes into the sitting position)", then saying "What would you like to do next? Head, arms, legs, hands, eyes, feet, or Tai Chi?", and finally passing control back to the conversational agent.

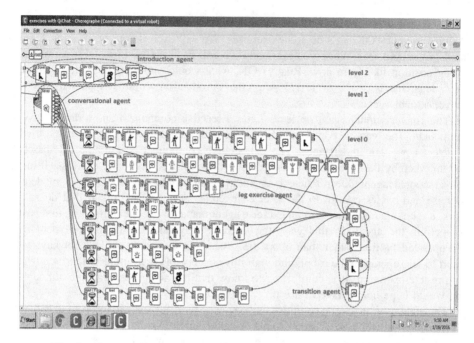

Fig. 2. Structure and implementation of the agents in the exercise routine [3]

4   Results of the Study

4.1   Preliminary Results of the Pilot Study

We designed three activities for the participants: (i) a warm-up routine with NAO, (ii) an exercise routine with NAO, and (iii) a session with Paro. Paro is robotic model of a baby harp seal and is used for pet therapy. The total amount of time spent with the participants was 36 min. Of those 36 min, the warm-up comprised 8 min, the exercise routine 15 min, and the session with Paro 9 min, totaling 32 min of interaction with the robots.

Each participant was scored on a scale of 1 to 10 for each observable discussed in Sect. 2.2 above. An additional category "Approval" was a subjective assessment by the authors. The "Total Engagement" score reflects the number of engagement points out of 100. Table 2 shows the preliminary results in raw form for the time spent interacting with the robots.

The average total engagement score of residents is 53 and the average score of staff is 71.4, suggesting that residents are slightly more than ambivalent in their approval of the robot, while nurses and administrators approve. Visual inspection shows that three of the residents highly approve, one is rather ambivalent, and two do not appear to approve – R1 and R5. We note that the low scores of R1 and R5 correlate with their mental and physical characteristics as shown in Table 1.

The evidence thus far invites a number of questions and suggests some lessons learned that will influence our future work. Inasmuch as the results show a stronger

Table 2. Preliminary results

	R1	R2	R3	R4	R5	R6	S1	S2	S3	S4	S5
Attentive	7	10	10	8	10	10	10	10	10	10	10
Smiling & laughter	1	9	10	2	3	10	9	9	10	8	8
Head-nodding	0	0	3	0	0	10	0	3	10	0	0
Leaning forward	0	0	0	0	0	6	0	0	7	0	0
Cooperation	2	9	10	6	4	9	8	8	8	7	7
Imitation	0	10	9	7	0	7	8	8	8	7	7
Initiation	0	3	2	9	0	9	8	8	8	0	7
Intensity	0	10	10	7	0	9	8	8	8	8	8
Participation	3	10	10	7	1	9	10	10	10	10	10
Approval	1	9	9	5	4	9	9	10	10	8	9
Total engagement	14	70	73	51	22	88	70	74	89	58	66

approval of the robots by staff than by residents, a number of potential factors causing the difference are possible. The difference may reflect one or more of the following kinds of differences separating the two groups: prior technological experience; experience with art forms such as movies involving robots; and the role differences in the Golden Oaks environment.

Based on the mapping of our observables into the design concepts, the first four items reflect pleasure and fun while the remaining items reflect actual work. With the exception of R1 and R5, one can see by visual inspection of the table that the participants were enjoying the engagement with the robot. One can see also that all participants scored favorably with respect to Cooperation, Imitation, Intensity, and Participation. These observations are encouraging.

4.2   An Analysis of Responsiveness by the Participants and by the Robot

In a second analysis of the video data we wished to examine more closely the responsiveness of the participants to the robot's directives and also the responsiveness of the robot to the participants' directives. For this study we considered the activities with NAO since Paro does not issue directives. During the warm-up routine and the exercise routine with NAO there were a total of 70 directives issued by the robot and 23 directives issued by the participants.

Table 3 shows the number of times when at least one participant responded to the robot's verbal directives. During the warm-up routine the robot asked the audience to ask him to do something, where each participant had a "cheat sheet" listing the things the robot could do: sing, recite a poem, tell a joke, wave, chill out, exercise, dance, or stop. During the exercise routine the robot asked the participants what exercise they would like to do next, also with a cheat sheet: head, arms, legs, Tai Chi, hands, eyes, feet, or stop. Sometimes a participant would respond immediately without prompting, sometimes only when prompted or encouraged by a nurse or administrator, and sometimes a participant would respond too quickly, i.e. before the robot finished a directive, in which case the robot did not hear the response. The numbers in Table 3 show that the HRI along this dimension was very good, where the participants improved from the first activity to the second. The fact that the participants responded too quickly 4 times during the exercise routine might suggest increased comfort with the robot.

Table 3. Participants' verbal response to robot directives during both routines (27 total)

	Warm-up routine	Exercise routine
Without prompting	7	14
With prompting	2	0
Too quickly	0	4
No response	0	0

For each exercise during the exercise routine, the robot issued additional directives instructing the participants to perform some sort of bodily behavior. We considered the response of the participants to be positive if the majority of them performed the behavior. For example, during the arm exercise the robot issued the following six directives (refer to Fig. 1): "Let's stand up with our arms down, like this" – "now let's raise our arms just a little and count like this 1 2 3 4 5 … good" – "now let them rest to the side again like this 1 2 3 4 5 … good" – "now raise them high and count to five like this 1 2 3 4 5 … good" – "now back to the side … beautiful" – "let's sit back and relax". Table 4 shows that the participants followed all exercise directives perfectly without prompting. This result is quite encouraging.

Table 4. Participants' exercise response to robot directives during the exercise routine (43 total)

	Warm-up routine	Exercise routine
Without prompting	N/A	43
With prompting	N/A	0
No response	N/A	0

Next we measured how the robot responded to directives from the participants, where sometimes the robot understood the speaker perfectly without the speaker having to repeat the directive, sometimes only when the speaker repeated a directive, and sometimes the robot entirely misunderstood the directive. For example, twice the robot misunderstood "feet" for "hands," once "legs" for "eyes", once "legs" for "hands," once "Tai Chi" for "arms," and once the robot heard "sing" without a directive at all. Table 5 shows that the robot understood a directive correctly without repetition roughly one third of the time, needed repeating a third of the time, and misunderstood the directive roughly a third of the time. Clearly the robot, the participants, and/or the environment need to be examined to improve the auditory capability of the robot. On the other hand we noted that participants laughed profusely when the robot misunderstood a directive, which may have contributed to the smiling/laughter item in our preliminary analysis.

Table 5. Robot's response to human directives during both routines (23 total)

	Warm-up routine	Exercise routine
Without repetition	3	5
With repetition	5	4
Misunderstood	1	5
No response	0	0

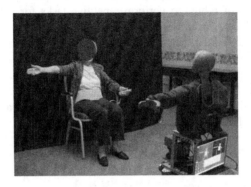

Fig. 3.   Coach-to-senior robot at USC [1]

Fig. 4.   RoboCoach [9]

4.3   Subsequent Focus Sessions with the Participants

The subsequent focus sessions with each group provided a wealth of suggestions to consider. The discussion among the nursing participants was fruitful, producing positive, enthusiastic, and creative discussion of the potential of robots on the premises. The group displayed sophistication concerning HRI, independently raising subjects and questions that one finds in HRI literature. Among the nurses, prior direct experience with a telepresence robot allowing hospital physicians to visit patients remotely was mentioned. A nurse's suggestion of robots possibly mimicking family members of Alzheimer residents provoked discussion or related ethical concerns. Another nurse participant introduced discussion of privacy issues, e.g. wondering whether robots watching residents would generate objections.

Ideas from administrative participants began with a marketing agent's speculation that the current generation of residents might show acceptance of robot presence rather different from the impending generation of new residents from the "baby boomer" generation. The marketing agent also commented on the psychological value of robots that could recognize individual residents and address them by name. The facility director and the marketing agent displayed further HRI sophistication in observations concerning the ability of robots to perform repetitive tasks for long periods of time without showing human frustration about wasting time. Referring to the facility's dining hall, the director added that a robot capable of taking orders at the tables and transmitting them wirelessly to the kitchen could save time and be "very helpful."

The resident group clearly enjoyed talking about the robot as well, evidenced by humor and laughter. Suggestions from the residents included talking to the robot, asking questions, and playing with it as one would play with a toy. One resident expressed a desire to have a robot to replace his wife.

5   Conclusion

The results of our first session of a humanoid robot leading exercises at a senior living community encourage us to continue our work. Plus, the results of other research groups on the same topic are motivating. Figure 3 shows the exercise coach developed at the University of Southern California (USC) in the United States and Fig. 4 shows the same by a research group at Ngee Ann Polytechnic (NAP) in Singapore.

The primary difference between the USC study and our study are (i) our study involves robot-to-group interaction whereas the USC study involves robot-to-individual-senior interaction one at a time, (ii) our robot communicates with participants verbally and physically in mutual interaction whereas the USC robot communicates via a remote clicker, and (iii) our study faces the complexities and dynamics of senior living in their real environment whereas the USC study is more of a lab experiment. Further, we question whether the USC robot satisfies the programmability and portability design concepts posited in our study. Nonetheless, the results of the USC study are largely consistent with those of our study. With respect to the NAP robot, Fig. 4 shows that the environment and the robot-to-group interaction is similar to that in our study. However, to our knowledge there is no published work that discusses the study in detail from a experimental/scientific point of view. For example, we cannot determine whether the robot is tele-operated or autonomous. In our study, the robot is completely autonomous.

For the next sessions at the facility we plan to (i) improve the conversational behavior of the robot to make it more personal [10], (ii) improve the responsiveness of the robot to the participants' directives, (iii) improve our study evaluation criteria, (iv) program the robot so that it can learn to attach names to faces and thus identify participants by name, (v) obtain more complete profile information on the participants in order to compare diverse population groups, (vi) add more exercise routines, e.g. a simple meditation exercise, and (vii) carefully select viable candidates with varying degrees of cognitive and physical abilities. For example, a better demographic profile of the participants (age, gender, level of education, previous technological experience) and clear selection criteria for the participants could make a good case for the integration of socially assistive robots into the lives of older people with varying cognitive and physical impairments. Future research will focus on these issues.

It is difficult to assess or predict the participants' motivation over time with multiple sessions based on the one experiment described here. Our future research will address this question. Future research will study also the effect of participants' characteristics with respect to personality, background, and individual preferences. We plan to tabulate and compare the response results in Tables 3, 4 and 5 for each participant. It is likely that reaction patterns will vary with each individual. If the pattern could be classified based on the type of personal characteristic, the result would be useful. Finally, our analysis focusses mostly on results for NAO. Further analysis will include results for PARO.

References

1. Juan Fasola, J., Matarić, M.: A socially assistive robot exercise coach for the elderly. J. Hum.-Robot Interact. **2**(2), 3–32 (2013)
2. www.goldenoaks.com. Accessed 19 Apr 2016
3. Lewis, L., Metzler, T., Cook, L.: Results of a pilot study with a robot instructor for group exercise at a senior living community. In: Bajo, J., et al. (eds.) PAAMS 2016 Workshops. CCIS, vol. 616, pp. 3–14. Springer, Heidelberg (2016). doi:10.1007/978-3-319-39387-2_1
4. Talaei-Khoei, A., Lewis, L., Talaei-Khoei, T., Ghapanchi, A.: Seniors' perspectives on perceived transfer effects of assistive robots in elderly care: a capability approach analysis. In: International Conference on Information Systems, Fort Worth, Texas (2015)
5. Checkland, P., Poulter, J.: Learning for Action: A Short Definitive Account of Soft Systems Methodology, and Its Use for Practitioners Teachers and Students. Wiley, Chichester (2007)
6. Checkland, P.: Soft systems methodology: a thirty year retrospective. Syst. Res. Behav. Sci. **17**(1), 11–58 (2000)
7. doc.aldebaran.com/2-1/family/robots/index_robots.html#all-robots. Accessed 20 May 2016
8. Brooks, R.: A robust layered control system for a mobile robot. IEEE J. Robot. Autom. **2**(1), 14–23 (1986)
9. www.np.edu.sg/sg50/events/Pages/20150101_npevents_robocoach.aspx. Accessed 19 Apr 2016
10. Lewis, L.: Using narrative with avatars and robots to enhance elder care (Chapter 14). In: Ghapanchi, A., Tavana, M., Talaei-Khoeif, A. (eds.) Healthcare Informatics and Analytics: Emerging Issues and Trends. IGI Global, Pennsylvania (2014)

Estimating the Effect of Dynamic Variable Resistance in Strength Training

Tomosuke Komiyama, Yoshiki Muramatsu, Takuya Hashimoto[✉],
and Hiroshi Kobayashi

Tokyo University of Science, Tokyo, Japan
tak@rs.tus.ac.jp

Abstract. Strength training is important for elderly to maintain their physical ability for independent walking. Meanwhile, the risks of high-intensity strength training for elderly should be considered: blood pressure elevation and arteriosclerosis. To tackle these problems, a multi-functional training machine which offers four kinds of training motions for lower limbs and controls training load dynamically have been developed. In this study, to estimate the effects of variable training load, measurements of muscle strength, blood pressure, and muscle fatigue are conducted in strength training experiments under various load conditions. The results indicate that the dynamic load conditions have potential to inhibit blood pressure elevation and muscle fatigue with increasing muscle strength comparing to constant load condition.

Keywords: Rehabilitation · Strength training · Dynamic load · Muscle strength · Blood pressure · Muscle fatigue

1 Introduction

The progress of aging society is a big social issue in Japan. The number of elderly population in Japan is expected to grow at 38.8 % with unprecedented rapidity by 2050 [1]. Physical and cognitive abilities are naturally impaired with aging. In particular, a decline in muscle strength which relates to daily activities such as maintaining a posture, standing, walking, and so forth seriously decreases the quality of life.

Strength trainings of lower limbs such as leg press, leg extension, hip adduction, and hip abduction contribute to improve and maintain muscle strength for independent living ability for elderly. However, various kinds of machines and a large space are required for various trainings because a conventional training machine is basically developed for a single training. Therefore, elderly have to go to specialized rehabilitation facilities twice or three times a week, and it is physical, psychological, and economical burden for their caregivers. Moreover, since high-intensity strength training has risks of blood pressure elevation [2] and arteriosclerosis [3], training intensity should be carefully determined for elderly.

To solve the problems mentioned above, we have developed a compact-type strength training machine for home use, named "Dynamically Variable Resistance Training Machine (DVR)", which can control training load dynamically and offer four kinds of muscle trainings to maintain muscle strength for independent living. We found

N. Kubota et al. (Eds.): ICIRA 2016, Part II, LNAI 9835, pp. 26–35, 2016.
DOI: 10.1007/978-3-319-43518-3_3

that variable resistance training could increase muscle strength with inhibiting blood pressure elevation in experiments using DVR. However, not only blood pressure elevation, but also muscle fatigue should be considered, because it is thought that elderly fatigue easily and recover slowly.

In this paper, we performed experiment of strength training to investigate influence of variable resistance on subject in terms of muscle fatigue in addition to blood pressure and muscle strength.

2 Dynamically Variable Resistance Training Machine: DVR

2.1 Overview of DVR

Figure 1 shows an overview of a training machine, named DVR (Dynamically Variable Resistance Training Machine), which applies training load dynamically while a conventional training machine can apply only a constant load. A characteristic of DVR is to offer four kinds of lower limb trainings to maintain walking muscle strength by utilizing only one machine as shown in Fig. 2: leg press, leg extension, hip adduction, and hip abduction. It can also control training load dynamically by a DC motor. The size is 952 mm in height, 460 mm in width, and 2000 mm (maximum) in depth. The weight is 98 kg which is less than conventional training machines.

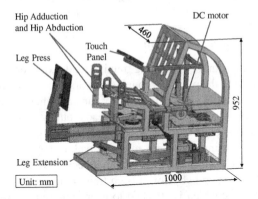

Fig. 1. Overview of dynamically variable resistance training machine: DVR

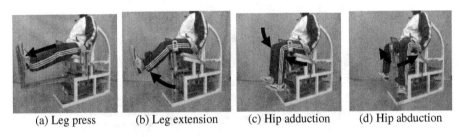

(a) Leg press   (b) Leg extension   (c) Hip adduction   (d) Hip abduction

Fig. 2. Four kinds of strength trainings of lower limb

2.2   System Configuration

Figure 3 shows the system configuration of DVR which consists of DC motor (13924G-900-10K36K, Tokushu Denso Co., Ltd.), regeneration brake driver (S20563, Tokushu Denso Co., Ltd.), control PC, USB module (USB-FSIO30, Km2Net Inc.), and rotary encoder (E6A2-CW3C, OMRON Co.). The control system controls motor torque depending on control commands sent to the regeneration brake driver from the control PC. Figure 4 shows the linear relationship between voltage signal and motor torque. DVR generates torque between 0.44 Nm and 9.6 Nm by voltage signal between 0.0 V and 3.0 V. The rotary encoder is used to measure movement length.

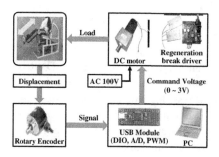

Fig. 3. System configuration

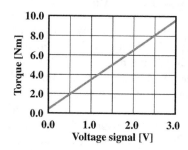

Fig. 4. The relationship between voltage signal and load torque

2.3   Mechanism for Load Generation

Figure 5 shows a mechanism for generating training load. Load torque generated by the DC motor is transmitted to the pulleys through the gears. Then load torque is applied to the end effector of each training unit. In this mechanism, when a pulley is rotated, its rotation doesn't interfere with others because each pulley has a one way crutch. Therefore load torque is applied independently to each training mechanism despite load torque is generated by using only one motor. Load is applied only to concentric contraction of main muscles in each training motion because eccentric contraction leads damage of muscle [4]. The end effector of each training unit returns to initial position by a constant force spring which is attached to every pulley (Fig. 6).

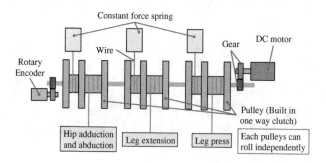

Fig. 5. Mechanical structure for generating training load

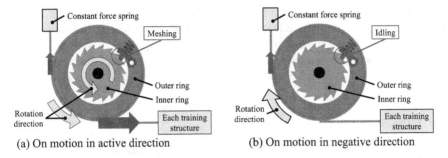

Fig. 6. Schematic of one way clutch pulley

3 Evaluation Experiment of DVR by Measuring Blood Pressure, Muscle Strength, and Muscle Fatigue

3.1 Objective and Evaluation Method

Maximum voluntary muscle force is generally changed depending on joint angle. For example, maximum voluntary force in leg press is generated at a knee angle of about 120°. Moreover, different kinds of muscle fibers respond to different intensities. Slow muscle fiber is mobilized to low muscle activity and fast muscle fiber responds to high intensity. Therefore, a variety of training loads may influence changes in blood pressure and muscle fatigue involved in strengthening muscle.

The objective of this study is to investigate the method which can inhibit blood pressure elevation and muscle fatigue with increasing muscle force by varying training load. Thus, we measured muscle force of lower limb in leg extension and estimated muscle fatigue by surface electromyogram (sEMG).

3.2 Load Patterns

Figure 7 shows load patterns used in experiments. In the constant load, called CL, training load is constant as a conventional training machine using weights. One of the dynamic load conditions provides an increasing load which is convex upward, called

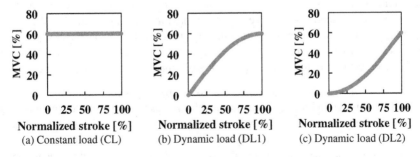

Fig. 7. Load patterns

DL1. The other dynamic load is an increasing load which is convex downward, called DL2. In each figure, normalized length with maximum movable range of subject is on the horizontal axis, and maximum voluntary contraction (MVC) of subject is on the vertical axis. The maximum load was set to 60 % MVC in this experiment as 60 % of maximum voluntary contraction (60 % MVC) is recommended load for elderly.

3.3   Subjects

Fifteen healthy 20's males, who are not used to exercise, participated in the experiment as subjects. Subjects are divided into 3 groups according to load conditions. First group has 5 subjects (age: 23.4 ± 1.0 yr, height: 172.2 ± 6.8 cm, weight: 72.6 ± 10.4 kg), who are labeled as Subjects A–C, participated in CL condition. Second group includes 5 subjects (age: 23.4 ± 1.5 yr, height: 170.0 ± 4.3 cm, weight: 64.4 ± 10.2 kg) experienced the strength training with DL1 condition and they were labeled as Subjects D–H. Third group also has 5 subjects (age: 23.1 ± 0.8 yr, height: 171.2 ± 1.7 cm, weight: 70.2 ± 9.6 kg) participated in DL2 condition and they were labeled as Subjects I–M.

3.4   Training Procedure

All subjects performed three sets of 10 repetitions of leg extension with one-minute break between sets in a day, three times per week. This training was undergone for five weeks. In total, all subjects performed the training fifteen times.

3.5   Evaluation Method

Measurement of Muscle Strength in Knee Extension. To estimate the increasing in muscle strength between before and after the training, muscle strength of quadriceps femoris muscles which mainly work in leg extension was measured by using the dynamometer of lower limb (MDKKS, Molten Co.) as shown in Fig. 8. The measurement was carried out before the experiment and once a week. In the measurement, each subject's muscle force was measured to calculate his average muscle force.

Measurement of Blood Pressure. To calculate the increasing amount in blood pressure in different load patterns, we measured maximal blood pressure (MBP) of each subject before and after the training by a manometer (HEM-762, OMRON Co.). Every subject was instructed to refrain from eating and drinking, strenuous exercise and sleeping from 90 min before the training to avoid influencing blood pressure.

Measurement of Muscle Fatigue. In order to estimate muscle fatigue, we performed surface electromyogram (sEMG) measurement for rectus femoris muscle which is a part of quadriceps femoris muscles. Figure 9 shows the measurement point. sEMG is known as a method to estimate muscle activity non-invasively and its frequency-domain features are useful for the assessment of muscle fatigue. In particular, mean power

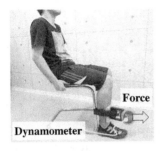

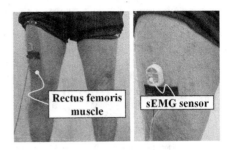

Fig. 8. Measurement method of muscle strength

Fig. 9. Measuring point

frequency (MPF) is widely used to estimate muscle fatigue [5]. In this study, we also used MPF obtained from FFT analysis of sEMG signal to assess muscle fatigue. Figure 10(a) shows an example of sEMG signal obtained from the experiment. In the analysis, we first extracted periods of movement from the signal and applied FFT analysis to each period. Then, we calculated MPF in each period as shown in Fig. 10(b). We used a multi-telemetry system (WEB7000, Nihon Kohden Co.) to measure sEMG signal. The sampling frequency was 250 Hz and the time constant was 0.1 s. In addition, high-pass filter with 30 Hz and low-pass filter with 500 Hz were applied to raw sEMG signal.

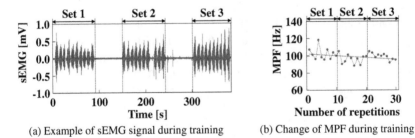

(a) Example of sEMG signal during training   (b) Change of MPF during training

Fig. 10. Example of measuring local muscle fatigue by sEMG

4 Experimental Results

4.1 Result of Muscle Force Measurement

Figures 11(a)–(c) show the increasing rates in muscle force of the subjects in three load conditions and Fig. 11(d) shows the average rates. The values of muscle strength of each subject are normalized by his initial muscle strength. Table 1 shows correlation coefficients between the number of weeks and the increasing rates in muscle force of each subject.

First, a one-way analysis of variance (ANOVA) was used to compare the final increasing rates in each load pattern based on the result of Fig. 11(d), and there were no

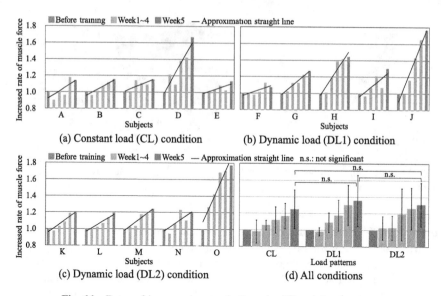

Fig. 11. Rates of increase in muscle force in different load patterns

Table 1. Correlation coefficient R between the number of weeks and the increasing rates in muscle force

	Subject					Avg.	S.D.
	Correlation coefficient R						
Constant load (CL)	A	B	C	D	E	0.844	0.084
	0.741	0.928	0.7989	0.932	0.820		
Dynamic load 1 (DL1)	F	G	H	I	J	0.899	0.103
	0.757	0.981	0.954	0.820	0.976		
Dynamic load 2 (DL2)	K	L	M	N	O	0.919	0.075
	0.932	0.954	0.954	0.785	0.961		

significant differences between load patterns. However, the average values were high in order of DL1, DL2, CL. Then, we performed correlation analysis to measure the strength of the association between the number of weeks and the increasing rates based on Pearson's correlation coefficient in each load pattern. As the results, there were significant positive correlations ($p < .05$) in three subjects (subjects B, D, and E) of CL condition, four subjects (subjects G, H, I, and J) of DL1 condition, and four subjects (subjects K, L, M, O) of DL2 condition. Therefore, every load pattern was considered as having potential in increasing muscle strength.

4.2   Result of Blood Pressure Measurement

Figures 12(a)–(c) show the average maximum blood pressures (MBPs) of the subjects before and after the training in three conditions and Fig. 12(d) shows the averages of increasing amount and standard deviations of MBPs in different conditions. Independent t-test was used to compare MBPs before and after the training, and there were significant differences (P < .05) in all subjects of CL condition, two subjects (G, H) of DL1 condition, and three subjects (K, M, O) of DL2 condition. ANOVA and Tukey's test were used to compare the increasing rates of blood pressure in three conditions based on Fig. 12(d). As the results, it is found that the dynamic load conditions can inhibit blood pressure elevation comparing to the constant load.

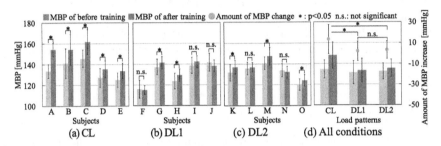

Fig. 12. MBP variation

4.3   Result of Muscle Fatigue Measurement

The average MPFs per set of 10 repetitions of the subjects in three conditions are shown in Figs. 13(a)–(c) and (d) shows the average MPFs. The MPFs of 2nd and 3rd sets are normalized by 1st set. Table 2 shows correlation coefficients between the number of sets and MPF.

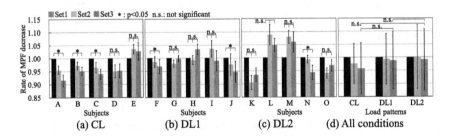

Fig. 13. Decreasing rate of MPF rate of each set

We performed correlation analysis to measure the association between the number of sets and the changes in MPFs based on Pearson's correlation coefficient. The results show that there were significant correlations (p < .05) in three subjects (A, B, C) of CL condition, two subjects (F, J) of DL1 condition, and one subject (N) of DL2 condition.

Table 2. Correlation coefficient R between the number of sets and MPF

	Subject					Avg.	S.D.
	Correlation coefficient R						
Constant load (CL)	A	B	C	D	E	0.914	0.109
	0.998	0.997	0.996	0.850	0.729		
Dynamic load 1 (DL1)	F	G	H	I	J	0.613	0.390
	0.999	0.083	0.769	0.217	0.999		
Dynamic load 2 (DL2)	K	L	M	N	O	0.674	0.143
	0.677	0.553	0.726	0.912	0.505		

Then, ANOVA was used to compare MPFs of final set in different conditions based on the result of Fig. 13(d), and there were no significant differences among different conditions. However, the decreasing rates of average MPFs of DL1 and DL2 conditions were smaller than that of CL condition. Therefore, the dynamic load conditions have potential to inhibit muscle fatigue comparing to the constant load condition.

5 Conclusion

We have developed a compact-type strength training machine, "Dynamically Variable Resistance Training Machine (DVR)", which control resistance dynamically as training load and offer four kinds of trainings of lower limb. In this paper, we performed experiment of strength training using DVR and evaluated the influence of variable training loads by measuring muscle strength, blood pressure, and muscle fatigue to investigate the method which can inhibit blood pressure elevation and muscle fatigue while increasing muscle force. As the results, it is found that the dynamic load conditions have potential for inhibiting blood pressure elevation and muscle fatigue with increasing muscle strength. The results will contribute to determine a strategy for strength training for elderly.

One of our future works is to estimate the effect of dynamic variable load in strength training in more detail by conducting experiments under various conditions: variety of trainings and subjects, and training duration.

References

1. Cabinet Office, Government of Japan: Situation on aging. In: Annual Report on the Aging Society: 2013 (Summary) (2013). http://www8.cao.go.jp/kourei/english/annualreport/2013/pdf/1-1.pdf. Accessed 15 Jan 2014
2. Roltsch, M.H., Mendez, T., Wilund, K.R., Hagberg, J.M.: Acute resistive exercise does not affect ambulatory blood pressure in young men and women. Med. Sci. Sports Exerc. **33**(6), 881–886 (2001)
3. Bertovic, D.A., Waddell, T.K., Gatzka, C.D., Cameron, J.D., Dart, A.M., Kingwell, B.A.: Muscular strength training is associated with low arterial compliance and high pulse pressure. Hypertension **33**(6), 1385–1391 (1999)

4. Ebbeling, Cara, Clarkson, B., Priscilla, M.: Exercise-induced muscle damage and adaptation. Sports Med. **7**(4), 207–234 (1989)
5. De Luca, C.J.: Myoelectrical manifestations of localized muscular fatigue in humans. Crit. Rev. Biomed. Eng. **11**(4), 251–279 (1984)

Smart Device Interlocked Robot Partner for Elderly Care

Takenori Obo[✉], Siqi Sun[✉], and Naoyuki Kubota[✉]

Department of System Design,
Tokyo Metropolitan University, Tokyo 191-0065, Japan
sun-siqi@ed.tmu.ac.jp, {takebo,kubota}@tmu.ac.jp

Abstract. This paper presents smart device interlocked robots partner for elderly care. Aging society in Japan can be a big serious problem. The number of caregivers is not enough in the current situation and is not expected to substantially increase in future. Hence, comprehensive care for elderly people should be provided by the local community. With the rapid development of ICT, it has been used to improve QOL and QOC for elderly people. Smart devices, especially, have become more familiar to us. In this paper, we introduce some applications for elderly care with smart device interlocked robot partners and discuss the scalability of the robot partner for the elderly care.

Keywords: Smart device interlocked robot partner · Healthcare system · Informationally structured space

1 Introduction

Aging society in Japan can be a big serious problem. According to the National Institute of Population and Social Security Research, the increase of elderly individuals living alone is estimated to reach into approximately 37.7 % overall by 2035, and to 44 % in Tokyo area. To address the problem, ideally, caregivers of elderly people in nursing home should create time regularly for supporting them. However, the number of caregivers is not enough in the current situation and is not expected to substantially increase in future. Elderly care currently is hence shifting from hospital care to community-based care and home care. However, this may lead to raise the burden on elderly's family members. To address the problems, comprehensive care for elderly people should be provide by the local community.

Recently, Information and Communication Technology (ICT) has been used to improve the quality of life (QOL) and the quality of community (QOC) [1–6]. With the rapid development of ICT, smart devices (e.g. smartphones and tablets) have become more familiar to us. The penetration of smartphone and personal computers (PC) in Japan respectively reached 64.2 % and 78.0 %, according to annual report published by Ministry of Internal Affairs and Communications. Until a few years ago, it is said that usage of PC is arduous for elderly people. It is true that there are many elderly persons that used to only basic functions of ICT, such as ICT device to control home appliances and to monitor the state of a living house. However, the internet population of seniors (aged 60 years of age and older) has been increasing steadily over time: the population

© Springer International Publishing Switzerland 2016
N. Kubota et al. (Eds.): ICIRA 2016, Part II, LNAI 9835, pp. 36–47, 2016.
DOI: 10.1007/978-3-319-43518-3_4

in their 60s reached nearly 75 % at the end of 2014. Thus, in near future, ICT devices could be necessities for their life if the devices would be more human-friendly for them.

Integration of smart devices and human-friendly robot partners can be a solution to realize the human-friendly system. Smart devices such as smartphones or tablet PCs provide us with the human-friendly interface and the easy access to every types of information such as (1) personal information, (2) environmental information, and (3) Internet information in addition to (4) people, (5) place, (6) objects, and (7) events. Furthermore, such smart devices are equipped with various sensors and a high-end CPU that is enough to be applied to a robot partner. In this study, we have developed smart device interlocked robot partners for elderly care. The robot partner can be a living partner to obtain useful and important information for the user from the Internet and sensor network devices in a house, and perform natural communication based on facial and gestural expression and verbal information. The robot partners are utilized from viewpoints of different types of interaction styles: physical robot partner, pocket robot partner, and virtual robot partner. Physical robot partner can interact with a person by using multi-modal communication like a human. Pocket robot partner can be easily brought as a portable smart device everywhere. The virtual robot partner is in the virtual space within the computer, but we can interact with it through avatars in the virtual space. The interaction style of these three robots is different, but they share the same personal database and interaction logs, and can interact with the person based on the same interaction rules independent from the style of interfaces. This functional flexibility leads to a comprehensive support for the users, e.g., lifelog measurement, user modeling, and daily exercise encouragement. The functionality of the robot partner corresponding to the system structure is presented in Table 1.

Table 1. Functionality of smart device interlocked robot partner for eldery care

System structure	Function	Application
Only smart device	Touch interaction	Information recommendation
	Verval communication	Navigation
	Outdoor behavior measurement	Outdoor lifelogging
Smart device with body part	Nonverbal communication (e.g. gesture)	Exercise encouragement
		Entertainment (e.g. Dance)
Smart device with sensor networks and DB	Indoor behavior measurement	Indoor lifelogging
		Abnormality detection
	User modeling	Preference analysis
		Health promotion

In this paper, we introduce some applications for elderly care with smart device interlocked robot partners and discuss the scalability of the robot partner. This paper is organized as follows. Section 2 explains the smart device interlocked robot partners and surrounding system for elderly people. Sections 3, 4 and 5 show respectively lifelogging system, health promotion system, and exercise encouragement system, as an application of the developed system. Finally, we summarize this paper, and mention the future direction of this research in Sect. 6.

2 Elderly Care System with Robot Partner

To realize the functional flexibility mentioned in Sect. 1, we have to consider the system integration from various points of view: robot technology (RT), network technology (NT), information technology (IT), and intelligence technology. In such system, the environment surrounding a person should have a structured platform for gathering, storing, transforming, and providing information. We call the shared cognitive environment Informationally Structured Space (ISS). This section explains the concept of ISS and introduces the developed robot partners and surrounding systems for elderly care.

2.1 Informationally Structured Space

ISS was proposed by Kubota et al. [7]. The emerging synthesis of IT, NT, and RT is one of the most promising approaches to realize a safe, secure, and comfortable society for the next generation. NT can provide the robot with computational capabilities based on various types of information outside of robots. Actually the robot directly receives environmental information through a local area network without the measurement by the robot itself. Wireless sensor networks realize to gather huge data on environments. However, it is very difficult to store all the huge data in real time. Furthermore, some features should be extracted from the gathered data to obtain the required information. Therefore, intelligent technology is required in wireless sensor networks.

Intelligence technology and information technology have been discussed from various points of view. Information resources and the accessibility within an environment are essential for both people and robots. Therefore, the environment surrounding people and robots should have a structured platform for gathering, storing, transforming, and providing information. The intelligent technology for the design and usage of the ISS should be discussed from various points of view such as information gathering of real environment and cyber space, structuralization, visualization and display of the gathered information. The structuralization of ISS realizes the quick update and access to valuable and useful information for people.

2.2 Robot Partner

In previous works, we have developed smart device interlocked robot partners called iPhonoid and iPadrone [8] presented in Fig. 1. In ISS, the smart device interlocked robot partners can adaptively vary functionality with the combination of the surrounding systems and devices. In the minimum configuration, the robot partners obtain environmental information from the own sensors, and interact with the user mainly through the touch interface. With wearable devices, wireless sensors and database server, the system easily realizes the user preference analysis and data mining. Moreover, the robots can perform nonverbal communication by adding the body actuators. This functional

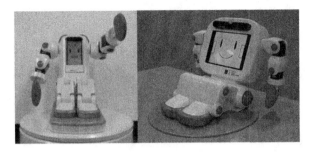

Fig. 1. Smart device interlocked robot partners: iPhonoid (left) and iPadrone (right).

flexibility is essential for supporting elderly people in order to serve the lifelog measurement, user modeling, and daily exercise encouragement. Figure 2 shows the overview of the framework in ISS.

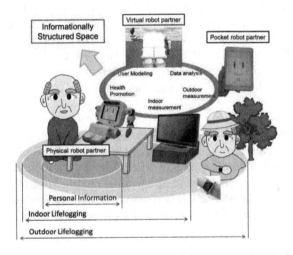

Fig. 2. Elderly care in ISS

2.3   Measurement System

In related works, various types of sensors have been applied to lifelogging systems: accelerometer, gyro sensor, illuminance sensor, pyroelectric infrared sensor, laser sensor, image sensor, and so on [9, 10]. The sensor devices should be selected according to objective human behavioral patterns, but it is difficult to predefine desired behavior patterns. To build such measurement system, a complementary measurement to comprehensively cover the measurement area should be performed by various sensors.

For example, laser range finder (LRF) and 3D image sensor can be applied to human and object tracking as a global measurement methodology. Figure 3(a) shows a LRF (URG-04LX) developed by HOKUYO. It has a contactless sensing system for measuring 2-dimensional distance based on time of flight principal by using laser. The sensor

(a)   URG-04LX   (b)   SunSPOT

Fig. 3. Sensors for human behavior measurement.

can measure precisely with measurement range up to 4095 [mm] in 682 different directions where the covering measurement range is 240 [deg]. 3D image sensor can also measure the distance from the device like a camera in real time. Kinect sensor, developed by Microsoft, is a typical 3D image sensor. The sensor includes a 3D image sensor, microphones, an RGB camera, an accelerometer, and a tilting-up mechanism, and it has a built-in processor to detect human posture and facial expression. Meanwhile, Fig. 3(b) illustrates a wireless sensor (SunSPOT) developed by Oracle Labs. The small, wireless, battery-powered device has three sensors, accelerometer, illuminance sensor, and temperature sensor. This type of the sensor can be used for constructing a local measurement system.

Furthermore, in recent years, many types of wearable smart devices have been developed and released rapidly [11–15]. Wearable smart devices do not only easily provide a biometric signal measurement but also have essential properties that serve as a fashion accessory and familiar necessity. Smart watch is a representative wearable smart device. Moreover, the device can be a familiar user interface to connect a person with the robot partner. Through the interaction with the user, the robot partner can obtain the detail information of user preference difficult to be measured by other sensor devices.

3   Lifelogging System in ISS

As shown in Fig. 2, the required lifelogging system should be discussed in terms of data types and measurement areas: personal information, indoor and outdoor measurement. For gathering the personal information, the system should present a human-friendly interface to interact with a person. Figure 4 shows a lifelogging system with smart watch. On the touch interface, 9 buttons on that an activity name is written are displayed. If the user pushes a button, the information is transferred to the database immediately. Such interface can easily obtain the labeled personal information without signal processing and data analysis. However, it can be a burdensome system to the user. On the other hand, in indoor and outdoor measurement, we can gather many kinds of sensor data corresponding to human behavioral pattern. In the indoor measurement, wireless sensor devices that attached to furniture and home electronics are applied to the measurement of motion patterns. In the outdoor measurement, the behavioral patterns are measured by internal sensors of portable devices. These measurements permit the natural usage

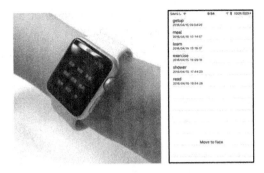

Fig. 4. Lifelogging system with smart watch.

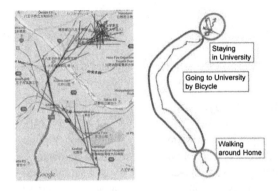

Fig. 5. Measurement GPS data of outdoor movement. In left figure, blue line indicates the vector norm of moving speed, and red line is represented as the trajectory of human movement. The history of human state is shown in right figure. (Color figure online)

of the portable devices and home electronics, however, the signal processing and pattern recognition are required to estimate the human states. Therefore, the contextual relation modeling between behavior pattern and personal information is important to produce a lifelogging system adapted to the user's lifestyle.

In [16], we have proposed a method for the contextual relation modeling, using Growing Neural Gas (GNG) and Spiking Neural Network (SNN). GNG is applied to the feature extraction of human behavior, and SNN is used to associate the features with event information labels.

Figure 5 presents a measured GPS dataset of outdoor movement. This is the data record of the movement of a university student in one day. In the data, the person went to school by bicycle in morning, after that, the student stayed at the university until evening and finally came back home. Figure 6 shows the feature extraction result of outdoor movement. From the figures, several clusters are produced depending on the procedure of GNG learning. At 5000 iteration steps, the trajectory data is abstracted so much that every data belongs to a same cluster. However, as the number of iteration steps increases, the number of clusters increases. Finally, at 20000 steps, 3 clusters are created. As a result, the clustering can represent 3 features of the movement corresponding to the actual human state shown in Fig. 5.

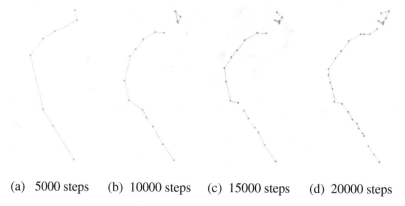

(a) 5000 steps   (b) 10000 steps   (c) 15000 steps   (d) 20000 steps

Fig. 6. Feature extraction of outdoor movement performed by GNG.

4   Health Promotion System

Various types of health promotion systems with ICT have been developed so far. In [17, 18], a smart device is applied as a telemonitoring system that can obtain some biological data-sets measured by external sensors, e.g., physical activity, heart rate, and blood pressure. Such systems aim to provide important information to caregivers in order to assist their timely healthcare decisions. On the other hand, health promotion is one of the most important tasks for supporting elderly people. The purpose is to lead elderly people to reassess their life-style. To heighten awareness about health, suitable knowledge of healthcare should be provided to elderly people improving their motivation. We have therefore presented an e-learning system for health promotion implemented into a smart device interlocked robot partner. In the system, the robot partner can give some quizzes to the user. The content of the quizzes is produced based on the transtheoretical model of health behavior change proposed by Prochaska [19].

In the system, the users are automatically classified into the five stages (Precontem-plation, Contemplation, Preparation, Action, and Maintenance) on the basis of their lifelog measured in ISS. To analysis the user model in each stage, the questionnaire is conducted through the touch interface on the smart device interlocked robot partners. Here, the robots provoke five questions to assess the degree of motivation for health. The probability model for quiz selection is generated based on the user property. The property is represented as a vector $\mathbf{u}$ (1×5 questions) given weighted values.

The quizzes are categorized into four groups based on the following objectives: (1) learning the pros and cons of exercise, (2) studying exercise caution, (3) considering various ways to realize new healthy behavior, and (4) gaining knowledge of various ways to exercise. The relationship between the quiz category and user property is expressed by predefined correlation values. If we assume that the feature vector of i-th category is defined as $\mathbf{q}_i$, the similarity S between the user property and the category is derived from a cosine distance as follows:

$$S_i = \cos \theta_i = \frac{\mathbf{u} \cdot \mathbf{q}_i}{\|\mathbf{u}\| \|\mathbf{q}_i\|} \tag{1}$$

Moreover, we apply stochastic method with Boltzmann selection for the quiz selection. The number of correct answers to each quiz is used to control the selection strength.

To discuss the effectiveness of the system, we have conducted an experiment to test the following hypotheses:

- The quiz category that has a high rating will be necessary knowledge for changing the user's daily behavior and habits.
- The user's self-efficacy can be improved by providing proper knowledge of health through the proposed system.

The experiment involves the following four steps: (1) preliminary questionnaire, (2) quiz selection and answer, (3) update of the probability model, and (4) post-survey of self-efficacy. First, the user needs to answer five questions for evaluating the degree of motivation for health through the touch interface. The robot uses the answers to the questions as the user property, and creates the probability model for selecting a quiz category. After the preliminary questionnaire, the robot partner provides quizzes to the user. To update the probability model, ten quizzes are presented in every trial. After completing the health promotion program, we have a questionnaire for the post-survey of self-efficacy to the user. Here, the purpose is to measure the ratio of the user's confidence to perform exercise routine regularly in the situations listed in Table 2.

Table 2. Self-efficacy questionnaire for Post-survey.

ID	Situations	Confidence rating
Q1	When I am feeling tired	1 · 2 · 3 · 4 · 5
Q2	When I am feeling depressed	1 · 2 · 3 · 4 · 5
Q3	When I have too much work to do at home	1 · 2 · 3 · 4 · 5
Q4	When I can afford to do anything	1 · 2 · 3 · 4 · 5
Q5	During bad weather	1 · 2 · 3 · 4 · 5

Figure 7 illustrates the result of self-efficacy questionnaire before and after use of the health promotion system. From the result, the proposed system increases self-efficacy in most of the cases. Especially, the health promotion is effective against barriers that interfere with the health activities: tiredness, overwork, and bad weather. In this experiment, over half of individuals were in the stage "Preparation". In fact, the system relatively chose quizzes and advices to make time for exercise. This indicates that the contents of quizzes provided by the robot partner may improve the motivation of some participants.

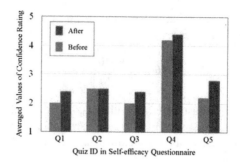

Fig. 7. The result of self-efficacy questionnaire.

5   Exercise Encouragement System

In [20], we have built an exercise encouragement system with robot partner. The exercise encouragement system aims to prevent age-related physical deconditioning. We applied a robot partner as an exercise instructor, and used a 3D image sensor to measure and evaluate the human motion during the exercise. To encourage an elderly person to do daily exercise, the robot partner needs to improve his/her motivation for the exercise. In this system, the robot partner scores the exercise performance out of 100. The user can enjoy the exercise as though it was a game, and it motivates them to great performance.

To assess performance of elderly's exercise, the instructor robot needs to share the body image with an elderly person. We have therefore built a virtual robot partner shown in Fig. 8(a) and proposed a method to transform the configuration space of human into that of the robot.

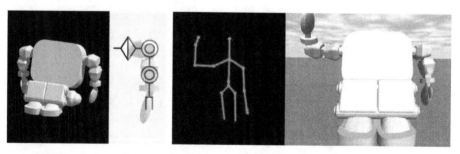

(a) Kinematic model   (b) Posture estimation

Fig. 8.   Virtual robot partner.

We applied a genetic algorithm as an optimization method to find a set of suitable joint angles for the virtual robot partner. The postural data **P** obtained from Kinect sensor is represented as:

$$\mathbf{P} = \left(\mathbf{p}_{1,1}, \mathbf{p}_{2,2}, \ldots, \mathbf{p}_{k,t}, \ldots, \mathbf{p}_{m,K,T_m} \right) \tag{2}$$

$$\mathbf{p}_{k,t} = \begin{bmatrix} x_{k,t} & y_{k,t} & z_{k,t} \end{bmatrix} \tag{3}$$

where $\mathbf{p}_{k,t}$ is the 3D position of the k-th joint at time t. $\mathbf{p}_{1,t}$ $\mathbf{p}_{2,t}$ and $\mathbf{p}_{3,t}$ are defined as the position of shoulder, elbow and hand of right arm, $\mathbf{p}_{4,t}$ $\mathbf{p}_{5,t}$ and $\mathbf{p}_{6,t}$ are defined as the position of shoulder, elbow and hand of left arm respectively.

The candidate solution is composed of numerical parameters corresponding to the joint angles as follows:

$$\mathbf{g}_{i,t} = \left(\theta_{i,1,t}, \dots, \theta_{i,j,t}, \dots, \theta_{i,M,t}, \gamma_{i,t} \right) \tag{4}$$

where M is the number of joints and $\theta_{i,j,t}$ is the j-th joint angle of the i-th candidate solution at time t. The objective of the genetic algorithm is to minimize the difference between actual human posture and posture of the robot model. To achieve the objective, the fitness value of candidate solution is calculated by

$$f_i = \sum_{k=1}^{K} \left\| \mathbf{p}^{GA}_{m,k,t} - \mathbf{p}^{*}_{m,k,t} \right\| \tag{5}$$

where $\mathbf{p}^{GA}_{m,k,t}$ is the position of body parts calculated by forward kinematics model using the candidate solution, $\mathbf{p}^{*}_{m,k,t}$ is the topologically nearest joint position measured by the Kinect sensor to the k-th joint position of the kinematic model at time t.

Furthermore, after the measurement of the exercise performance, the robot has to evaluate it and provide some advices to motivate the elderly person. Some persons can enjoy the exercise with the robot partner, and the others may be discouraged due to their bad scores. Hence, we have developed a human-robot communication model for improving a feeling of self-efficacy [22] of elderly people. In the system, the robot partner can choose a user level of exercise according to the past performance, and the robot can make comments to accept the blame for the instruction ability when the score is not so good. The effectiveness of communication model has been confirmed through some experiments [21].

6 Conclusion

This paper presents the smart device interlocked robot partners and their application for elderly care. As you can see from the above introduction, it can be discussed from various viewpoints. If the smart device interlocked robot partners in ISS have the functional flexibility and expandability, we can realize a comprehensive elderly care.

First, in this paper, we explained the essence of the functional flexibility corresponding to the system architecture. Next, we showed the lifelogging system, health promotion system, and exercise encouragement system in order. In the lifelogging system, the robot can be a pocket robot partner for outdoor behavior measurement and connect to the external smart devices and sensors to extend the functionality of indoor measurement. In the health promotion system, the robot partner can regularly provide quizzes for lifelong learning of healthcare. Moreover, the exercise encouragement system uses the virtual robot partner for the human motion analysis and utilizes the bodily architecture to demonstrate the exercise.

In the future works, we intend to build a platform to integrate the systems for measurement, data collection, motion analysis, persona update, persona analysis, and health promotion program. It can be called Computational Systems Care (CSC). Moreover, we will conduct long-time experiments in real environment, in order to examine the effectiveness of the developed system.

References

1. Clark, M., Lim, J., Tewolde, G., Kwon, J.: Affordable remote health monitoring system for the elderly using smart mobile device. Sens. Transducers **184**(1), 77–83 (2015)
2. Haluza, D., Jungwirth, D.: ICT and the future of health care: aspects of health promotion. Int. J. Med. Inform. **84**, 48–57 (2015)
3. Anguita, D., Ghio, A., Oneto, L., Parra, X., Reyes-Ortiz, J.L.: Human activity recognition on smartphones using a multiclass hardware-friendly support vector machine. In: Bravo, J., Hervás, R., Rodríguez, M. (eds.) IWAAL 2012. LNCS, vol. 7657, pp. 216–223. Springer, Heidelberg (2012)
4. Boulos, M.N.K., Wheeler, S., Tavares, C., Jones, R.: How smartphones are changing the face of mobile and participatory healthcare: an overview, with example from eCAALYX. Biomed. Eng. Online **10**(24) (2011)
5. Blaya, J., Fraser, H., Holt, B.: E-Health technologies show promise in developing countries. Health Aff. **29**(2), 244–251 (2010)
6. Street, R.L., Gold, W.R., Manning, T.: Health promotion and interactive technology: theoretical applications and future directions. Routledge, London (2013)
7. Kubota, N., Tang, D., Obo, T., Wakisaka, S.: Localization of human based on fuzzy spiking neural network in informationally structured space. In: IEEE World Congress on Computational Intelligence (WCCI), pp. 2209–2214 (2010)
8. Woo, J., Botzheim, J., Kubota, N.: Verbal conversation system for a socially embedded robot partner using emotional model. In: 24th IEEE International Symposium on Robot and Human Interactive Communication (RO-MAN), pp. 37–42 (2015)
9. Aztiria, A., Izaguirre, A., Augusto, J.C.: Learning patterns in ambient intelligence environments: a survey. Artif. Intell. Rev. **34**(1), 35–51 (2010)
10. Chan, M., Esteve, D., Escriba, C., Campo, E.: A review of smart homes—present state and future challenges. Comput. Methods Programs Biomed. **91**(1), 55–81 (2008)
11. Lutze, R., Waldhör, K.: A smartwatch software architecture for health hazard handling for elderly people. In: 2015 International Conference on Healthcare Informatics (ICHI), pp. 356–361 (2015)
12. Maglogiannis, I., Ioannou, C., Spyroglou, G., Tsanakas, P.: Fall detection using commodity smart watch and smart phone. In: Maglogiannis, I.L., Papadopoulos, H., Sioutas, S., Makris, C. (eds.) Artificial Intelligence Applications and Innovations, pp. 70–78. Springer, Heidelberg (2014)
13. Waldhör, K., Baldauf, R.: Recognizing drinking ADLs in real time using smartwatches and data mining. In: 6th RapidMiner Wisdom (2015)
14. Bieber, G., Fernholz, N., Gaerber, M.: Smart watches for home interaction services. In: Stephanidis, C. (ed.) HCII 2013, Part I. CCIS, vol. 373, pp. 293–297. Springer, Heidelberg (2013)

15. Boletsis, C., McCallum, S., Landmark, B.F.: The use of smartwatches for health monitoring in home-based dementia care. In: Zhou, J., Salvendy, G. (eds.) ITAP 2015. LNCS, vol. 9194, pp. 15–26. Springer, Heidelberg (2015)
16. Obo, T., Kakudi, H., Loo, C.K., Kubota, N.: Behavior pattern extraction based on growing neural networks for informationally structured space. In: IEEE Symposium Series on Computational Intelligence, pp. 138–144 (2015)
17. Kratzke, C., Cox, C.: Smartphone technology and apps: rapidly changing health promotion. Int. Electron. J. Health Educ. **15**, 72–82 (2012)
18. Giacometti, M., Gualano, M.R., Siliquini, R.: Smartphones and health promotion: a review of the evidence. J. Med. Syst. **38** (1) (2014)
19. Prochaska, J.O., Velicer, W.F.: The transtheoretical model of health behavior change. Am. J. Health Promot. **12**(1), 38–48 (1977)
20. Obo, T., Loo, C.K., Kubota, N.: Imitation learning for daily exercise support with robot partner. In: 24th IEEE International Symposium on Robot and Human Interactive Communication (RO-MAN), pp. 752–757 (2015)
21. Ono, S., Obo, T., Loo, C.K., Kubota, N.: Robot communication based on relational trust model. In: 41st Annual Conference of the IEEE Industrial Electronics Society (IECON), YF-025941 (2015)
22. Bandura, A.: Self-efficacy mechanism in human agency. Am. Psychol. **37**(2), 122–147 (1982)

Simplified Standing Function Evaluation System for Fall Prevention

Mami Sakata[1(✉)], Keisuke Shima[1], Koji Shimatani[2], and Hiroyuki Izumi[3]

[1] Yokohama National University,
79-5 Tokiwadai, Hodogaya-ku, Yokohama 240-8501, Japan
sakata-mami-dz@ynu.jp, shima@ynu.ac.jp
[2] Prefectural University of Hiroshima, 1-1 Gakuen-cho, Mihara 723-0053, Japan
shimatani@pu-hiroshima.ac.jp
[3] University of Occupational and Environmental Health,
1-1 Iseigaoka, Yahatanishi-ku, Kitakyushu 807-8555, Japan
izumi-h@med.uoeh-u.ac.jp

Abstract. This paper proposes a simplified standing-function evaluation system based on virtual light touch contact using a Wii balance board and a wearable VLTC device. In this system, a virtual partition by VLTC is first created from measurements of the subject's trunk and finger positions. Random on/off control of virtual forces from the partition enables evaluation of standing function based on the presence or absence of somatic sensory stimulation to the fingers. Evaluation and comparison experiments were conducted with 35 healthy male subjects. The results suggest that the proposed system allows to evaluate the standing function evaluation quantitavively.

Keywords: Virtual light touch contact · Evaluation system · Fall prevention

1 Introduction

Just over 25 % of Japan's population is at least 65 years old, and the number of elderly people in the country is projected to rise. Individuals in this older segment of the population tend to suffer more falls and other accidents than others, and are more commonly diagnosed with Parkinson's disease and other conditions. Falls are often fatal to elderly people; in 2010, they were the fourth most common issue resulting in a requirement for nursing care in this segment [1]. This situation gives rise to the need for a novel system to address the issue of falls in the older population.

Canes, walking frames and other physical support devices are commonly used by elderly people to prevent falls. However, as these solutions are designed specifically to support ambulation only, further consideration for both standing and walking functions is required. These solutions may be ineffective if the device cannot be chosen and applied optimally based on the individual's motor function

N. Kubota et al. (Eds.): ICIRA 2016, Part II, LNAI 9835, pp. 48–57, 2016.
DOI: 10.1007/978-3-319-43518-3_5

and walking ability. Against this background, a technique for precise evaluation of the wide-ranging motor functionality observed in standing and ambulation is needed toward the development of a comprehensive approach to fall prevention.

In previous work, Niino et al. investigated fall factors with focus on the situation and frequency of falls [2], and Kobayashi et al. worked on the development of a toe-function evaluation and training method for standing individuals to help prevent falls [3]. In addition, Shimada et al. performed a study to elucidate fall mechanisms via stimulation experiments involving stumbling, slipping and other fall factors [4]. However, these methods place a heavy physical and mental burden on the subject due to the numerous tests required for function evaluation. Accordingly, a new method for simple and quantitative evaluation without such burdens is needed.

Some researchers have previously sought to quantify standing function based on stimulation of the human senses. Examples of research in this area include studies to investigate three sensory modalities in human standing based on stimulation of individual senses (i.e., visual, vestibular and proprioceptive modalities) [5,6], and examination of a method to evaluate differences between people who had previously suffered falls and others who had not based on the application of disturbance stimuli [7]. The authors of this paper also previously proposed a novel standing-function evaluation method in which virtual light touch contact (VLTC) is utilized for somatosensory stimulation [8,9]. However, this evaluation method involves the use of force plates to determine the center of pressure (COP) while the subject is standing, and therefore requires simplification for clinical use. The results of evaluation regarding the virtual partition also indicated that the VLTC method has the potential for simplification to support acceleration sensor usage [10].

This paper outlines a novel system for simple and quantitative standing-function evaluation based on the previously proposed evaluation method, and discusses the results of evaluation experiments performed to verify its efficacy and applicability.

In the rest of this paper, Sect. 2 outlines the proposed method, Sect. 3 describes experimental validation of the proposed system, and Sect. 4 presents the conclusion and a discussion of future work.

2   Evaluation System Using Virtual Light Touch Contact

Postural sway is known to be mitigated in humans through light fingertip contact (less than 1 N) with physical objects such as cloth curtains (Jeka [11], 1994). A variety of studies on this phenomenon, which is known as LTC (light touch contact) have been reported. On the other hand, the authors previously proposed a novel concept in standing support called VLTC, which provides the effects of LTC virtually without the use of a physical partition [12,13]. Figure 1 gives an overview of the VLTC system. VLTC can be used to control stimulation of the subject's fingertips by changing the parameters of the virtual partition.

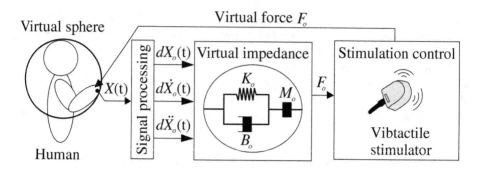

Fig. 1. Structure of virtual light touch contact (VLTC) [8]

The authors also previously proposed a novel standing-function evaluation system in which VLTC is used for somatosensory stimulation [8], and additionally reported on a VLTC-based wearable system incorporating an acceleration sensor (referred to as acceleration-based VLTC, or AVLTC) with which postural sway can be mitigated [10]. This paper proposes a system for simple and quantitative standing-function evaluation using AVLTC (Fig. 2). In the proposed system, a Microsoft Kinect sensor and Wii Balance Board are utilized to determine the center of mass (COM) and the center of pressure (COP). A vibrotactile stimulator is utilized for virtual force feedback.

During evaluation, the subjects were asked to retain a standing posture on a Wii Balance Board under contact with a surrounding virtual partition created by the system. The partition's properties were then changed from a force application state (in which forces produced by touching the partition are fed back to the subject) to a no-force state (in which no feedback is provided if the fingertips come into contact with the wall while the subject is in a standing state). The details of the system are further outlined below.

2.1   Estimation of Virtual Forces

In VLTC [8], M_o, B_o and K_o represent the virtual inertia, viscosity and stiffness matrices associated with the sphere, respectively. the virtual force $F_o(t)$ exerted from the virtual partitionto the fingertip is defined as follows:

$$\boldsymbol{F}_o(t) = -(M_o d\ddot{\boldsymbol{X}}_o(t) + B_o d\dot{\boldsymbol{X}}_o(t) + K_o d\boldsymbol{X}_o(t)) \tag{1}$$

$d\dot{\boldsymbol{X}}_o(t)$ is the normal vector from the surface of the internal sphere to the finger. The authors previously demonstrated that the effects of VLTC can be provided simply using an acceleration sensor attached to one of the user's fingertips [10]. In the proposed system, virtual forces to be fed to the user are approximately calculated from the user's fingertip acceleration $a(t) \in \Re^3$.

$$\boldsymbol{F}'_o(t) = -M_o d\ddot{\boldsymbol{X}}_o(t) \tag{2}$$

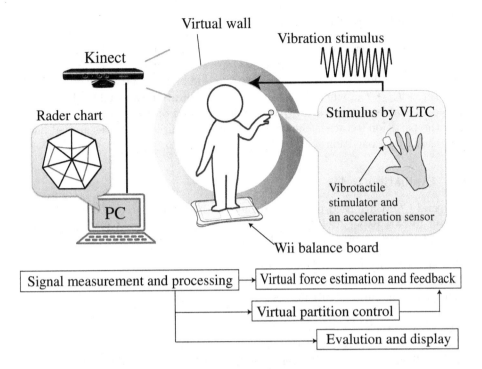

Fig. 2. Overview of the proposed simplified evaluations system

$$d\ddot{\boldsymbol{X}}_o(t) \cong \begin{cases} |\boldsymbol{a}(t)| - a_{\text{th}} & (|\boldsymbol{a}(t)| > a_{\text{th}}) \\ 0 & (|\boldsymbol{a}(t)| \leq a_{\text{th}}) \end{cases} \tag{3}$$

Here, the threshold a_{th} for acceleration is set for determination of contact with the partition.

2.2 Force Feedback Using Vibrotactile Stimulation

For the VLTC system, an approximate method for applying virtual force using a compact vibrotactile stimulator containing a small voice coil motor was developed. Tactile stimulators are devices that generate tactile sensations against the skin of the user, and can be controlled to apply vibration of any amplitude up to a maximum value of A_{max} (vibrational frequency: f_l Hz). Here, f_l is a constant value, and signal amplitudes $A_m(t)$ are computed according to the norm of the virtual force vector $\boldsymbol{F}'_o(t)$ as follows:

$$A_m(t) = \begin{cases} k|\boldsymbol{F}'_o(t)| & (k|\boldsymbol{F}'_o(t)| < A_{\text{max}}) \\ A_{\text{max}} & (k|\boldsymbol{F}'_o(t)| \geq A_{\text{max}}) \end{cases} \tag{4}$$

where k is the constant gain and $A_m(t)$ is set as $A_m(t) = A_{\text{max}}$ if $k|\boldsymbol{F}'_o(t)|$ exceeds the maximum amplitude. This allows the user to generally sense virtual forces through contact with the virtual partition in the air.

2.3 Virtual Force Control

In the proposed method, parameters M_o, B_o and K_o are defined as having the state $P_n = \{M_o^n, B_o^n, K_o^n\}$, and N states are set in advance. $P_n(n = 1, \ldots, N)$ are changed randomly $(N-1)$ times over a period of T seconds. The duration of Pn is set as $T_n \geq T_{th}$, where T_{th} is the minimum duration threshold. The system is used to evaluate postural sway from stimuli based on the difference observed after the partition's state is changed. By way of example, no forces are fed back from the virtual partition if $P_n = \{0.0, 0.0, 0.0\}$.

2.4 Evaluation Index

COP and COM are used to evaluate the standing-function computing of some indices. In this paper, five COP evaluation indices and four COM indices are defined based on previous observations as follows:

- Total trajectory length of COP: L_{COP}
- Rectangle area of COP: S_{rect}
- Outer peripheral area of COP: S_{peri}
- Average velocity of COP: $\overline{v_{COP}}$
- Average vector of COP: $\overline{L}$
- Total trajectory length of COM: L_{COM}
- Index differences observed after a change in the virtual partition's state
- Average velocity of COM: $\overline{v_{COMx}}, \overline{v_{COMy}}, \overline{v_{COMz}}$

L_{COP}, S_{rect} and S_{peri} are utilized to evaluate the instability of body sway. Specifically, the indices of the COP area indicated significant differences between people who had previously suffered falls and those who had not when the subjects retained a standing state and when vibration stimulation was applied to their feet [7]. $\overline{v_{COP}}$ is the total trajectory length per unit time. As larger values are known to indicate greater postural sway in standing [14], this metric is utilized to evaluate postural instability. The average vector of COP is often used for assessment of vestibular disorders and other conditions [14]. COM changes linearly with ankle joint torque [15]. Accordingly, the system uses the average vector of COP to evaluate the effect of a virtual partition for mitigation of body sway. In evaluation, disturbances are applied to the user by changing the state of the virtual partition in the proposed system, which then evaluates postural sway based on stimuli from the difference observed after the partition state change. Thus, d_{10-12} are differences observed in the rectangle area after a virtual-partition state change. The index values are standardized with reference to a control subject group of young people. Standardization indices (I_l) are calculated using the mean (μ^l) and standard deviation (σ^l) of standard-subject data for each index value.

$$I_l = \frac{d_l - \mu^l}{\sigma^l} \tag{5}$$

I_l can be used to evaluate differences between control subject group data and measurement data.

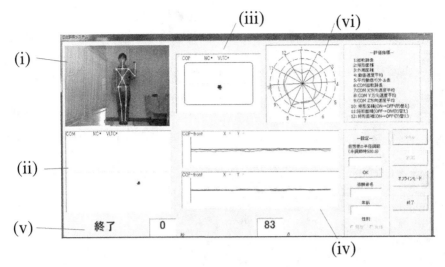

(i) Image of subject   (ii) COM   (iii) COP

(iv) Time series waveforms of COP   (v) Countdown   (vi) Radar chart

Fig. 3. System screen

2.5   Graphic Display

Measured data and index values were plotted as shown in Fig. 3, in which (i) shows an image of a subject from Kinect. During evaluation, the system displays the results of skeleton tracking, the location of each joint and the position of COM. The trajectories of COM from Kinect and COP are displayed in Fig. 3 (ii) and (iii), respectively. The time-series variations of COP and COM in the X and Y directions are plotted in (iv), and (v) shows the duration of the state of virtual force (P_n). Figure 3 (vi) shows 12 evaluation indices in radar chart form based on measurement over a duration of T seconds. The system's functions include footage playback, measurement data display and scalability, thereby supporting its use as an electronic medical chart.

3   Experiments

3.1   Method

Experiments were conducted to verify the proposed method with 35 healthy male subjects aged from 20 to 60. The experimental setup (Fig. 4) involved the use of Kinect (Microsoft; sampling frequency: 30 Hz) and a Wii Balance Board (Nintendo Co., Ltd.; sampling frequency: 100 Hz). Preliminary experiments were first conducted to verify the effectiveness of AVLTC with five healthy subjects. The subject was asked to retain a tandem (heel-to-toe) stance with his eyes

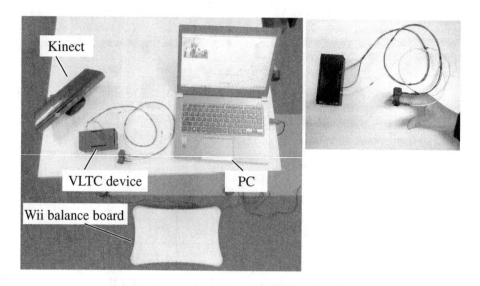

Fig. 4. Prototype of the simplified standing function evaluation system

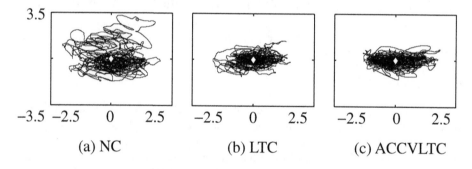

(a) NC   (b) LTC   (c) ACCVLTC

Fig. 5. Examples of COP measured during each task [10]

d_1 : Total trajectory length of COP d_4 : Average velocity of COP
d_2 : Rectangle area of COP d_5 : Average vector of COP
d_3 : Outer peripheral area of COP d_6-d_8: Index difference before and after
                                          the change of Virtual partition

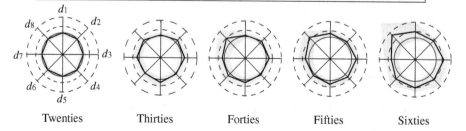

Twenties   Thirties   Forties   Fifties   Sixties

Fig. 6. Results of index value standardization

closed and were instructed to perform three tasks: (i) no contact (NC) anywhere; (ii) LTC with a piece of paper hanging to the subject's right; and (iii) acceleration-based VLTC (AVLTC). The location of the paper used for LTC was also adjusted to the optimal position for contact as indicated by the subject. The number of trials was 10, and the duration of each task exceeded 60 s.

Next, standing-function evaluation experiments were conducted. OpenNI (PrimeSense) was used as a driver for Kinect and NiTE was used as middleware in order to extract body link information on the subject and compute COM. The subjects were asked to retain a standing posture with their eyes closed and their legs together for 60 s ($T = 60$). The parameter of virtual forces (P_n) was $n = 4$, and each value was set $P_1 = P_3 = \{1.0, 10.0, 50.0\}$, $P_2 = P_4 = \{0.0, 0.0, 0.0\}$. The investigation was approved by the local ethics committee, and written informed consent was obtained from all subjects.

3.2 Verification of VLTC with an Acceleration Sensor

Examples of COP data obtained during each task over periods of 60 s are shown in Fig. 5. These results indicate that postural sway can be mitigated by AVLTC as well as by LTC. This results show that the effectiveness of the AVLTC as same as the previous work [10]. Here, it is known that postural sway can be mitigated by touching a hard surface object lightly besides a paper [11]. Thus the result show that AVLTC can mitigate postural sway as well as touching the object. Thus, the authors concluded that AVLTC can be used as a tool to mitigate body sway, and that the standing state of the user can be changed by switching the stimulation state from AVLTC to NC.

3.3 Evaluation of Standing Function

The total index values of each virtual partition state were standardized with reference to data from six subjects in their twenties. The results are shown in Fig. 6. Here, since the indices from d_9 to d_{12} have large variation for each subject, the figure plots the incides of COP (d_1–d_8) measured from a Wii Balance Board were used for evaluation in this paper. Radar charts made using standardized index results are shown in Fig. 6, where the solid lines indicate the average values for young subjects and the dotted lines show double and quintuple the standard deviation (2SD, 5SD). It can be seen that the radar for subjects in their twenties is almost perfectly octagonal, and that their values are close to the mean. The standardization values increase with the subjects' age. It is particularly notable that the rectangle areas are larger after the virtual-partition change (d_6–d_8) in line with the average ages. This indicates that changing the virtual forces for a subject in a standing state significantly facilitates understanding of standing function based on the subject's age. The system can thus be used to control stability in a standing state based on virtual-partition changes, and enables both visual and index-based evaluation of standing function. It also supports quantification of standing function with radar charts. As this study focused exclusively on subjects who had not suffered falls previously, consideration of the relationships between this group and those who have suffered falls, as well as evaluation of the proposed method, remain incomplete. Accordingly, the authors plan to conduct a more detailed evaluation and further consider body sway observed before and after the state of the virtual partition is changed.

4 Conclusion

This paper proposes a novel standing-function evaluation method involving the use of VLTC for somatosensory stimulation. In the proposed method, forces from a virtual partition created around the subject are changed randomly via on/off control. This enables quantitative evaluation of standing function based on changes in somatosensory stimulation of the fingertips using the VLTC system. Experiments were conducted to evaluate posture controllability for individuals in a standing state using external stimulation based on whether the subject touched the virtual wall or not. The following results were observed:

- Acceleration-based VLTC (AVLTC) can be used to mitigate the body sway of a subject in a standing state.
- The standing state of the subject can be controlled using force control from the virtual partition based on AVLTC.
- Standing function can be quantitatively evaluated using a simplified system with 8 indices and a radar chart.

The results suggest that the proposed method is applicable to consideration of the relationship between standing function and age. In future work, the approach described here will be tested with a larger subject group. Alternative indices potentially applicable to the clarification of standing-function differences between healthy individuals and people with various conditions will also be examined.

Acknowledgments. The authors would like to cordially acknowledge and express appreciation to Dr. T. Yokota and Mr. S. Yamashita for his assistance in the implementation of this research. This work was supported by JSPS Kakenhi Grant Number 25870242 and 16H05914.

References

1. Ministry of Health, Labour, Welfare, Comprehensive Survey of Living Conditions: 2013 in Japanese (2013)
2. Niino, N., Tsuzuku, S., Ando, F., Shimokata, H.: Frequencies and circumstances offalls in the National Institute for Longevity Sciences, Longitudinal Study of Aging(NILS-LSA). J. Epidemiol **10**(1), S90–S94 (2000)
3. Kobayashi, R., et al.: Effects of toe grasp training for the aged on spontaneous postural sway. J. Phys. Ther. Sci. **11**(1), 31–34 (1999)
4. Shimada, H., et al.: New intervention program for preventing falls among frail elderly people: the effects of perturbed walking exercise using a bilateral separated treadmill. Am. J. Phys. Med. Rehabil. **83**(7), 493–499 (2004)

5. Day, B.L., Severac Cauquil, A., Bartolomei, L., Pastor, M.A., Lyon, I.N.: Human body-segment tilts induced by galvanic stimulation: a vestibularly driven balance protection mechanism. J. Physiol. **500**(3), 661–672 (1997)
6. Ivanenko, Y.P., Grasso, R., Lacquaniti, F.: Effect of gaze on postural responses to neck proprioceptive and vestibular stimulation in humans. J. Physiol. **519**(1), 301–314 (1999)
7. Izumi, K., Makimoto, K., Kato, M., Hiramatsu, T.: Prospective study of fall risk assessment among institutionalized elderly in Japan. Nurs. Health Sci. **4**(4), 141–147 (2002)
8. Sakata, M., Shima, K., Shimatani, K.: A standing-function evaluation method based on virtual light touch contact. In: Proceedings of the IEEE Biomedical Circuits and Systems Conference (BioCAS), pp. 153–156, 22–24 October 2015
9. Sakata, M., Shima, K., Shimatani, K.: Novel concept for evaluation of standing function based on virtual light touch contact. In: Proceedings of 37th Annual International Conference of the IEEE Engineering in Medicine and Biology Society (EMBC2015), 12.LB1 (2015)
10. Sakata, M., Shima, K., Shimatani, K.: Investigation of relationships between body sway and fingertip vibrotactile stimulation based on virtual light touch contact. In: Proceedings of International Symposium on Artificial Life and Robotics (AROB 2016), pp. 280–283 (2016)
11. Jeka, J.J.: Light touch contact as a balance aid. Phys. Ther. **77**(5), 476–487 (1997)
12. Shima, K., Shimatani, K., Sugie, A., Kurita, Y., Kohno, R., Tsuji, T.: Virtual light touch contact: a novel concept for mitigation of body sway. In: Proceedings of 7th International Symposium on Medical Information and Communication Technology (ISMICT), pp. 108–111 (2013)
13. Shima, K., Shimatani, K., Kurita, Y., Tsuji, T.: Novel method for mitigation of body sway and preliminary results for tandem standing. In: Proceedings of 2013 International Symposium on Micro-NanoMechatronics and Human Science (MHS), pp. 71–73 (2013)
14. Kouzaki, M., Masani, K.: Reduced postural sway during quiet standing by light touch is due to finger tactile feedback but not mechanical support. Exp. Brain Res. **157**(3), 153–158 (2008)
15. Winter, D.A., Patla, A.E., Prince, F., Ishac, M., Gielo-Perczak, K.: Stiffness control of balance in quiet standing. J. Neurophysiol. **80**(3), 1211–1221 (1998)

Psychographic Profiling for Use in Assistive Social Robots for the Elderly

Kayo Sakamoto[✉], Sue-Ann Lee Ching Fern, Lin Han,
and Joo Chuan Tong

Institute of High Performance Computing,
Singapore Agency, Science, Technology, and Research,
1 Fusionopolis Way, #16-16 Connexis, Singapore 138632, Singapore
{sakamotok,leesa,linhan,tongjc}@ihpc.a-star.edu.sg

Abstract. In this study, we proposed two psychographic profiling models that can be applied in Socially Assistive Robots (SARs) for the elderly. We designed those models based on Self-Determination Theory (SDT), and those models can acquire user model about the domain of needs and the motivation styles respectively. We demonstrated the model can be implemented in interactive dialogue systems, and the psychographic profiling helps SARs deliver more meaningful and personalised service to the elderlies.

Keywords: Social robot · User model · Profiling · Psychographics · Ageing · Motivation · Psychological needs

1 Introduction

Due to a combination of declining fertility rates and decreased mortality, countries all across the world are facing the headwinds of a rapidly ageing population. According to the UN's "World Population Ageing 2013" report, not only is the global population of elderly aged above 60 expected to almost triple by 2050, the already low old-age support ratios in many countries will likely continue to plummet, adding fiscal and social pressures on society.

In keeping with the extended life expectancies, incidence of non-communicable diseases, disabilities and cognitive declines will increase, especially given the projected global trends of obesity, diabetes, and neurodegenerative diseases such as Alzheimer's disease. To overcome this massive challenge, the healthcare sector has to boost the number of trained healthcare workers, increase development of therapeutic programmes extensively, and raise accessibility and quality of healthcare services. It is already apparent that countries need even more trained healthcare workers than previously projected [1].

A promising solution to the problem of trained manpower shortages is emerging: assistive technology, in the form of socially assistive robots (SAR) that provide assistances to people through social interaction. SAR can be equipped with the capabilities to support the elderly across a range of needs: help with cognitive tasks, daily living activities, enhancement of psychological well-being, and increase of social participation [2]. If successfully deployed, SAR increases the likelihood that the elderly

N. Kubota et al. (Eds.): ICIRA 2016, Part II, LNAI 9835, pp. 58–67, 2016.
DOI: 10.1007/978-3-319-43518-3_6

can continue living at home, even with physical or cognitive disabilities, and stave off the need for institutionalized care. It relieves the economic, and social pressures, and the physical burden of their caregivers.

The elderly population reflects the heterogeneity of the human population as a whole, in that it presents a wide spectrum of needs depending on their socio-economic status, physical, mental and cognitive health, and personality. To design and deploy SAR best suited to each individual's needs, it is essential to profile and understand the elderly users, and prioritize their problems and needs.

Although this can be achieved through interviews with social workers and allied healthcare professionals, the number of elderly who can benefit from this service will be limited by the schedule and workload of interviewers, and by the geographical area the interviewers are active in. Further, manual screening of the elderly will result in non-integrated pockets of data residing on different and possibly non-compatible systems. The lack of a common database of the psychographic and functional needs data of the elderly will make it difficult to conduct insights analysis, and to follow up with an evaluation of the efficacy of the assistance given.

Automating this task via a dialog system will allow for effective outreach to potentially the entire elderly community. The dialog system can be embedded as chatterbots in officially approved websites, or in strategically located nation-wide self-help kiosks. By removing human-based errors and biases from the decision making process, such systems will also help ensure consistent and objective assessment of the elderly's needs.

In this study, we will introduce two psychographic profiling models (user modeling) that can be applied in SARs for the elderly. In Sect. 2, we review background theories for the psychographic profiling models specifically focussed on an integration of the self-determination theory (SDT). In Sect. 3, we describe the details of our profiling models. In Sect. 4, we demonstrate cases of profiling applying the proposed models to several elderly personas in Singapore.

2 Theoretical Background

Our proposed psychographic framework that underpins such a dialog system is based on an integration of the self-determination theory (SDT) [3] and Maslow's Hierarchy of Needs [4]. The model forms the basis for assessing and prioritizing needs for elderly individuals.

Briefly, SDT describes the concept that an individual's psychological health relies on the fulfilment of these basic psychological needs: autonomy (feeling behaviors are self-chosen and self-controlled), relatedness (feeling of being connected with others) and competence (feeling of effective and competent). Numerous studies have attested that fulfillment of different psychological needs is conducive to well-being.

Maslow's model establishes the idea of a hierarchy of needs ranging from physiological (most fundamental) to safety, belonging, esteem and self-actualization (least fundamental). In theory, an individual instinctively prioritizes the fulfilment of more fundamental (lower) needs before thinking of higher needs. However, real world data showing that people can still gain subjective well-being by fulfilment of social needs (belonging and esteem), even when their physiological needs are not met, suggests that this hierarchy can be fluid [5].

Ageing is inherently a difficult time in an individual's life; as the leading risk factor for most human diseases, the elderly is often besieged on multiple fronts by physical, psychological and cognitive frailty. Accordingly, successful ageing has been proposed to encompass low incidence of disease and related disability, maintenance of good physical and cognitive health, and active participation in living [6]. Because of diverse physical conditions, life experiences, long established habits and preferences, and expectancy for the rest of the life, elderly tend to possess a combination of needs and varying levels of motivation at any one time. Physical, psychological and cognitive health are so closely intertwined that any disruption to one often results in downstream effects on the other two. Understanding where the elderly lie within the continuum of motivation and needs will be essential to help them achieve subjective well-being and the overall goal of successful ageing.

In-depth ethnographic research in Singapore has led to broad categorization of elderly individuals into the 3 SDT motivators [7]. Elderly whose behavior is related to a lack of Autonomy are prone to passiveness and belief in their lack of control over their lives due to external factors. This in turn leads to reduced expression of the other motivations of Relatedness and Competence. For example, elderly who are sick or immobile might rely solely on the initiative of friends and relatives for help with medication, activities of daily living and socialization, without which they might simply pass time alone by watching television at home.

In fact, many of the elderly are socially isolated, putting them at higher risk of physical and cognitive decline [8]. Such elderly might be estranged from their family, or living alone after the death of their spouse; they have trouble making social connections to other people for various reasons. There is a vicious cycle of declining health and social isolation. Successfully helping these elderly build their social network in an organic and self-sustainable manner requires an understanding of individuals' reasons for isolation.

Lack of Competence often manifests in risk-adverse behavior in the elderly and an unwillingness to seek new challenges in their lives. Retirees might be reluctant to seek reemployment or engage in social activities out of fear that they are not up to date with societal and technological advancements. Idleness and boredom sets in, leading to dissatisfaction at the lack of a meaning or purpose in their lives. Successful ageing for these elderly becomes a problem as they are liable to be less active in daily life and at higher risk for depression and other health problems [9].

Seeing those findings and mechanisms of ageing issues, it can be said that psychological profiling of the elderly becomes essential to understand the primary motivation driving their needs and satisfy the most urgent need at the point of the psychological assessment. Subsequent assessments can be conducted if and when necessary until the elderly achieve their best outcome for life. In this way, government, community and individual resources can be most effectively utilized to raise their quality of life.

3 Psychographic User Models and Acquisition Processes

We will demonstrate two user models and the acquisition processes for which the different aspects of Self-determination theory (SDT) are applied. These user models can be used for SAR for the elderly.

3.1 Domain of Needs

This user model is based on the three basic psychological needs in STD [3]: *Competence, Relatedness*, and *Autonomy*. According to the original theory, these needs can be salient and relevant across universal contexts and situations.

We apply this framework to specify the user's most important domain of service needs. In elderly care, we can assume specific contexts in which specific psychological needs are more likely to be salient, and certain types of services are relevant to the context:

For Competence, we assume "aspiration" contexts, in which an elderly person seeks opportunities to express their talents or skills, to desire to feel that they are useful and capable. An example of elderly service for this context is a volunteer recruitment agency that provides senior volunteers with opportunities to serve the community with their talent.

For Relatedness, we assume "socialization" contexts, in which an elderly person seeks interaction with other people to feel emotionally connected to or cared for, as well as to care for others. A befriending service for elderly will be relevant to this context.

For Autonomy, we assume "care" contexts, in which an elderly person receives aid or assistance so that they can gain control over things they are struggling at. For example, community healthcare and financial assistance are the relevant ones to this context.

Figure 1 is a proposed acquisition process for this user model. A user model is obtained through interaction with the potential assistive social robot. The process begins with obtaining non-psychographic data, which includes items shown in Table 1.

Profiles are used to specify a potential domain of needs. We set heuristic rules for this specification such as IF Income = "Low" OR Health Status = "Struggling", THEN

Table 1. Non-psychographic information used in the acquisition process of "Domain of Needs" user model.

Item	Levels
Age	"Young", "Middle", "Old"
Living arrangement	"Alone", "With Spouse", "With Family"
Health status	"Healthy", "Under Control", "Struggling"
Income	"High", "Middle", "Low"
Education	"High", "Middle", "Low"
(Past) occupation	"Professional/Manager", "Service/Administration worker", "Manual Labour/unemployed"

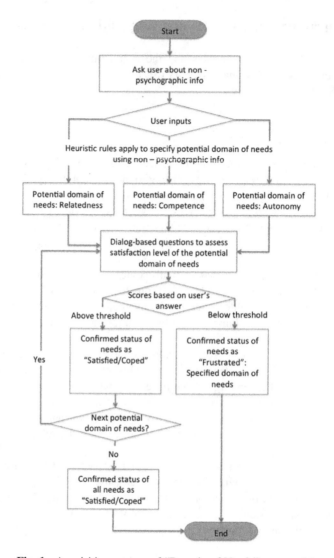

Fig. 1. Acquisition process of "Domain of Needs" user model.

Potential Domain = "Autonomy". After specifying a potential domain of needs, the process assesses the elderly client's satisfaction level for the potential domain. This assessment is conducted by dialogues based questions asking clients to rate their basic psychological needs on a scale (e.g., [10]). We then score user's answers to those dialog-based questions. The satisfaction level ("Satisfied/Coped" vs. "Frustrated") is determined by the score.

When the potential domain of need turns "Frustrated", a need is confirmed as the in-need domain for the user. Otherwise, the process continues to evaluate other potential needs until an in-need domain is determined.

In cases where the satisfaction levels of all three needs are "Satisfied/Coped", the system classifies the user as someone with a high aspiration, suitable for activities with higher level purpose such as the self-actualisation [4].

3.2 Motivation Style

3.2.1 User Model and the Usage

Another sub-theory under SDT provides the framework for our user model of motivation style. This profile enhances the performance of assistive social robots by encouraging, teaching, and managing elderly users' behavior such as exercising, dietary practice, or participation to social activities. While many versions of the motivation framework have been proposed over time (e.g., [3, 11, 12]), in this study, we apply a simpler version of the framework of motivational states for better usability (e.g., [13]): *intrinsically motivated, motivated by self-determined extrinsic motivation, motivated by nonself-determined extrinsic motivation*, and *amotivated*.

When a user is intrinsically motivated, they are simply rewarded by the internal experience of fulfilling in the behaviour (e.g., interest). The robot or system does not have to make extra effort to drive the user to the behavior; it only needs to inform them how to perform an existing behavior in better ways. Take, as an example, an exercise recommendation to a user who already believes that exercise is fun: this exercising behaviour is voluntarily performed without any external rewards or punishments. The user is also highly likely to exercise again.

When a user is driven by self-determined extrinsic motivation, behaviour is prompted by external factors (e.g., fame, money, etc.). Such users determine that performing a given behavior brings an external reward that they value. Because they agree on the contingency, a user in this profile can have strong motivation override other emotional impulsiveness. Such behavior is also likely to be voluntarily performed, and robust. For example, an elderly user who regularly exercises may not enjoy exercising itself, but they believe they can stay charming by exercising. She consciously choses regular exercise despite some impulse to skip it. For this profile, effective engagement involves the systems or robots enhancing the chance that users perform better in the behavior (e.g., by providing social feedback), or reassuring the user that the expected self-image is recognized through repeating the relevant behavior.

When a user is motivated by nonself-determined extrinsic motivation, the behaviour is not as robust or sustainable. Besides motivation, behavior must also regulated by external factors. In other words, behaviour is not chosen by the user voluntarily, and the reward/punishment is also not under the user's control but others. For example, an elderly exercises because he was told to do so by a doctor. In this case, they are indifferent to potential rewards, such as better health, but are forced by the doctor's potential criticism. An effective engagement for potential systems or robots is to introduce and induce alternative rewards that can be regulated by the user themselves.

When a user is unmotivated, they cannot associate specific behaviors with intrinsic or extrinsic reward/punishment. For example, an elderly user who refuses to exercise because they do not see any point to do so. This profile will be challenging to engage. Systems or robots need to start by enable the awareness of the user about the potential relevance of specific behaviours.

3.2.2   User Model Acquisition

Figure 2 shows the acquisition process through dialog-based interaction between a user and the conversational social robot. This dialog based questionnaire is designed from scales which measure self-regulatory state (e.g., [14, 15]). The basic structure of these dialogs is to get information about the user and the target behavior, why they do it (or don't). The question "Why do/don't you do it?" is answered by selecting answer options given in the dialog such as "Is it because you are supposed to do it?", "Is it because you chose it for your own good?", or "Is it because it is just fun?". This process, which has a simple and clear structure, can be used to specify a profile. However, in order to enhance reliability of the profile, this probably question needs to be asked and answered using varied dialogues for the same option.

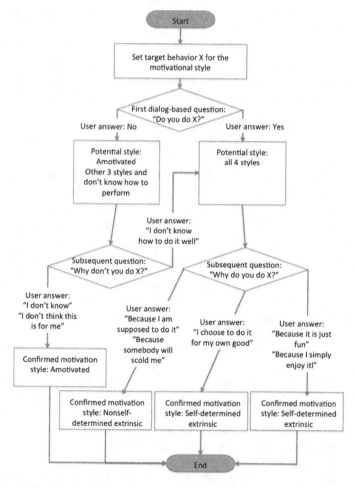

Fig. 2. Acquisition process of "Motivation Style" user model.

4 Case Illustrations

We illustrate actual profiling cases using the proposed dialog-based user modeling frameworks. We selected and simplified 3 personas out of 12 representative Singapore's elderly personas, which was developed by Singapore Design Council based on fieldwork [7]. Table 2 shows the 3 selected personas' basic profiles. Figure 3 is a dialogue illustrating someone with "The Enclosed" persona being profiled through the Domain of needs model. From his non-psychographic profiles, the model first specifies his potential domain of needs as Relatedness, and subsequent interaction confirms he is frustrated in this specific domain. Figure 4 illustrates dialogue with persons with personas "The Dependent" and "The Denier" being profiled through the Motivation Style model. This reveals that even though their profiles are similar–their health status is "struggling with illness", and their domain of needs is probably Autonomy–they have distinct motivation styles. The Motivation Style model will suggest different engagement approaches for these users. In other words, a combination of the two models will produce a stepwise identification such that interaction strategies could be narrowed down first and then differentiated at a personal level.

Table 2. Personas of Singapore's elderlies. Sourced from [7] and modified.

The Enclosed	
With the passing of his spouse and friends, he suddenly finds himself alone in an uncertain situation. Due to his introversion, he rarely meets other positive elderly role models that can contradict his beliefs.	
Age: Old	Living arrangement: Alone
Health: Healthy	Education: Middle
Income: Low	Occupation: Service/Administration worker

The Dependent	
She requires assistance to achieve her daily needs and relies on other to provide support. Her life is shaped around her caregiver's schedule, as she is unable to do most tasks without her caregiver.	
Age: Old	Living arrangement: With family
Health: Struggling	Education: Middle
Income: Middle	Occupation: Unemployed

The Denier	
She perceives the implications of her illness as serious, but accepting it requires lifestyle changes. Therefore, she finds it easier to deny her illness. She feels overwhelmed and lacks directions in her life.	
Age: Old	Living arrangement: Alone
Health: Struggling	Education: Low
Income: Low	Occupation: Manual Labour

(From user's non-psychological profiles, Age: Old, Health: Healthy, and Living Arrangement: Alone, the model specifies his potential domain of needs is Relatedness domain.)

[Model] Hi [Enclosed user]! Do you have anybody you often contact with?
[User] No, I'm totally alone. Even my children seldom visit me.
[Model] Oh, don't you feel connected with people who   care   for   you, and for whom you care?
[User] Not at all.
[Model] Well, you mean you have impression that they dislike you?
[User] I don't think so, but...
[Model] Then, you mean relationship you have with them is superficial?
[User] Somewhat, I think so.
[Model] All right! Don't worry, I can understand you...

(Model confirmed that the user's domain of needs is Relatedness. He is Frustrated in this domain.)

Fig. 3. Supposed dialogue with the domain of needs model.

(Target behavior: Regular walking)

[Model] Hello [Dependent user]! Do you take a walk?
[User 1] Sometimes. With my daughter.
[Model] Ok... Why do you take a walk? Is it because it is fun for you?
[User 1] Well.... I don't know.
[Model] Then is it because you believe it is good for your health?
[User 1] Uh... Maybe...
[Model] Is it because somebody tells you to take a walk?
[User 1] Yes. My daughter says that's for my good.
[Model] All right! I understand you...

(Model confirmed her motivation style is Nonself-determined extrinsic)

[Model] Hello [Denial user]! Do you take a walk?
[User 2] No. Not at all.
[Model] Ok... Why don't you take a walk? Is it because you are tired?
[User 2] No.
[Model] Hmmmm... Then is it because you cannot see the point?
[User 2] Ya lah. I'm too old, waiting to die already. No point.
[Model] All right! I understand you...

(Model confirmed her motivation style is Unmotivated)

Fig. 4. Two dialogues with the motivation style model.

5 Conclusion

In this study, we proposed two psychographic profiling models that can be applied in SARs for the elderly. We demonstrated the model can be implemented in interactive dialogue systems, and the profiling result informs SARs with better understanding of the elderly users in terms of their area of needs, and motivational style. Hence, psychographic profiling helps SARs serve more meaningful personalised service according to elderly's individual psychological needs.

References

1. Fujisawa, R., Columbo, F.: The long-term care workforce: overview and strategies to adapt supply to a growing demand. OECD Health Working Papers No. 44, OECD Publishing, Paris (2009)
2. Broekens, J., Heerink, M., Rosendal, H.: Assistive social robots in elderly care: a review. Gerontechnology 8(2), 94–103 (2009)
3. Deci, E.L., Ryan, R.M.: Intrinsic Motivation and Self-determination in Human Behaviour. Springer, New York (1985)
4. Maslow, A.H.: A theory of human motivation. Psych. Rev. 50(4), 370–396 (1943)
5. Tay, L., Diener, E.: Needs and subjective well-being around the world. J. Pers. Soc. Psychol. 101(2), 354–365 (2011)
6. Rowe, J.W., Kahn, R.L.: Successful aging. Gerontogist 37(4), 433–440 (1998)
7. Design Singapore Council Asian Insights and Design Innovation unit. https://www. designsingapore.org/for_enterprises/Design_Research/DesignforAgeingGracefully.aspx. Accessed 14 Apr 2016
8. Hawton, A., Green, C., Dickens, A.P., Richards, S.H., Taylor, R.S., Edwards, R., Greaves, C.J., Campbell, J.L.: The impact of social isolation on the health status and health-related quality of life of older people. Qual. Life Res. 20(1), 1–11 (2011)
9. Moreno, R.L., Godoy-Izquierdo, D., Perez, M.L.V., Garcia, A.P., Serrano, F.A., Garcia, J.F. G.: Multidimensional psychosocial profiles in the elderly and happiness: a cluster-based identification. Aging Ment. Health 18(4), 489–503 (2014)
10. Campbell, R., Vansteenkiste, M., Delesie, L., Mariman, A., Soenens, B., Tobback, E., Van der Kaap-Deeder, J., Vogelaers, D.: Examining the role of psychological need satisfaction in sleep: a Self-Determination Theory perspective. Personality Individ. Differ. 77, 199–204 (2015)
11. Ryan, R.M., Deci, E.L.: Self-determination theory and the facilitation of intrinsic motivation, social development, and well-being. Am. Psychol. 55(1), 68–78 (2000)
12. Deci, E.L., Ryan, R.M.: Facilitating optimal motivation and psychological well-being across life's domains. Can. Psychol. 49(1), 14–23 (2008)
13. Vallerand, R.J., O'Connor, B.P., Hamel, M.: Motivation in later life: theory and assessment. Int. J. Aging Hum. Dev. 41(3), 221–238 (1995)
14. Bober, S., Grolnick, W.S.: Motivational factors related to differences in self-schemas. Motiv. Emot. 19, 307–327 (1995)
15. Levesque, C.S., Williams, G.C., Elliot, D., Pickering, M.A., Bodenhamer, B., Finley, P.J.: Validating the theoretical structure of the treatment self-regulation questionnaire (TSRQ) across three different health behaviors. Health Educ. Res. 21, 691–702 (2007)

Human Support Robotics

Interface Design Proposal of Card-Type Programming Tool

Naoki Tetsumura[1]($\boxtimes$), Toru Oshima[1], Ken'ichi Koyanagi[1], Hiroyuki Masuta[1],
Tatsuo Motoyoshi[1], and Hiroshi Kawakami[2]

[1] Department of Intelligent System Design Engineering,
Toyama Prefectural University, 5180 Kurokawa, Imizu, Toyama, Japan
t654010@st.pu-toyama.ac.jp,
{oshima,koyanagi,masuta,motoyosh}@pu-toyama.ac.jp
[2] Unit of Design, Kyoto University, 1 Yoshidanakaadachicho, Sakyoku, Kyoto, Japan
kawakami@design.kyoto-u.ac.jp

Abstract. We developed a card-type programming tool "Pro-Tan" as
an education tool for beginners in programming. This tool was designed
to be usable without any instruction on its operations. In the experiment,
we did not tell the users how to use any of the tools. The subjects were
instructed to create a program by themselves. We found problems of
Pro-Tan's design in the experiment. In this paper, we report the system
configuration of Pro-Tan and problem of Pro-Tan's design.

Keywords: Programming education tool · Tangible interface · Beginners · Active learning · RFID system

1 Introduction

Recently, studying programming is of increasing importance on an international
level [1]. In Japan, programming education is introduced into the education system at the secondary-education level [2]. The tool to learn programming is a
PC. Making a source code according to the rules of a computer language is
one method of programming. Some programming tools which it is easy to learn
programming use a graphical, mouse-driven programming environment [3,4].
Programming education using the PC have problem that using such programming tools compels users to become accustomed to operating a keyboard and a
mouse. Adopting a tangible user interface (TUI) [5] is one of the solutions for
these problems [6,7,9]. We have been developing a card-type programming tool
called Pro-Tan which uses an RFID system to detect the structure of a program.
Pro-Tan is designed to be usable by users on their own accord; therefore, it is
considered that users can use the tool even if no explanation is given on how
to use it. We conducted an experiment to evaluate Pro-Tan. The purpose of the
experiment is to evaluate whether user can create a program without instructions. In the experiment, we did not tell the users how to use any of the tools.
The subjects were instructed to create a program by themselves. We understood

N. Kubota et al. (Eds.): ICIRA 2016, Part II, LNAI 9835, pp. 71–77, 2016.
DOI: 10.1007/978-3-319-43518-3_7

problems of Pro-Tan's design in the experiment. The remainder of this paper is organized as follows. Pro-Tan is described in Sect. 2. Section 3 presents programming tool using TUI. Section 4 describes the problem of Pro-Tan's design. Section 5 presents a discussion and draws a conclusion.

2 Pro-Tan

2.1 System Concept

Figure 1 shows the concept of Pro-Tan. Pro-Tan consists of a controlled object, programming cards, a program panel, and a PC for data transfer. A user can make a program for controlling a controlled object by attaching programming cards onto the program panel. The programming cards corresponds to various programming elements such as the controlled object's motion, while loop and sequential branch the sensors for controlling a controlled object. Users control the object by positioning the programming cards on the program panel. With this system, users are able to create three types of programs: sequential OPEN, sequential LOOP, and conditional branch. Pro-Tan is intended to teach the fundamental programming structures (sequences, branches and loops) used in beginner-level programming. We designed the tool so that it is easy for beginners to operate. Thus, we think that younger students can use and enjoy the system while learning programming. We consider that the methodology of this system will help users learn fundamental programming concepts.

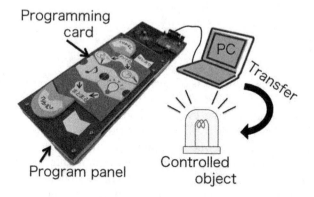

Fig. 1. Overview of the system.

2.2 Programming Cards

There are four types of programming cards: Motion, Timer, IF, and LOOP. Each programming card has a RFID tag to identify the card type, and has a stainless steel sheet attached to the obverse face for attachment to the panel. The programming cards are made from expanded polystyrene board. All the cards fit each other if users create a correct program. Also, each programming card has an illustration on the surface that enables understanding of the function of each card.

Motion Cards. Figure 2 shows examples of Motion cards: Beep buzzer and Light up. Motion cards give performance instructions to the controlled object. The illustrations on the top surface of the card indicate the object's action. As for the example of the object's action, turn on light and make a sound.

Timer Cards. The Timer cards set the movement duration of the controlled object on a scale of one to four seconds. Users have to place a Timer card to the right of a Motion card. The illustrations on the top surface of the cards have timepiece shapes.

IF Cards. The IF cards correspond to the branch functions of each sensor on the controlled object. The IF cards consist of a Separate card and a Joint card. By using IF cards, the user can create conditional branch programs. The illustrations on the top surface of the cards have action for conditional branch. Letters to express a meaning is written on the top surface of the cards.

LOOP Cards. The LOOP cards correspond to the while loop functions of a program. The controlled object repeats the movements of cards between a pair of LOOP cards. The LOOP cards consist of a Begin card and an End card. The illustrations on the top surface of the cards have arrow. Key word to express a meaning is written on the top surface of the cards.

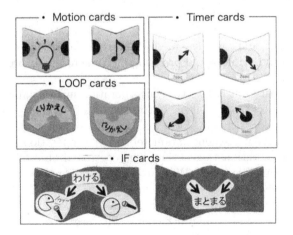

Fig. 2. Programming cards.

2.3   Program Panel

Figure 3 show the program panel system. The program panel has 10 cells to which programming cards can be attached, and is equipped with RFID readers under each cell. Key word to understand place where firstly put a card is written on the top line of panel. The card type information is obtained from the RFID tags of the programming cards and is transmitted by the PC. After that, the information is conveyed to the controlled object.

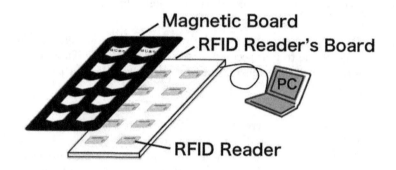

Fig. 3. Program panel system.

3   Programming Tool Using TUI

E-Block [6], Algo Block [7] and Tern [9] are able to be classified into one of programming education tool using TUI. E-Block consists of several types of blocks using wireless technology for teaching the basics of programming to younger learners. Several blocks contains an LED, a battery, an infrared receiver, an infrared transmitter, and a wireless module. The interface that users touch equips precision instruments. However, a precision instrument might break when users create programs.

Algo Block consists of blocks that incorporate a microcomputer and a PC, and controls an object on a monitor. In this system, users control the submarine by arranging the blocks. Each block has an input switch that can be used to change the submarine movement parameters, such as operational distance, rotational angle, and conditional execution. The controlled object is not an actual machine such as a mobile robot. We consider that it is preferable to control an actual machine when learning programming.

Tern consists of programming bricks made from wood with AR marker, a mobile robot, a PC, and a digital camera. The user assembles bricks in the manner of a jigsaw puzzle into the structure of a sequential program to control a mobile robot. The placement of assembled bricks are analyzed by a digital camera. Then, information of the placement bricks is transported to the mobile robot by digital camera. The creation of the program area is limited because of the use of a digital camera.

We consider that those tool need explanation of how to use. Pro-Tan is designed so that user can create a program without instructions. The tool is easy to make trial and error for users. We consider that users can understand the function of program elements by not instructions but operation experiences of the tool.

4   Problem of Pro-Tan's Design

We conducted an experiment for evaluation of Pro-Tan. Pro-Tan is designed so that users can easily understand how it works. The purpose of the experiment is to evaluate whether a user can create a program without instructions. In the experiment, we did not tell the users how to use any of the tools. The subjects were instructed to create a program by themselves. The experimental subjects were five male and five female participants between 20 and 24 years of age, and all were inexperienced in programming. We evaluated how well subjects completed three programming tasks. The contents of tasks are as follows:

Task 1 Light up the LED for one second.
Task 2 Light up the LED for one second. Then, beep the buzzer for two seconds.
        Repeat the sequence.
Task 3 Beep the buzzer if a sound is heard and otherwise light up the LED.

Four subjects completed the Task 1 at the first trial. Other four subjects completed the Task 2 without any trials. There are no subjects who completed the Task 3 at the first trial. We instructed how to use Pro-Tan to four subjects at the Task 3. Six subjects were able to complete all tasks without requiring any explanations of how to use Pro-Tan.

We observed mistakes made by subjects in the experiment. Also, we discovered problems of Pro-Tan's design in the experiment. Typical types of mistake are shown in Fig. 4. The most typical mistake was that the Motion card and the Timer card were arranged in tandem. The design of the card should be modified

Fig. 4. Typical types of mistakes.

in order to be understood that the Timer cards have to be placed to the right side of the Motion cards. Other type of mistake is using only one LOOP card, not using a set of LOOP cards. We consider that the design of the LOOP cards should be modified in order to be understood that the LOOP cards are used in pairs. Another type of mistake is only using one IF card, not using a set of IF cards. We consider that the design of the IF cards should be modified in order to be understood that cards are used in pairs.

After the experiment, we conducted a questionnaire survey to all subjects. Subjects pointed out problems of the interface design that they did not understand how to use the cards. We asked subjects to comment about easiness of the usage of programming cards.

- Some subjects could not understand that the cards have to be positioned from the top to the bottom of the panel.
- Some subjects could not understand how to use a set of IF cards and LOOP cards.
- Some subjects could not understand the place of the card which should be placed at the beginning position of the panel.

We got opinions like the consideration from mistakes. The design of the panel should be modified in order that it be readily understood by users that the cards have to be attached to the top line of panel.

5   Discussion and Conclusion

In this paper, we introduced the Pro-Tan, which is used as an education tool for beginners in programming. The tool uses an RFID system to detect the structure of a program and is designed so that user can create a program without instructions. The tool is easy to make trial and error for users. We consider that users can understand the function of program elements by not instructions but operation experiences of the tool. To identify the characteristic of the Pro-Tan, we conducted the experiment which programming beginners make programs using Pro-Tan. In the experiment, all subject were able to create a program using Pro-Tan, but some subjects got wrong operations. After the experiment, we conducted a questionnaire survey, and clarified the type of the programming card which is hard to imagine the usage of the card intuitively, that is, we should improve the shape and the interface design of the card to be understood that IF cards and LOOP cards should be used as a set. Furthermore, we will verify the learning effect of the Pro-Tan.

References

1. http://www.soumu.go.jp/maincontent/000361430.pdf
2. http://www.mext.go.jp/a_menu/shotou/new-cs/news/080216/003.pdf
3. https://scratch.mit.edu

4. Harada, Y., Potter, R.: Fuzzy rewriting soft program semantics for children. In: Human Centric Computing Languages and Environments, vol. 1, pp. 39–46. IEEE (2003)
5. Ishi, H., Ulmer, B.: Tangible bit: toward seamless interface between people, bits and atoms. In: Proceedings of Conference on Human Factors in Computing Systems, CHI 1997, pp. 234–241 (1997)
6. Wang, D., Zhang, Y., Chen, S.: E-block: a tangible programming tool with graphical blocks. In: Proceedings of Mathematical Problems in Engineering, MPE 2013, vol. 2013 (2013). Article ID 598547
7. Hideyuki, S., Hiroshi, K.: Interaction level support for collaborative learning: Algo-Block an open programming language. In: Proceedings of the First International Conference on Computer Support for Collaborative Learning, CSCL 1995, pp. 349–355 (1995)
8. Kai, K., Kimuro, Y., Sakaguchi, Y., Yasuura, H.: An educational method of computer principles for elementary and junior high school students. In: Proceedings of the Information Processing Society of Japan, IPSJ 2002, pp. 1121–1131 (2002)
9. Horn, M.S., Solovey, E.T., Crouser, R.J., Jacob, R.J.K.: Comparing the use of tangible and graphical programming languages for informal science education. In: Proceedings of the SIGCHI Conference on Human Factors in Computing Systems, CHI 2009, pp. 975–984 (2009)

Direct Perception of Easily Visible Information for Unknown Object Grasping

Hiroyuki Masuta[1(✉)], Tatsuo Motoyoshi[1], Ken'ichi Koyanagi[1],
Toru Oshima[1], and Hun-ok Lim[2]

[1] Toyama Prefectural University, Imizu, Japan
{masuta,motoyosh,koyanagi,oshima}@pu-toyama.ac.jp
[2] Kanagawa University, Yokohama, Japan
holim@kanagawa-u.ac.jp

Abstract. This paper discusses the estimation of the easily visible information for deciding action of unknown object grasping without all of predefined knowledge in a real situation. To estimate the easily visible information, active perception is an important concept. A conventional active perception method estimates an optimum sensing position to measure the accurate object information. However, a robot has to move all around the object to measure the accurate data beforehand. This is not realistic for a robot operation. On the other hand, a robot wants to know the most easily visible sensing position in the neighborhood of a current position to make a decision as soon as possible from a viewpoint of a robot action. As a first step, we propose an estimation of an easily visible information using a point cloud data of an object which is limited data. Now, we apply the unknown object detection method based on plane detection. And, we propose the graspability and the easily visible information estimation method from the point cloud data of the object. We verify the proposed method in a multiple object environment. The effectiveness of our proposal are discussed for grasping an unknown object.

Keywords: 3D-range camera · Unknown object detection · Service robot · Robot vision

1 Introduction

Recently, various types of intelligent robots have been developed to work in our life for the next human society. An intelligent robot is expected to work in a familiar environment such as homes and commercial and public facilities [1]. An intelligent robot should remain in action in order to fulfill specific tasks in real time, even in an unknown environment. A robot perception to work in unknown environment is the important element.

To perceive an unknown object, depth sensors are usually used to perceive 3D space in an environment. A depth sensor delivers a 3D point cloud including distance and optical image. There are many previous studies that a robot recognizes an environment using 3D point cloud data [2]. For example, 3D template matching is a typical method to detect a specific object [3]. A robot can obtain a specific object, if accurate 3D templates are provided by operators. However, it is impossible to provide all of 3D

N. Kubota et al. (Eds.): ICIRA 2016, Part II, LNAI 9835, pp. 78–89, 2016.
DOI: 10.1007/978-3-319-43518-3_8

template data by operator in real unknown environment. Therefore, 3D template matching is difficult to apply when a robot is the first contact with the real unknown environment. On the other hand, the sample consensus and segmentation methods are popular methods to detect an unknown object from point cloud data using RGB-D sensor. Random sample consensus (RANSAC) and Randomized Hough Transform (RHT) are a popular segmentation method for 3D space processing [4, 5]. However, these methods require the accurate measurement data, namely the methods are combined with a 3D reconstruction method. To make an accurate 3D map by 3D reconstruction method, a robot should move a sensor in order to get invisible side. As a related study of moving sensor, there are active perception [6–9]. Active perception estimates an optimum sensing position to measure the accurate object information. Actually, a robot may moves all around the object to measure the accurate data beforehand, but it is not realistic for a robot operation. Moreover, the estimated optimum position by active perception method is not always suitable position to grasp an object. A robot wants to know the most easily visible sensing position in the neighborhood of a current position to make a decision as soon as possible from a viewpoint of a robot action. Therefore, a robot should decide an action by using limited information, for example a measured data is only one side.

We have proposed a perceptual system for an unknown object detection from point cloud data without predefined knowledge in order to execute a flexible action. To detect an unknown object, we have proposed a plane detection based object detection method [10]. Our plane detection method can detect small planes accurately and computational cost is less than previous methods drastically. An object is detected by combining planes. Our research target is that a robot makes a decision of an action using a point cloud data of an object. However, the accuracy of a point cloud data of an object is influenced a great deal by a sensing position. A robot should move a sensor in order to get easily visible information for decision making of an action. An easily visible information is affected by the characteristic of the device, the sensing position, environmental condition and others. It is desirable to move a sensor to most easily visible information for grasping an unknown object. So, a robot should recognize a current visible state considering with a grasping action. As a first step, we propose an estimation of an easily visible information using a point cloud data of an object which is limited data.

This paper is organized as follows. Section 2 explains our robot system to measure the unknown environment. Section 3 explains the plane detection method and object extraction from 3D point cloud data. Section 4 shows the index parameter for the easily visible value in the facing situation, Sect. 5 shows experimental results. Finally, Sect. 6 concludes this paper.

2 The Intelligent Robot System for a Clearing Table

2.1 Outline of the Service Robot System

Figure 1 shows an overview of our intelligent robot system for clearing a table. The intelligent robot system consists of a service robot, an interactive robot, and an intelligent space server. In this system, the service robot can clear the table automatically by collaboration with the interaction robot and the intelligent space server [11]. The object information is provided beforehand by a human operator using RFID tags. The service robot cannot pick up an object, if there are unknown objects which are unregistered objects.

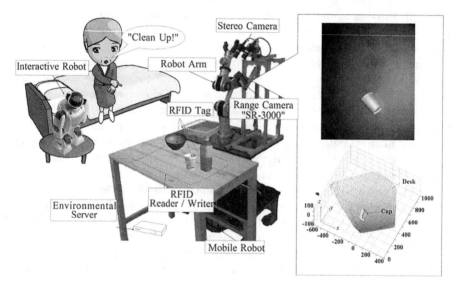

Fig. 1. The robot system overview for clearing table

The robot arm "Katana" is installed on the service robot. Katana has 5 degree-of-freedom (5DOF) structure and 6 motors, including a gripper. Katana movement is restricted to approach from the front side. The robot action is restricted by the embodiment of robotic structure.

To detect an unknown object, we installed a 3D depth sensor "SR-3000" to the service robot.

2.2   Point Cloud Data of Object on the Table

We use SR-3000 as the 3D depth sensor to measure the environment. SR-3000 is a Time-Of-Flight range image sensor made by the MESA imaging AG that can measure 3D distance and 2D luminance. The range of this sensor is a maximum of 7.5 [m][12], and the output signal is a quarter common intermediate format (QCIF 176 * 144 pixels). Therefore, this sensor can measure a 25,344-directional distance information simultaneously. The maximum frame rate of this sensor is 50 [fps], but we set a fixed frame rate of 20 [fps] because the sensing stability and real time sensing are considered.

SR-3000 is located at the right side of the robot arm. SR-3000 can be moved pan-tilt and vertical direction. The moving range of pan-tilt direction is −30 to 30° from the front direction. The range of vertical movement is 600 [mm] height from the base of the robot arm.

Figure 1 right side shows a snapshot of the distance and amplitude image of a paper cup. We define the measuring point of the image pixel as the w-h axis of the coordinate. Moreover, the measuring point of the real world is defined as the x-y-z axis of the robot coordinate, where the origin is the base of the robot arm.

A cup is recognized from a 3D distance data by observation. However, the point cloud data of the round surface is not complete. One side of the cup is not measured by occlusion, and the surface boundary is smoothed with a smoothing filter. Furthermore, the edge of the measured field is curved because of lens distortion. Therefore, all the aforementioned conditions should be considered in order to perceive unexpected situations. To detect an object to grasp from this sensing data, we focus on the plane detection method.

3   Unknown Object Detection Based on Plane Detection

This section explains an unknown object extraction method to extract a point cloud group of an unknown object. We have proposed the Simplified Plane Detection based SEgmentation of a point cloud into object's clusters (SPD-SE) [10]. Figure 2 shows the processing flow of the proposed method with SPD-SE. The SPD-SE consists of the particle swarm optimization (PSO) [13], the simplified plane detection (SPD) with region growing (RG) and the plane integrating object extraction.

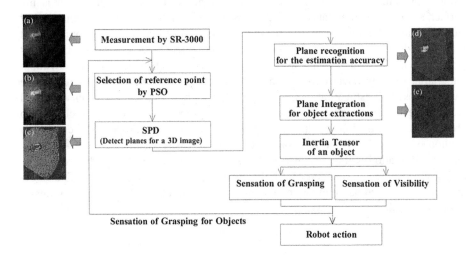

Fig. 2. Flowchart of an unknown object detection method

The PSO for the seeded RG is the first step to detect planes which is composed of object from point cloud data. Figure 2(a), (b) and (c) shows a snapshot of SR-3000 image, distribution of particles of PSO and a result of simplified plane detection, respectively. The particles of PSO are updated to gather around detected small planes. As a result, small planes are detected preferentially [10].

However, it is difficult to detect a stable plane because of sensing noise. We applied the estimation accuracy parameter to consider the changing time series. As a result, a stable plane detection result is obtained like Fig. 2(d). There are many planes on the object position in the case of a round object and multi-object situation. Therefore, the planes are integrated to one group as an object based on geometric invariance of each plane, like Fig. 2(e). The largest detected plane as a table plane is eliminated from the detected plane. After that, we calculated the parameter of the sensation of grasping which means the graspability in the facing situation [15].

The experimental result of unknown object detection was shown that the SPD-SE can reduce the computational cost drastically compared with the Point Cloud Library, and can extract unknown objects accurately [10]. Moreover, the decision of grasping is made simply [15]. However, a detected object has to detect by a proper balance of multiple planes to obtain better estimation results of graspability.

In this paper, we propose the sensation of visibility as follows.

4   Estimation of Graspability and Easily Visible Information

4.1   Inertia Tensor for a Detected Object

A gravity position of an extracted object is calculated from set of point cloud data. However, it is difficult to recognize accurate object position, posture, shape and size without predefined knowledge. Because, the point cloud data of the object is not perfect, such as the point data on back side is not exist and the edge of an object is not sharp by smoothing filter. The accurate object size, posture and shape are not important if a person decides to grasp an object [14]. There is a direct perception concept. A direct perception insists that relevant information for action is perceived directly without integrating some physical information such as size, posture and so on. Therefore, we have proposed an estimation method of graspability based on an inertia tensor model [15]. However, a detected object has to detect by a proper balance of multiple planes to obtain better estimation results of graspability. Hence, we propose an estimation method of an easily visible information to move to easily visible position. An easily visible information will be changed by the specification of the sensing device and the facing environmental situation like light intensity. Basically, an easily visible infor-mation is calculated based on an inertia tensor of a point cloud of an object. An inertia tensor is calculated as follows.

$$
\mathbf{T}_{P_k} =
\begin{bmatrix}
I_{x'} & I_{y'x'} & I_{z'x'} \\
I_{x'y'} & I_{y'} & I_{z'y'} \\
I_{x'z'} & I_{y'z'} & I_{z'}
\end{bmatrix}
$$

$$
=
\begin{bmatrix}
\frac{1}{N}\sum_{n=1}^{N} m(y_n'^2 + z_n'^2) & \frac{1}{N}\sum_{n=1}^{N}(-m \cdot y_n' \cdot x_n') & \frac{1}{N}\sum_{n=1}^{N}(-m \cdot z_n' \cdot x_n') \\
\frac{1}{N}\sum_{n=1}^{N}(-m \cdot x_n' \cdot y_n') & \frac{1}{N}\sum_{n=1}^{N} m(x_n'^2 + z_n'^2) & \frac{1}{N}\sum_{n=1}^{N}(-m \cdot z_n' \cdot y_n') \\
\frac{1}{N}\sum_{n=1}^{N}(-m \cdot x_n' \cdot z_n') & \frac{1}{N}\sum_{n=1}^{N}(-m \cdot y_n' \cdot z_n') & \frac{1}{N}\sum_{n=1}^{N} m(x_n'^2 + y_n'^2)
\end{bmatrix}
\tag{1}
$$

where $\mathbf{T}_{P_k}$ is an inertia tensor of the k-th object. x_n', y_n' and z_n' are measure points transformed into an object gravity center coordinate shown by Fig. 3. m is a unit weight, N is the number of the point cloud on an object.

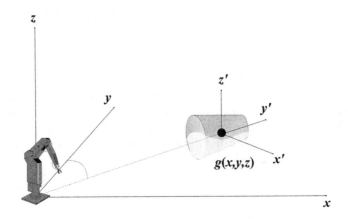

Fig. 3. Control architecture for a service robot based on human interaction

4.2 The Estimation of Graspability

The value of principal moment of inertia indicates the size of an extracted object. For example, the value of principal moment of inertia around the y-axis means the invariant of a width size in robot view. Especially, the inertia moment around y' and z' axis should be smaller than the object size by grasping of robot hand, because the robot arm can only reach from the front side. Therefore, the combination of principal moments of inertia is explained the invariant of grasping. The estimation value of graspability is calculated as follows based on simplified fuzzy inference [16].

$$
\mu_i = \prod_{j=1}^{n} \mu_{A_{i,j}}(x_j)
\tag{2}
$$

$$S = \frac{\sum\limits_{i=1}^{r} \mu_i \cdot y_i}{\sum\limits_{i=1}^{r} \mu_i} \tag{3}$$

where μ_i is the fitness value of the i-th rule applying membership function $A_{i,j}$ of input x_j; r is the number of rules. S is the output, which shows the estimation value of graspability. The membership function shows Fig. 4, SM, MM and LM are shown small, middle and large of inertia moment, respectively. The size of membership function is decided in consideration of the embodiment of the robot which considered a size of gripper. Table 1 shows the rules of fuzzy inference. The consequent part function means the sensation of grasping given by singleton. There are 19 rules, 1.0 means a most likely graspability situation. In this case, the rule is decided by heuristics. The rules consist of the principal moment of inertia, because the movement of Katana can not change the coordinate of x'_n, y'_n and z'_n. Fuzzy inference calculates a value considered a size and posture. Finally, the invariant perceptual information value A is calculated as Eq. (4) which consider the distance between object position and robot position.

$$A(\mathbf{I}) = S(\mathbf{I}) \cdot e^{-\frac{(L-c)^2}{2\zeta^2}} \tag{4}$$

where ζ is standard deviation, L is the distance between robot position and the gravity position of the object, c is a constant value. A robot can decide to grasp an object when the invariant has high value. Therefore, the invariant affords the possibility of action to a robot directly without inference from object's property such as size, posture and shape.

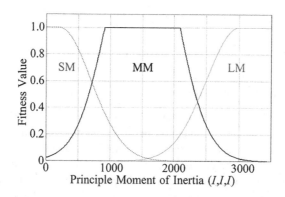

Fig. 4. The membership function for each principal moment

Table 1. Fuzzy rules for estimating graspability

Ix'				SM				
Iy'	SM			MM			LM		
Iz'	SM	MM	LM	SM	MM	LM	SM	MM	LM
yf	0.0		0.0		0.0		0.0		0.0

Ix'				MM				
Iy'	SM			MM			LM		
Iz'	SM	MM	LM	SM	MM	LM	SM	MM	LM
yf		0.0		0.0	1.0	0.7		0.2	0.1

Ix'				LM				
Iy'	SM			MM			LM		
Iz'	SM	MM	LM	SM	MM	LM	SM	MM	LM
yf	0.0		0.0		0.7	1.0	0.0	1.0	0.0

4.3 Estimation of Easily Visible Information

The relationship with principal moment of inertia only considers the position relation and variance of a point cloud of an object. The estimation value of graspability is not suitable for the facing situation in case of a detected plane is in the absence of a part of the surface or there are some differences of measurement data by noisy input. A robot wants to know the appropriateness of a detected object. The acceptable result of an object detection is integrated multiple planes are well balanced. Therefore, we focus on the principal axis of inertia. An inertia tensor $\mathbf{T}_{P_k}$ is transformed into different axis $\mathbf{T}'_{P_k}$ based on rotation matrix. A rotation matrix $\mathbf{R}$ is shown based on Euler Angle as follows.

$$\mathbf{R} = \begin{bmatrix} \cos\phi\cos\theta\cos\psi - \sin\phi\sin\psi & -\cos\phi\cos\theta\sin\psi - \sin\phi\sin\psi & \cos\phi\sin\theta \\ \sin\phi\cos\theta\cos\psi + \cos\phi\sin\psi & \cos\phi\cos\theta\cos\psi + \cos\phi\cos\psi & \sin\phi\sin\theta \\ -\sin\theta\cos\psi & \sin\theta\sin\psi & \cos\theta \end{bmatrix}$$

$$(5)$$

$$\mathbf{T}'_{P_k} = \mathbf{R}\mathbf{T}_{P_k}\mathbf{R}^{-1} = \begin{bmatrix} I'_{x'} & I'_{y'x'} & I'_{z'x'} \\ I'_{x'y'} & I'_{y'} & I'_{z'y'} \\ I'_{x'z'} & I'_{y'z'} & I'_{z'} \end{bmatrix}$$

$$(6)$$

where ϕ, θ and ψ are a rotation angle of z-axis, x-axis and z-axis, respectively. There is the principal axis of inertia has $\mathbf{R}$ which product of inertia is 0. Therefore, we search $\mathbf{R}$ to get the smallest product of inertia using the following equation.

$$F = \left(I'_{y'x'}\right)^2 + \left(I'_{z'x'}\right)^2 + \left(I'_{z'y'}\right)^2$$

$$(7)$$

$$V = \underset{\phi,\theta,\psi}{\arg\min}\, F(\phi, \theta, \psi)$$

$$(8)$$

where V is an estimation value of easily visible information. This paper is a first step to estimate an easily visible information. So, we apply an all combination search which ϕ, θ and ψ are changed by the 1° step from –45° to 45°, respectively. If the V is nearly equal 0, the distribution of point cloud data of the detected object is well-balanced around the center of gravity of an object. Therefore, the V shows a degree of easily visible information.

5 Experiment

5.1 Experimental Condition

Experiments were conducted using the proposed method under certain conditions. The experimental condition is shown by Fig. 5(a). There are 6 various types of objects on the table, and the robot position is fixed. The position of SR-3000 is fixed on the 500 [mm] height to fit objects into the sensor range. We perform the offline calculation using acquired measurement data beforehand, because the all combination search required a lot of computational cost. As a result of object detection, Fig. 5(b) upper side is a snapshot of SR-3000 image, Fig. 5(b) lower side is the object detection result. Figure 6 shows the point cloud data that have overwritten the object detection result. Each detected object is named A to F, respectively. In this specific environment, the SPD-SE can stably detect the planes and the objects. The object A, B, C and D are detected multiple planes, but the object E and F are detected only one plane.

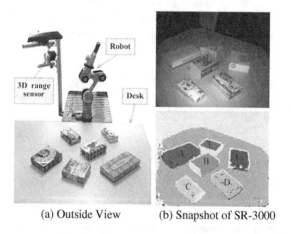

(a) Outside View   (b) Snapshot of SR-3000

Fig. 5. The experimental condition

5.2 Experimental Results

The experimental result of a graspability value is shown in Fig. 6. The bars on each object are the output of A in Eq. (4). In the object B has high value of the output A. This means that the posture of object B is appropriate to grasp and the distance to the object is also appropriate. As a result, a robot can make decision to grasp an object

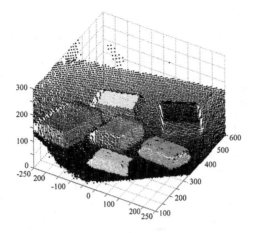

Fig. 6. The result of object detection shown by 3D view

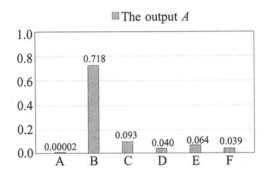

Fig. 7. The experimental result of the graspability

based on the graspability value, directly. The grasping of an object has an over 80 % success rate when the graspability value is higher than 0.70 in preliminary experiments. Thus, the graspability value can afford a specific action, even if the measurement is imperfect data (Fig. 8).

Next, the experimental result of an easily visible information is shown in Fig. 7. The point cloud data of a detected object is fluctuated with sensing noise, therefore Fig. 7 shows average value, standard deviation, maximum value and minimum value. The vertical axis shows the output V. The object A, B and C have smaller value, on the other hand the object D, E and F have larger value. The object A, B and C are detected multiple planes which are top and side plane, so these results are understandable. However, the object D has large value even if multiple planes are detected. The reasons why the side planes of the object D looks flat plane, but it's actually curved because the position of object D is on the edge of sensor angle. Hence, the proposed method of an easily visible information expresses not only the accuracy of the plane detection, but also a feature of the sensing device. A robot may decide grasping action accurately, if it gets the low easily visible information and high graspability value.

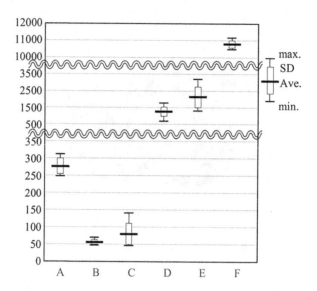

Fig. 8. The experimental result of the easily visible information

6 Conclusion

This paper discusses the estimation of the easily visible information for deciding action of unknown object grasping without all of predefined knowledge in a real situation.

To estimate the easily visible information, active perception is an important concept. A conventional active perception method estimates an optimum sensing position to measure the accurate object information. However, a robot has to moves all around the object to measure the accurate data beforehand. This is not realistic for a robot operation. Moreover, the estimated optimum position by active perception is not always suitable position to grasp an object. A robot wants to know the most easily visible sensing position in the neighborhood of a current position to make a decision as soon as possible from a viewpoint of a robot action. As a first step, we propose an estimation of an easily visible information using a point cloud data of an object which is limited data.

Previously, we have proposed a perceptual system which is composed of an online processable object extraction method "SPD-SE" based on plane detection. The SPD-SE is collaborated with PSO-based SPD with RG and the object detection method based on geometric invariance. By using SPD-SE, the point cloud data of the object are obtained in real time. Now, we propose the graspability and the easily visible information estimation method from the point cloud data of the object. The graspability is explained by the principal moment of the inertia tensor and fuzzy inference, the easily visible information is explained by the product of inertia of the inertia tensor.

We verify the proposed method in a multiple object environment. As a result, the graspability value can afford a specific action, even if the measurement is imperfect data. And the easily visible information is shown not only the accuracy of the plane detection, but also a feature of the sensing device.

As future work, we will apply the high-speed optimization algorithm to search the easily visible information. Moreover, we will propose the easily visible position inference method to obtain better perception.

References

1. Mitsunaga, N., Miyashita, Z., Shinozawa, K., Miyashita, T., Ishiguro, H., Hagita, N.: What makes people accept a robot in a social environment. In: International Conference on Intelligent Robots and Systems, pp. 3336–3343 (2008)
2. Zhang, Z.: Microsoft kinect sensor and its effect. IEEE Multimedia **19**(2), 4–10 (2012)
3. Jurie, F., Dhome, M.: Real time 3D template matching. In: Proceedings of the 2001 IEEE Computer Society Conference on Computer Vision and Pattern Recognition, vol. 1, pp. I-791–I-796 (2001)
4. Fischler, M.A., Bolles, R.C.: Random sample consensus: a paradigm for model fitting with applications to image analysis and automated cartography. Commun. ACM **24**(6), 381–395 (1981)
5. Xu, L., Oja, E.: Randomized hough transform (RHT): basic mechanisms, algorithms, and computational complexities. CVGIP Image Underst. **57**(2), 131–154 (1993)
6. Lyudmila, M., Tine, L., Herman, B., Klaas, G., Joris, D.S.: A comparison of decision making criteria and optimization methods for active robotic sensing. In: Dimov, I.T., Lirkov, I., Margenov, S., Zlatev, Z. (eds.) NMA 2002. LNCS, vol. 2542, pp. 316–324. Springer, Heidelberg (2003)
7. Gupta, S., Arbelaez, P., Girshick, R., Malik, J.: Inferring 3D object pose in RGB-D images. Comput. Vis. Pattern Recogn. (2015). arXiv preprint arXiv:1502.04652
8. Alenya, G., Foix, S., Torras, C.: ToF cameras for active vision in robotics. Sens. Actuators, A **218**, 10–22 (2014)
9. Atanasov, N., Sankaran, B., Pappas, G.J., Daniilidis, K.: Nonmyopic view planning for active object classification and pose estimation. IEEE Trans. Rob. **30**(5), 1078–1090 (2014)
10. Masuta, H., Makino, S., Lim, H., Motoyoshi, T., Koyanagi, K., Oshima, T.: Unknown object extraction for robot partner using depth sensor. In: 4th International Conference on Informatics, Electronics &Vision (2015)
11. Masuta, H., Hiwada, E., Kubota, N.: Control architecture for human friendly robots based on interacting with human. In: Jeschke, S., Liu, H., Schilberg, D. (eds.) ICIRA 2011, Part II. LNCS, vol. 7102, pp. 210–219. Springer, Heidelberg (2011)
12. Oggier, T., Lehmann, M., Kaufmannn, R., Schweizer, M., Richter, M., Metzler, P., Lang, G., Lustenberger, F., Blanc, N.: An all-solid-state optical range camera for 3D-real-time imaging with sub-centimeter depth-resolution (SwissRanger). Proc. SPIE **5249**, 534–545 (2003)
13. Kennedy, J., Eberhart, R.C.: Particle swarm optimization. In: Proceedings of the 1995 IEEE International Conference on Neural Networks, vol. 4, pp. 1942–1948 (1995)
14. Carello, C., Turvey, M.T.: Rotational invariants and dynamic touch. In: Touch, Representation and Blindness, pp. 27–66 (2000)
15. Masuta, H., Lim, H., Motoyoshi, T., Koyanagi, K., Oshima, T.: Invariant perception for grasping an unknown object using 3D depth sensor. In: 2015 IEEE Symposium Series on Computational Intelligence, pp. 122–129 (2015)
16. Takagi, T., Sugeno, M.: Derivation of fuzzy control rules from human operator's control actions. In: Proceedings of the IFAC Symposium on Fuzzy Information, Knowledge Representation and Secision Analysis, vol. 6, pp. 55–60 (1983)

Multi-layer Situation Map Based on a Spring Model for Robot Interaction

Satoshi Yokoi$^{(\boxtimes)}$, Hiroyuki Masuta, Toru Oshima,
Ken'ichi Koyanagi, and Tatsuo Motoyoshi

Toyama Prefectural University, Imizu, Japan
t554017@st.pu-toyama.ac.jp,
{masuta,oshima,koyanagi,motoyosh}@pu-toyama.ac.jp

Abstract. This paper discusses the Multi-Layer Situation Map to estimate the relationship between the objects and the storage places for robot interaction. An interaction robot is required the many rules to take suitable action. But, It is difficult to provide rules based on all situations. Now, we focus on a cleanup service by a robot, and are developing an interaction robot which estimates the storage place where a robot should carry objects. In this system, an interaction robot has to know the relationship of the objects and the storage places in dynamically changing situation. Therefore, we propose the Multi-Layer Situation Map, and express the relationship between the object and the storage place. The multi-layer situation map can move each component using a spring model. To verify the effectivity of the multi-layer situation map, we have performed the simulation experiment. We discuss about the dynamically situation change when the same objects are exist on the table.

Keywords: Interaction robot · Spring model · Integration systems

1 Introduction

In recent years, interaction robots have been developed to home or nursing facilities. These robots should be more familiar with human being. In the future, these interaction robots will be used in various situations to provide services to human [1,2]. As a related study of interaction robots, a robot is not only making decision of action by own measurement, but also it collaborates with an environmental information which manages on the fixed servers [3–6].

In these studies, an environmental server measures and manages about the all measured data of objects, persons and robots in the room. Therefore, a robot has to understand a situation, then to decide action by collaborating with environmental server for providing a suitable service. Furthermore, these interaction robot is decided actions based on the rule based model. Many rules which is considered with the various situation are provided by operators.

Now, we focus on a cleanup service by a robot, and are developing an interaction robot which estimates the storage place where a robot should carry objects.

© Springer International Publishing Switzerland 2016
N. Kubota et al. (Eds.): ICIRA 2016, Part II, LNAI 9835, pp. 90–101, 2016.
DOI: 10.1007/978-3-319-43518-3_9

First, we have developed the robot integration system which installed various sensors in the room, and an interaction robot to communicate with human. An interaction robot talks a cleanup service, if a storage place is decided only one place. On the other hand, an interaction robot has to communicate with human about cleanup service in fuzzy situations. For example, a storage place can be selected multiple place when there is an unfinished drink on the table. If a drink is lower, a robot can clear a cup by own decision. On the other hand, if a drink is left almost half, a robot should ask options of storage places. In those cases, a rule based model is difficult to provide the all rules which considers all situation. Therefore, a robot has to know the relationship of the objects and the storage places in dynamically changing situation. Fuzzy inference is the one of the solutions for deciding fuzzy situations [7,8]. This method is used to convert conceptual and fuzzy situations into concrete values. However, the membership functions and the rules are complicated when the consideration of the situation is increased. Clustering methods like a Support Vector Machine (SVM) and Self-Organized Map (SOM) are applied to recognize the complicated situations [4,9,10]. An interaction robot should manage the relationship with multiple clusters when a robot communicates with human in fuzzy situations. However, traditional clustering methods cannot express the relationship with multiple clusters expressly. Because, SVM can identify the current situation to the specific cluster, but the relationship with each cluster is unclear. Also, SOM can express clusters on the visual map, but the relationship with each cluster is unclear. Therefore, we propose the Multi-Layer Situation Map to estimate the relationship between the objects and the storage places, which is used to the decision making of a interaction robot. Multi-layer situation map expresses the relationship of all objects and place without applying a clustering method. Especially, multi-layer situation map is applied a simple spring model to express the relationship between all objects and places. It is expected that an interaction robot communicates with human appropriately in fuzzy situations by learning of spring model parameters. As a first step, we consider the 2 layers which are an object layer and a place layer. This paper explains the outline of Multi-Layer Situation Map, and discuss the experimental simulations. This paper is organized as follows: Sect. 2 explains the hardware and software construction of our robot integration system, Sect. 3 explains the Multi-Layer situation map, Sect. 4 explains an experiment. Finally, Sect. 5 concludes this paper.

2 Storage Place Estimation System

2.1 The Target Environment for the Proposed System

We assume the dining room after a meal at home to install our robot integration system. Furthermore, the situation of the object is changed for example there are a leftovers on a dish, or an unused dish. The specification of the storage place estimation system in above environment is shown as follows.

- The robot should clear the table.
- The robot integration system measures information of all objects.
- Human can instruct a desired storage place arbitrarily.
- The database server updates the object's information whenever the object's state is changed.
- An interaction robot talks the appropriate storage place of the object by considering the environmental situation.

2.2 The Base Rules for Clearing the Table

We have developed the rule set for clearing table in various situations. Table 1 shows the rule set. This rules are a basic knowledge based on general awareness. There are 5 points which are a type, a usage condition, an owner, weight and a rest time to decide the storage place. The example as follows, if the plate's weight is light, the output is that the storage place is sink. If its weight is middle or heavy, the storage place is changed according to other point such as the staying time.

2.3 The Systems Overview

Our robot integration system is composed of vision sensor, force plate and RFID reader/writer. Figure 1 shows the overview of the robot integration system. RFID tags are installed to all objects. There is a force plate with RFID reader/writer which can acquire the weight and tag information of all objects on the table. The RFID tag is registered an object information by a human operator beforehand. An object information is an object ID, a type, an owner, an usage condition (used or unused) and weight. A vision sensor is mounted on a ceiling. The role of vision sensor is to recognize the object position and a type by image processing. Finally, an interaction robot speaks the action for clearing the environment.

We apply the OpenRTM-aist (RTM) to develop the robot integration system [11]. By OpenRTM-aist the robot system can be constructed by making the program of each functional module and connecting each other module. Therefore, various modules developed by other university and company can connect easily. A software module of RTM is called RT component (RTC). There are 5 RTCs in our integration system. The image processing RTC detects the object type and the position on the table, which are outputted to the environmental server. The weight measurement RTC detects the weight and the center of gravity of objects, which are outputted to the environment server. The RFID control RTC detects the tag information on the object, which are outputted to the environment server. The environment server RTC integrates and manages the object information using of all object information. The Estimate RTC estimates the storage position of a target object. We will explain the detail of all RTCs.

- The Image Processing RTC
  The image processing RTC detects the object type and the position on the table. A vision sensor is mounted on a ceiling [12]. We have used the OpenCV

Table 1. The base rules for clearing the table

Object	Usage condition	Owner	Weight	Staying time	Storage place	No.
Rice bowl	Unused	Owner A	-	-	Shelf A	01
		Owner B	-	-	Shelf B	02
		Owner C	-	-	Shelf C	03
		Sharing	-	-	Shelf D	04
	Used	-	Light	-	Sink	05
		-	Middle	Less than 60 min	Refrigerator	06
		-		Over 60 min and more	Sink	07
		-	Heavy	Less than 60 min	Refrigerator	08
		-		Over 60 min and more	Sink	09
Plate	Unused	Owner A	-	-	Shelf A	10
		Owner B	-	-	Shelf B	11
		Owner C	-	-	Shelf C	12
		Sharing	-	-	Shelf D	13
	Used	-	Light	-	Sink	14
		-	Middle	Less than 60 min	Refrigerator	15
		-		Over 60 min and more	Sink	16
		-	Heavy	Less than 60 min	Refrigerator	17
		-		Over 60 min and more	Sink	18
Cup	Unused	Owner A	-	-	Shelf A	19
		Owner B	-	-	Shelf B	20
		Owner C	-	-	Shelf C	21
		Sharing	-	-	Shelf D	22
	Used	-	Light	-	Sink	23
		-	Middle	-	Sink	24
		-	Heavy	Less than 60 min	Refrigerator	25
		-		Over 60 min and more	Sink	26
Plastic bottle	Unused	Owner A	-	-	Shelf A	27
		Owner B	-	-	Shelf B	28
		Owner C	-	-	Shelf C	29
		Sharing	-	-	Refrigerator	30
	Used	-	Light	-	Trash can	31
		-	Middle	Less than 60 min	Refrigerator	32
		-		Over 60 min and more	Sink	33
		-	Heavy	Less than 60 min	Refrigerator	34
		-		Over 60 min and more	Sink	35

library for image processing. The objects in the meal like a rice bowl and a cup are almost round shape. Therefore, we have used the ellipse detection method in the OpenCV library for the object detection method. Furthermore, the object types are detected from the difference of ellipse size.
- The Weight Measurement RTC
  The weight measurement RTC detects the weight and the center of gravity of objects. A commercial released force plate can't measure a few grams of change like a light dish weight and a leftovers on a dish, because the application is

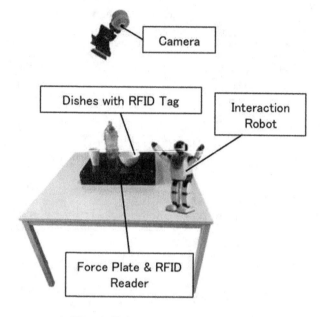

Fig. 1. Robot integration system

limited. Therefore, we have developed the force plate sensor to measure the change of a few grams like a light dish weight and a leftovers on a dish. The force plate is installed the 4 high-sensitivity load cells which can measure 1.2 g.
– The Environment Server RTC
The environment server RTC integrates and manages the object information using of all object information. We have used an open source database system "MySQL" [13]. The environment server RTC integrates an overlapping information from multiple inputs of the same object. For example, the image processing RTC and the weight measurement RTC output an object position, the image processing RTC and the RFID control RTC output a type of object. Furthermore, the environment server RTC estimates the change of a leftover on a dish from the change of weight. The weight of a leftover is categorized to light, middle and heavy. The timestamp is registered at the same time. By using the registered timestamp, the staying time of an object on the table is updated when an object state is changed. Hence, the environment server RTC is not just registered the measured information, but integrates objects to useful for deciding what an interaction talks.

3   Multi-layer Situation Model for the Relation Estimation

The Estimate RTC estimates the storage position of a target object by considering with the relationship between the object and the place. In this paper, we

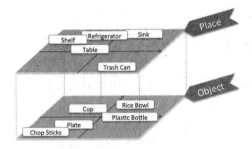

Fig. 2. Outline of multi-layer situation map

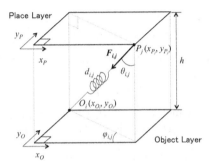

Fig. 3. Multi-layer spring model

propose a multi-layer situation map to explain the relationship of all object and place. In this paper, the relationship is defined that it is not only the relationship between an object and a storage place, but also the change of an object's situation gives an influence to all object and storage place.

3.1   The Multi-layer Situation Map

In related works, the keyword map was proposed to show the relationship of keywords [14–16]. The keyword map is that many keywords on a website are set on a two-dimensional place. Keywords are connected each other by virtual spring, and a tensile force is generated to the spring according to the detection frequency of web search. The distance of keywords is shown the strength of a relationship of keywords. However, we want to consider the relationship of objects and places, separately. Therefore, the keyword map can't explain the relationship of different categories. We propose a multi-layer situation map which prepared category layer like object and place.

Figure 2 shows an example of component layout. Each component is distributed to the corresponding layer. Each component between a neighboring layer are connected by virtual spring. All components are moved according to a spring model, when the environmental change. The difference from a place component to a connected object component is changed when an object property is changed.

The multi-layer situation map shows the relationship of each component from a projection of a component position. An tensile force is provided when an object property is changed.

3.2 Algorithms of the Multi-layer Situation Map

Firstly, the movement of a component is restricted to a place. Now, when a component is moved, other layered component is fixed. Figure 3 shows a multi-layer spring model. An object component position is $O_i (i = 1, 2, \cdots, n)$ and a place component position is $P_j (j = 1, 2, \cdots, m)$. A direct distance between O_i and P_j is $d_{i,j}$, a height between layer is h.

$$d_{i,j} = \sqrt{(x_{P_i} - x_{O_j})^2 + (y_{P_i} - y_{O_j})^2 + h^2} \tag{1}$$

$k_{i,j}$, $l_{i,j}$ and $F_{i,j}$ are a spring rate, a natural length of spring and a tensile force, respectively.

$$F_{i,j} = -k_{i,j} \times (d_{i,j} - l_{i,j}) + U_{i,j} \tag{2}$$

$U_{i,j}$ is an tensile force which is provided when an object property is changed. α is an attenuation factor of the tensile force. S is an arbitrary constant.

$$U_{i,j}(t+1) = \alpha U_{i,j}(t) + S \tag{3}$$

F_{P_j} is a total force for P_j.

$$F_{P_j} = \sum_{i=1}^{n} (F_{i,j} \sin \theta_{i,j}) \tag{4}$$

$F_{P_j x}$, $F_{P_j y}$ are divided F_{P_j} into two direction.

$$\boldsymbol{F_{P_j}} = \begin{bmatrix} F_{P_j x} \\ F_{P_j y} \end{bmatrix} = \begin{bmatrix} F_{P_j} \cos \phi_{i,j} \\ F_{P_j} \sin \phi_{i,j} \end{bmatrix} \tag{5}$$

M_{P_j} is mass, the motion equation is follows,

$$M_{P_j} \ddot{\boldsymbol{s}} = \boldsymbol{F_{P_j}} \tag{6}$$

A displacement is follows,

$$\boldsymbol{s} = \frac{1}{2M_{P_j}} \boldsymbol{F_{P_j}} t^2 + C \tag{7}$$

$$P_j = \begin{bmatrix} x_{P_j}(t+1) \\ y_{P_j}(t+1) \end{bmatrix} = \begin{bmatrix} \frac{1}{2M_{P_j}} F_{x P_j} \Delta t^2 + x_{P_j}(t) \\ \frac{1}{2M_{P_j}} F_{y P_j} \Delta t^2 + x_{P_j}(t) \end{bmatrix} \tag{8}$$

As above, the component movement of a multi-layer situation map is calculated. Moreover, the tensile force is reduced step by step, therefore each component will move away by spring model.

4   Experimental Simulations

4.1   Scenarios and Experimental Conditions

We developed the simulator to verify the Multi-Layer Situation Map. We have performed an experiment by using the simulator that simulates a dynamically changing situation when the same type object cup A and B exist on the table. Figure 4 shows the time flow of situation change in the simulation. The numerals within brackets in the Fig. 4 are the numbers of Table 1 that are corresponding to the change of an object property. After initialization, the usage condition of Cup A, Cup B and Plastic Bottle is set to "unused", "unused" and "used", respectively. The usage condition of Cup A and B is changed to "used" when the water is poured into the cup. First, 3 objects of cup A, cup B and plastic bottle are put on a table at 0 [s]. Then, the water in the plastic bottle is poured to cup A and B at 10 [s] and 20 [s], respectively. The plastic bottle is emptied after pouring the water at 20 [s]. The water in cup A is drunk to half at 30 [s]. Cup A is emptied at 50 [s]. There are 3 components that are cup A, cup B and plastic bottle in the object layer. There are 5 components that are table, shelf, sink, trash can and refrigerator in the place layer. The experimental conditions and the parameters of simulator are shown as follows.

- The simulation is executed for 60 [s] and the sampling time is 0.1 [s].
- Each layer has a width and a depth of 500, and the height distance between 2 layers is 100.
- The mass of each component is fixed 1.0.
- Parameters of natural lengths of the springs are 50 to 400 based on the heuristics. Table 2 shows the natural lengths of spring between each of components.
- Spring rate parameters are set the normal random numbers (center: 4.0, deviation: 1.0).
- Initial positions of each of components are set randomly.

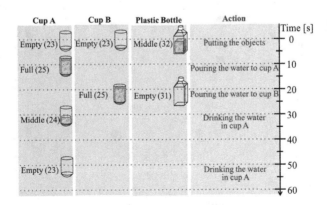

Fig. 4. Scenarios of simulation

Table 2. Natural length

Place / Object	Table	Shelf	Sink	Trash Can	Refrigerator
Cup	100	300	50	350	400
Rice Bowl	170	150	50	300	200
Plastic Bottle	240	200	160	140	100
Plate	50	100	50	400	200

– Initial state is that all object components are fixed and the tensile forces is generated by the position relation. The initializer routine keeps until the place components are balanced with each other.
– If an object state fits the rule of Table 1, the tensile force is provided between an considering object component and place component.
– The provided tensile force is set to 1000.

4.2   Experimental Results of the Simulations

We show the experimental result of a dynamically situation change when the same objects of cup A and B are exist on the table.

Figure 5 shows the position change of all components every 10 [s]. For easy to read, the object layer and the place layer are shown by same map. The circle markers indicate object component, and the squared markers indicate place component. Firstly, cup A, cup B and Table component are approaching until 10 [s]. Refrigerator component is approached to cup A component at 20 [s], because cup A has full of water. Moreover, trash can component is approached when cup A is emptied. Therefore, the relationship between an object and a place is shown by distance on the map.

Figure 6 show the distance change of time series between all places and cup A, plastic bottle and cup B, respectively. In Fig. 6(a), the distance between cup

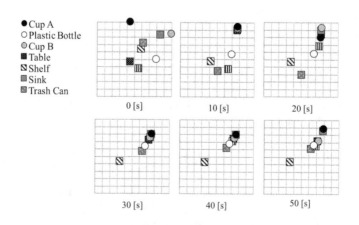

Fig. 5. Position of components

A and the refrigerator is the shortest from 28 [s]. On the other hand, the distance between cup B and Table is the shortest from 35 [s] at Fig. 6(b).

This means that even if 2 components are a same type, multi-layer situation map can express different relationships of the components. Moreover, the approaching speed of refrigerator and cup A, and cup B is different after a cup

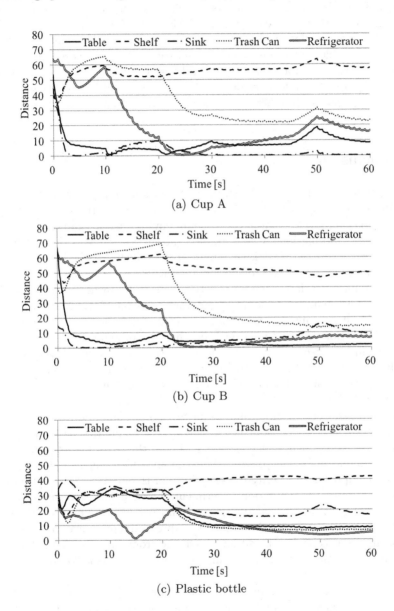

Fig. 6. Distance between object and place

has full of water. The multi-layer situation map can consider the situation change from the past.

Therefore, it is expected that an interaction robot can talk suitable contents according to the fuzzy situations. For example, if a robot starts cleaning up a table at 60 [s], an interaction robot decide that cup A moves to the sink, because cup A and sink is very closer than other places. The plastic bottle is close to the refrigerator, the trash can and the table. Therefore, an interaction robot can ask the storage place to human based on the 3 candidate places. Cup B is closest to the table, but the refrigerator and the sink is also closer. In this case, an intelligent robot can ask the major candidate to human, for example it talks "Cup B keeps on the table, doesn't it?". By using the multi-layer situation map, it is expected that an interaction robot can decide various action based on the current situation.

5   Conclusion

This paper has discussed the Multi-Layer Situation Map to estimate the relationship between the objects and the storage places for robot interaction. A recent interaction robot collaborates with an environmental management server to get various information. A conventional interaction robot is decided actions based on the rule based model. But, many rules which are considered with the various situations should be provided by operators.

Now, we have developed the robot integration system which installed various sensors in the room, and a robot to interact with humans. The environment server of this system integrates and manages the object information using of all sensor information. The interaction robot has to communicate with a human about cleanup service in fuzzy situations. In this case, a rule based model is difficult to provide all the rules which considers all situations. Therefore, a robot has to know the relationship of the objects and the storage places in dynamically changing situation.

Therefore, we propose the Multi-Layer Situation Map, and express the relationship between the object and the storage place. The multi-layer situation map can move each component using a spring model. If the environmental situation is changed, the tensile force is provided between an considering object component and place component. To verify the effectivity of the multi-layer situation map, we have performed the simulation experiment. The simulation assumes dynamic situation changes when the same objects of cup A and B are exist on the table. As a result, multi-layer situation map can express different relationships of the components, and consider the situation change from the past. Therefore, it is expected that an interaction robot can decide various actions based on the current situation.

As future work, we are going to install the multi-layer situation map to the real environmental system, and an interaction robot communicates with human.

References

1. Masuta, H., Matsuo, Y., Kubota, N., Lim, H.: Robot-human interaction to encourage voluntary action. In: Proceedings of 2014 IEEE International Conference on Fuzzy Systems, pp. 1006–1012 (2014)
2. Banda, N., Engelbrecht, A., Robinson, P.: Feature reduction for dimensional emotion recognition in human-robot interaction. In: Proceedings of 2015 IEEE Symposium Series on Computational Intelligence, pp. 803–810 (2015)
3. Gross, H., Schroeter, C., Mueller, S., Volkhardt, M., Einhorn, E., Bley, A., Martin, C., Langner, T., Merten, M.: I'll keep an eye on you: home robot companion for elderly people with cognitive impairment. In: Proceedings of 2011 IEEE International Conference on Systems, Man, and Cybernetics, pp. 2481–2488 (2011)
4. Masuta, H., Tamura, Y., Lim, H.: Self-organized map based learning system for estimating the specific task by simple instructions. J. Adv. Comput. Intell. Intell. Inf. 17(3), 450–458 (2013)
5. Hasegawa, T., Kurazume, R., Kimuro, Y.: A structured environment with sensor networks for intelligent robots. Proc. IEEE Sens. 2008, 705–708 (2008)
6. Kubota, N., Tang, D., Obo, T., Wkisaka, S.: Localization of human based on fuzzy spiking neural network in informationally structured space. In: Proceedings of 2010 IEEE International Conference Fuzzy Systems, pp. 1–6 (2010)
7. Zadeh, L.A.: Fuzzy logic and approximate reasoning. Synthese 30(3), 407–428 (1975)
8. Fukuda, T., Kubota, N.: An intelligent robotic system based on a fuzzy approach. Proc. IEEE 87(9), 1448–1470 (1999)
9. Ben-Hur, A., Horn, D., Siegelmann, H.T., Vapnik, V.: Support vector clustering. J. Mach. Learn. Res. 2, 125–137 (2001)
10. Kohonen, T.: Self-organized formation of topologically correct feature maps. Biol. Cybern. 43(1), 59–69 (1982)
11. Ando, N., Suehiro, T., Kitagaki, K., Kotoku, T., Yoon, W.: RT-middleware: distributed component middleware for RT (robot technology). In: Proceedings of 2005 IEEE/RSJ International Conference on Intelligent Robots and Systems, pp. 3933–3938 (2005)
12. Otsuka, Y., Hu, J., Inoue, T.: Tabletop dish recommendation system for social dining: group FDT design based on the investigations of dish recommendation. Int. J. Inf. Process. 21(1), 100–108 (2013)
13. Williams, H.E., Lane, D.: Web Database Applications with PHP and MySQL. O'Reilly Media, Sebastopol (2004)
14. Kajinami, T., Makihara, T., Takama, Y.: Interactive visualization system for decision making support in online shopping. In: Chawla, S., Washio, T., Minato, S., Tsumoto, S., Onoda, T., Yamada, S., Inokuchi, A. (eds.) PAKDD 2008. LNCS, vol. 5433, pp. 193–202. Springer, Heidelberg (2009)
15. Sabo, S., Kovarova, A., Navrat, P.: Multiple developing news stories identified and tracked by social insects and visualized using the new galactic streams and concurrent streams metaphors. Int. J. Hybrid Intell. Syst. 12(1), 27–39 (2015)
16. Srinivasa Rao, P., Krishna Prasad, M.H.M., Thammi Reddy, K.: An efficient semantic ranked keyword search of big data using map reduce. Int. J. Database Theor. Appl. 8(6), 47–56 (2015)

Service Robot Development with Design Thinking

Takeo Ainoya[1(✉)], Keiko Kasamatsu[1], Kouhei Shimizu[2], and Akio Tomita[3]

[1] Faculty of System Design, Tokyo Metropolitan University, Tokyo, Japan
ainoya@vds.tokyo
[2] Musashino Art University, Tokyo, Japan
[3] Misawa Homes Institute of Research and Development Co., Ltd., Tokyo, Japan

Abstract. This paper described about design method using design thinking by collaboration with designer and engineer for the service robot development. A service robot does not have the substitution function what does the thing a person can't do in concerning with a person. It's necessary for a service robot to create to be able to do because it's a robot. This research aimed at the light in outdoor. This made the watch function as well as the role as the illumination last, and the feedback which isn't robot-like was shown by the illumination pattern. The relation between an expression of illuminations and the feeling was examined using psychophysical method to express the function as an expression of illuminations by this illumination pattern. Developed illuminations could mount a function expression as watch and offer more solutions by data and a sensing.

Keywords: Service robot · Design thinking · Expression · Illumination · Collaboration

1 Introduction

Various problems of an environmental issue, an aging theme and an accident rehabilitation point are complicated in present Japan. It's necessary to return a seeds of a university to society as the role of the university. There is a movement which promotes an industry-academia cooperation study in a university. However the cooperation study doesn't often accomplish an enough outcome. The "co-creation" which connects a university and the industrial world, sticks the side skewer and promotes cooperation is necessary for an industry-academia cooperation project. It's necessary to establish methodology for co-creation to achieve true "co-creation". The human resources and space where practices the way is needed.

An incubation hub as "serBOTinQ" [1] was founded as the mechanism to return study results about a service robot to society by faculty of system design, Tokyo metropolitan university. This Hub combines the technology in on campus and the office related to a service robot and aims at the product development which premised on sale. The purpose in Hub is to create a prototype including upbringing of human resources necessary for the coming robot society. What kind of needs are there in the society for a service robot? What can be made a robot in the area and the house?

This paper introduces the service robotic development utilized collaboration with a designer and a researcher with a different specialized field, and design thinking as

N. Kubota et al. (Eds.): ICIRA 2016, Part II, LNAI 9835, pp. 102–109, 2016.
DOI: 10.1007/978-3-319-43518-3_10

methodology for co-creation. The illumination product developed by Ainoya et al. was taken up as a case in this research [2].

2 Service Robot

It's often used to classify as a robot for an industrial robot and a person with a service robot. However, it's defined as a robot with the function to offer the value a person can't do by this research. Not just to take the service a person was doing so far over in a robot, but technology has to argue what is made for a person. And we're thinking there is service which can be done because it's a robot, for a robot to coexist with a person while thinking what kind of future will create it. Because it's such robot, the service which can be done is designed.

3 The Process for Service Robot Development by Design Thinking

This research used design thinking to catch various problems in crossing way. It includes the following for a way of thinking (Fig. 1). What kinds of needs are there in the society, and what kind of thing can a robot do in the area and the house? What kind of service robot is considered while combining the technology in the university and the company? Idea with the design is introduced to a business and service to invent new innovation in IDEO [3]. There is d-school [4] as an education project using design thinking. The visible shape, the block which touches and the prototype model who works were made with a macro viewpoint and a micro viewpoint, hypothesis consideration and prototyping. It was being developed while sharing the part where couldn't be understand and conscious by creating the prototype.

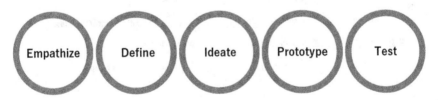

Fig. 1. The development process

4 Empathize

The designers and researchers piled up a discussion and planned for conscious sharing to illumination about the light in outdoor as a step of empathy of this research. We constructed the value by the robot as follows. Electrical appliances with the function of the plural of existence were integrated. The new system integration by this was constructed. It has been developed as a robot with the function as "illuminations + street light + watch sensing unit + alarm system" based on such way of thinking by this research. A concept decided to nestle close to a person, not a just machine as the new value by the design thinking. To make them feel to nestle close to a person, an expression was added to a machine. In other words, it was designed as the robot to which not as a humanoid, but the response from the person is feedback. We measured a person and considered the design of light with emotionally engaging to mount the feedback. The movement a person can sense psychologically was achieved as the function by adding an expression to illumination in this research.

5 Define

The proposed product applied the sensing technology having as seeds of existence, in cooperation with big data and home energy management system (HEMS), can provide information as Fig. 2. In addition, the feedback, including the cooperation with the community as illumination was defined as Fig. 3.

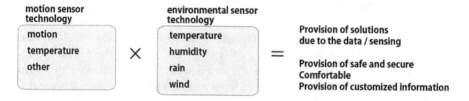

Fig. 2. The functions of proposed product

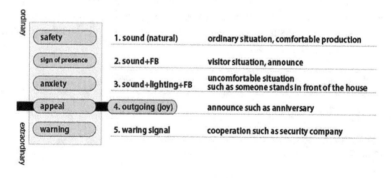

Fig. 3. Feeling and feedback

6 Ideate

The ideation to achieve it as a product in detail using defined technology was performed. This report proposed new lighting with a fusion of these features. The illumination products, not only light up at all times illumination for safety and security features. This utilizes the sensing technology, has the light-emitting pattern and a motion in the equipment, and is fused features on street lights and outdoor lights.

The concept of developed illumination product is to respond the human motion by sensing technology and to have the watching illumination as function. This lighting has the functions which feel reassurance like illumination, sensing the strange motion and intruder by motion sensors and do emission in accordance with the abnormal state of the attention to the warning.

The system concepts are the following two points.

- Light-emitting pattern and the illumination equipment by sensing device are operating, lighting range is changed.
- Provide information to the town security system, and operate to have a function as a security sensor in the street.

7 Prototype (1)

It's necessary for proposed illumination to add the function which makes relief or danger feel. We decided to express by the luminous pattern of the illumination as its function. Therefore a frequency of a threshold with the luminous pattern of the illumination of relief and danger was considered first.

7.1 Methods

A threshold with the luminous pattern of the relief and the danger has been decided by up-and-down method. The luminous pattern was shown on the display using a sin wave. The luminous pattern was shown on the display using a sin wave. The case by which the pattern felt to be relieved is judged changed from an early flash into a slow flash. The case by which the pattern felt to be dangerous is judged changed from a slow flash into an early flash. The participants evaluated about these patterns. Number of evaluations was each 10 times of each pattern. The participants were ten (seven males and three females, average age: 22.4 years old).

7.2 Threshold with the Luminous Pattern

A threshold of relief and danger in the luminous pattern of the illumination was 1.21 ± 0.135 Hz. The luminous pattern of nine conditions was made based on this threshold.

8 Test

8.1 Purpose

The luminous pattern of nine conditions to express a sense of security, attention and danger was prepared using the threshold with the luminous pattern decided at Sect. 7. The physiological reaction when a person judged its pattern by this research, was checked in the respective conditions, and impression evaluation was performed.

8.2 Methods

The presentation conditions were nine; three comfortable conditions (C1-C3), three attention conditions (A1-A3), and three warning conditions (W1-W3). Presentation time was a minute for comfortable conditions, thirty seconds for attention conditions, and fifteen seconds for warning conditions. However, experimenter informed consent to the participants. The experimenter told the participants to close eyes when the participant feel "feel bad" or "eye hurts", and decided to interrupt the experiment.

The measurement indices were galvanic skin response (GSR) and electrocardiogram (ECG) as physiological responses. The integral value of every second was calculated for GSR, and LF/HF (Low Frequency/High Frequency) was calculated by frequency analysis for ECG. LF/HF is the stress index which measures a balance of the autonomic nerve function. LF is the range of the LF ingredient (from 0.05 Hz to 0.15 Hz) of power spectrum, and HF is the range of the HF ingredient (from 0.15 Hz to 0.40 Hz). Impression was evaluated using 8 items by visual analog scale (VAS). Table 1 shows impression items.

Table 1. Impression evaluation items

Warm	Feel relief
Strong	Bright
Humanness	Natural
Scary	Feel danger

The experimental procedure was rest for a minute, presentation of a condition and evaluation. This procedure was repeated for nine conditions. At last, the experimenter interviewed to participants (Fig. 4).

The participants were ten (seven males and three females). The average age was 22.4 years old.

8.3 Results

The results of physiological indices were showed as follows. The presentation time for warning conditions was fifteen seconds, therefore the first fifteen seconds for comfortable and attention conditions were analyzed.

As for GSR, the ANOVA for conditions was performed to examine the difference between conditions. There was no significant difference on nine conditions, however

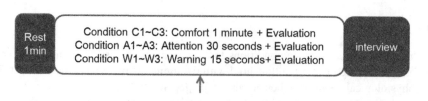

Fig. 4. Experimental procedure

integral value for GSR was high on attention conditions and the sympathetic nerve tended to activate (Fig. 5).

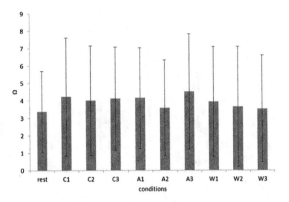

Fig. 5. Results on GSR

Next is LF/HF, the ANOVA for conditions was performed to examine the difference between conditions. There was no significant difference on 9 conditions, however LF/HF on C2 of comfortable condition was low, in other words, the parasympathetic nervous was activated (Fig. 6).

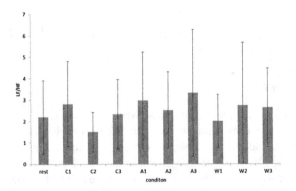

Fig. 6. Result on LF/HF

Figure 7 shows the results of VAS evaluation on eight items. As the results of ANOVA for conditions, there were significant differences on conditions with the exception of "humanness". Therefore, multiple comparison was occured to examine the differences between conditions. There were significant differences between comfortable and warning conditions.

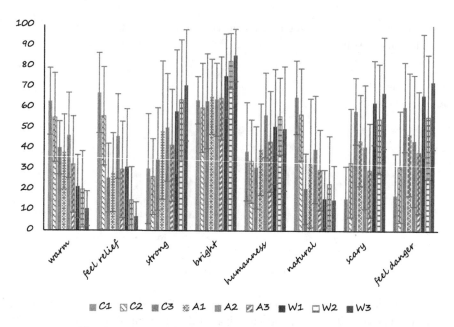

Fig. 7. Impression evaluation on nine presentation conditions

9   Prototype (2)

A prototype was produced based on a result of above mentioned test. Produced proto-types were the following 2 models.

9.1   Design Model A

The design model A is the lighting equipment with the function as fence and with light-emitting pattern for security with motion (the left on Fig. 8). This model reacts to person's movement, and when a person is approaching, the luminous pattern changes by a distance sensor. When it's delaying there in the state a person approached, it'll be judged as the abnormal state and be the luminous pattern to which warning is expressed.

9.2   Design Model B

The design model B is the lighting equipment with the function as foot light and with light-emitting pattern for security (the right on Fig. 8). When a person approaches, a part of illumination rises and has the function which lights up a wider area by a distance sensor.

A type   **B type**

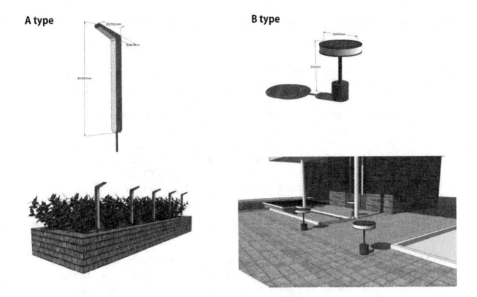

Fig. 8. Design model A and B

10   Conclusion and Future Works

The illumination with the luminous pattern which isn't a picture and the function as the crime prevention sensing with a movement was proposed by this research. We aimed as realization of creation of the case that it's possible because it was a robot, not to change that a person can't illuminate this time. This illumination proposed the watch system to construct the integrated new system. This watch system integrated the function of the plural as "illuminations + street light + watch sensing unit + alarm system", and it was made the value by the robot. On the other hand, it was expressed that the new value by the design thinking nestles close to a person, not a just machine. Therefore the luminous pattern emotionally engaging was considered and that was mounted using the luminous pattern, change of color, and movement as a feedback. The product with integrated system could be proposed by design thinking. In the future, new warm community construction will become possible by realization of sensing in the whole town. However, it's future's problem about how to treat of the behavior data and security of a network collected from sensors.

References

1. serBotinQ. http://www.comp.sd.tmu.ac.jp/serbotinq/index.html
2. Ainoya, T., Kasamatsu, K., Tomita, A.: Proposal of new lighting which combined functionality of street light and outdoor light. In: Yamamoto, S., Abbott, A.A. (eds.) HIMI 2015. LNCS, vol. 9172, pp. 491–499. Springer, Heidelberg (2015). doi:10.1007/978-3-319-20612-7_47
3. Brown, T.: Design thinking. Harvard Bus. Rev. **86**, 84–92 (2008)
4. d.school. http://dschool.stanford.edu

Advancement in the EEG-Based Chinese Spelling Systems

Minghui Shi[1,2(✉)], Changle Zhou[1,2(✉)], Min Jiang[1,2], Qingyang Hong[1,2], Fei Chao[1,2], Jun Xie[3], Weifeng Ren[1,2], Dajun Zhou[1,2], Tianyu Yang[1,2], and Xiangqian Liu[1,2]

[1] Department of Cognitive Science, School of Information Science and Engineering, Xiamen University, Xiamen 361005, Fujian, China
{smh,dozero,minjiang,qyhong,fchao}@xmu.edu.cn
[2] Fujian Provincial Key Laboratory of Brain-like Intelligent Systems, Xiamen University, Xiamen 361005, Fujian, China
[3] School of Mechanical Engineering, Xi'an Jiaotong University, Xi'an 710049, Shaanxi, China
xiejun@mail.xjtu.edu.cn

Abstract. EEG-based spelling systems have practical utilities, not only for spelling letters but also for "spelling" commands to control machines, such as robots, directly by analyzing brain signals. In recent years, EEG-based English Speller (EEGES) has been widely studied. However, only a few researches focused on EEG-based Chinese Speller (EEGCS), which is more difficult to be developed than EEGES. This paper introduced the current methods for EEGES, presented the advancement of methods (Shape-based and Phonetic-based) employed in the current EEGCS systems, discussed the existing problems, highlighted the future research direction, and showed that EEGCS would be a promising research field with rapid development in the future.

Keywords: Brain computer interface (BCI) · Electroencephalography (EEG) · Chinese speller · English speller

1 Introduction

A brain computer interface (BCI) can allow users to communicate with surroundings by means of their brain signals, without the involvement of peripheral nerves and muscles, and particularly benefit individuals suffering from severe disabilities such as amyotrophic lateral sclerosis (ALS) [1]. BCI techniques are also studied in the non-medical fields such as gaming and entertainment [2, 3].

The first BCI system was proposed by Vidal [4]. Presently, BCI is beginning to move from the proof-of-concept and emulation stages into maturity [5]. For noninvasive BCIs, Electroencephalogram (EEG) is favored over magnetoencephalography (MEG), functional near infrared spectroscopy (fNIRS), and functional magnetic resonance imaging (fMRI), due to its better properties such as high time resolution, low cost and portable devices [6, 7].

One of popular applications of EEG-based BCI is for spelling systems, which allow users to select or construct characters from a user interface based on their EEG patterns. Therefore, they enabled human beings to express their intentions directly by brain.

© Springer International Publishing Switzerland 2016
N. Kubota et al. (Eds.): ICIRA 2016, Part II, LNAI 9835, pp. 110–117, 2016.
DOI: 10.1007/978-3-319-43518-3_11

Furthermore, they can also be used for controlling mechanic systems, such as robots, by commands represented by characters.

Present-day methods employed in EEG-based spelling systems, according to the EEG patterns employed, can be roughly divided into three catalogs, i.e., ERP (Event-Related Potential)-based speller, SSVEP (Steady State Evoked Potential)-based speller, and MI (Motor Imager)-based speller.

The original ERP-based Speller is a P300-based speller proposed by Farwell and Donchin [8] in 1988. Thus far, P300 can be evoked by visual [9], auditory [10–12] or tactile stimulations [11, 13]. Most of P300-based spellers, however, exploit visual stimulations with various configurations such as different matrix size [14], novel stimulus patterns [15, 16], novel visual stimulus types [17, 18], predictive word model [6, 19–22], adaptive subject-specific methods [23, 24], dynamic stopping [9, 25–27], etc.

Another kind of ERP-based speller is N200-based speller. Hong et al. [28] and Zhang [29] proposed the motion-onset paradigms of N200-based speller with low contrast and luminance tolerance, and achieved a comparable performance with that of a P300-based speller.

The ERP-based Spellers generally has low information transfer rate (ITR) due to the need of repetitive stimulation sequences to average recorded signals. To address this issue, SSVEP-based BCIs have received increased attention [30, 31–35], since they do not require to average repetitive recorded signals.

The methods of the EEG-based spellers mentioned above depend on externally stimulus. The MI-based speller, however, has been realized in an internally self-paced mode. Several MI-based spelling devices have been proposed [36–40].

Thus far, most of EEG-based spellers are EEG-based English Spellers (EEGESs) developed for English spelling. By contrast, for Chinese spelling, EEG-based Chinese Speller (EEGCS) has not been studied until 2009 [41]. Since the number of sinograms (Chinese characters) are more than 11,000 [42], developing EEGCS is more difficult than developing EEGES.

Nevertheless, various aspects about EEGCS have been addressed: (1) both traditional Chinese characters (used in Taiwan and Hongkong) [43, 44] and the simplified Chinese characters (used in mainland China) [45] have been discussed; (2) two types of phonetic symbols[1], i.e., Pinyin (used in mainland China) [46] and Zhuyin (used in Taiwan) [47] have also been investigated; (3) special user interfaces [43, 47], Chinese language model [48], and novel applications [45, 46] have also been involved.

In this paper, we present the advancement in EEGCS, provide good references for either EEGCS or EEGES, and highlight the future research on EEGCS.

The remainder of this paper is organized as follows. The current EEGCS systems are investigated and discussed in Sects. 2 and 3, respectively. Section 4 concludes the paper.

---

[1] The phonetic symbols for a sinogram represent its pronunciation. There are two different Chinese phonetic symbol systems, i.e., Pinyin and Zhuyin, respectively adopted by mainland China and Taiwan. Pinyin symbols are also English letters but with different pronunciations, while Zhuyin symbols use parts of Chinese characters rather than English letters.

2 Advancement in EEGCS

2.1 Shape-Based Method

Shape-based methods allow users only to know the form of sinograms (with no requirement to know the pronunciation), and can be divided into the stroke-based method and the segment-based method.

Stroke-Based Method. The first P300 EEGCS was developed by Wu et al. [41] in 2009, using five basic strokes (—, |, /, ヽ, ¬). A user could input a sinogram by three steps with several EEG-based-choices, which is a concept we proposed for describing the process of choosing target items by analyzing EEG signals. The items, either strokes or sinograms (depending on the trial phases) flicked one by one. When the target item flicked, the user counted numbers to incur stronger P300 in his/her EEG singnals. Although the experiment did not provide a practicable EEGCS, it proved the potentiality.

In 2010, Jin et al. [45] also developed a EEGCS based on strokes, for sending messages on a cell phone. The user interface was referenced to the T9 stroke input system and included a four-by-four matrix, which contained five strokes, eight numbers (1–8) and three other control items.

Their goal [45] was to improve the accuracy rather than the speed. Their unique methods are the choice of electrodes by Particle Swarm Optimization (PSO) algorithm, and they employed a classification algorithm called Bayesian Linear Discriminate Analysis (BLDA).

Another novel stroke-by-stroke approach, based on SSVEP rather than P300, was proposed in 2012 by Zhao [50]. The limitation of their system is the limited number of frequencies used to evoke SSVEP. In fact, this system is a prototype to show the possibility to design a SSVEP-based EEGCS.

Segment-Based Method. To improve efficiency of stroke-based methods, several segment-based EEGCS systems have been proposed. In 2010, Minnet et al. [44] proposed a segment-based EEGCS. in 2012, the same research group [43] further proposed a EEGCS system using a novel segment-based method called First-Last, or FLAST. FLAST encoded 7,072 Chinese characters, and required three EEG-based-choices for selecting one sinogram: by the first two EEG-based-choices, FLAST selected the first and the last segments respectively, and produced several optional sinograms; by the third EEG-based-choice, FLAST selected the target sinogram.

Their experiment results showed that, when the input speed was set to one sinogram per 107 s, the mean accuracy of segments selections reached 82.8 %, delivering 12.93 bits per minute.

2.2 Phonetic Symbol-Based Method

Phonetic symbol-based EEGCS systems are complementary to the shape-based EEGCS systems, since they allow users to input sinograms without the need to know their forms. One problem is: generally, the same phonetic symbols may represent several different sinograms, which results in many optional sinograms, and more EEG-based-choices are required in a phonetic symbol-based EEGCS than in a shape-based EEGCS, to select the target sinogram from the optional ones. Nevertheless, many people prefer phonetic symbol-based method since it appears to be a more natural way than shape-based method.

In 2011, Sun et al. [49] developed a phonetic symbol-based EEGCS employing Zhuyin symbols. A three-by-four matrix, containing 11 groups of phonetic symbol and one control item, was presented to the user. The items were intensified in a similar way adopted by Firewell [8], i.e., the items in one row or one column were intensified at the same time. Their experiments showed that the target intensifications evoked significantly larger P300 components than the non-target intensifications. The results showed that the average accuracy was up to 90 %, and the average speed was up to one sinogram per 130 s.

In 2013, Huang et al. [47] further proposed an improved EEGCS based on phonetic symbols (Zhuyin), which were arranged into a matrix as their former approach in [49]. They developed a hybrid BCI employed both P300 and N200. To elicit the N200 component, beneath each phonetic symbol, there is a small rectangular, in which there is a bar moving at the coded time. The phonetic symbols, however, are not intensified. During the experiment, both P300 and N200 were used to classify the target ones, which contain the maximum N2P3-value, an amplitude difference between P300 and N200. The result showed that the accuracy was up to 95.8 % and the speed was up to 136.6 bits/min.

Xu et al. [48] reported another P300 EEGCS based on phonetic symbols (Pinyin) with the following characteristics: (1) the optional sinograms were produced by the system, and they did not replace the phonetic symbols in the user interface, instead they were presented around the phonetic symbols, to avoid the switch between the feature (phonetic symbols) area and candidate (optional sinograms) area; (2) the identifications flashed in a item-by-item way, a phonetic symbol item or a sinogram item; (3) the classification algorithm used was the support vector machine (SVM); (4) to improve speed, a statistical language model was employed to predict target sinograms. The results showed that the average speed was up to 0.90 sinogram per minute when the statistical language model was not used, and was improved to 1.13 sinogram per minute when the statistical language model was used.

In 2013, Chen et al. [46] proposed an EEGCS system based on motor imagery, without the need for exterior stimulus. Their system combined Chinese and English input in a common interface, and implemented an asynchronous Chinese-English BCI speller with a 2-D cursor control strategy. The letters, along with some control items, were arranged in three layers of Oct-o-spell interfaces. To choose a block of items, the user should perform three kind of motor imageries, i.e., moving his/her left hand, right hand or feet, to control a cursor initially located at the center of the interface until it hit the circle of the target block [46]. Their results showed that the mean accuracy was 96.33 % and the speed was 30.86 bits/min and 5.53 letters per minute.

3   Discussion

Among the shape-based methods, segment-based method is more efficient than stroke-based method. Among the phonetic symbol based methods, however, Pinyin-based methods and Zhuyin-based methods are complementary.

The EEGCS systems investigated above include the earlier shape-based methods (stroke-based or segment-based) and the recent phonetic symbol-based methods (Pinyin-based or Zhuyin-based), with diverse user interfaces, stimuli and applications.

Thus far, different components, such as P300, N200, SSVEP and MI, have been used to develop EEGCS systems. Researchers have also realized the potentiality to employ hybrid components, such as the combination of P300 with N200, to overcome the disadvantages of a single component.

Compared with EEGES, EEGCS has been far less fully investigated, and there are several problems in the current research: (1) although various EEG components have been employed in the current EEGCS research, other components except P300 have not been investigated as much as in the EEGES research. In fact, many methods now widely used in the EEGES systems can also be introduced into the EEGCS systems; (2) except for the hybrid components of P300/N200, other hybrid BCIs, such as P300/SSVEP BCIs [51–53] could also be introduced to develop the EEGCS systems; (3) most of the current EEGCS systems are tentative to explore the feasibility of their methods with only a small number of sinograms; (4) most of the experiments were carried out with few healthy subjects less than 20 subjects; (5) the accuracy and the speed reported in different systems may have less significance for comparing with each other.

In the further research on EEGCS, the following aspects are worthy of being concerned: (1) target user: the systems should be comfortable for the disabled people; (2) evaluation metric: the performance evaluation such as accuracy and speed should be comparable among different systems and approaches; (3) assistant techniques: some assistant technologies may greatly improve performance, i.e., a natural language model may predict the subsequent sinograms [6, 48]; (4) applications: some applications of the current EEGES systems, such as BCI-based web browser [54] and explorer [55], may be introduced into the EEGCS systems.

4   Conclusion

EEG-based spellers can be used not only for spelling letters but also for "spelling" commands to control machines, such as robots, directly by brain. Compared with the EEGES systems, developing the EEGCS systems is much more difficult. However, many methods employed in the current EEGES systems may be good references for developing EEGCS systems. This has not been fully done yet. Therefore, it can be expected that the research on EEGCS would prove a rapid development in the near future.

Acknowledgements. This work is supported by the State Foundation for Studying Abroad of China, the National Natural Science Foundation of China (Grant No. 61203336).

References

1. Nicolas-Alonso, L.F., Gomez-Gil, J.: Brain computer interfaces, a review. Sensors **12**(2), 1211–1279 (2012)
2. Nijholt, A.: BCI for games: a 'state of the art' survey. In: Stevens, S.M., Saldamarco, S.J. (eds.) ICEC 2008. LNCS, vol. 5309, pp. 225–228. Springer, Heidelberg (2008)

3. Blankertz, B., Tangermann, M., Vidaurre, C., Fazli, S., Sannelli, C., Haufe, S., Maeder, C., Ramsey, L., Sturm, I., Curio, G.: The Berlin brain–computer interface: non-medical uses of BCI technology. Front. Neurosci. **4**, 1–2 (2010)
4. Vidal, J.-J.: Toward direct brain-computer communication. Ann. Rev. Biophys. Bioeng. **2**(1), 157–180 (1973)
5. Allison, B.Z., Dunne, S., Leeb, R., Millán, J.D.R., Nijholt, A.: Towards practical brain-computer interfaces: bridging the gap from research to real-world applications. Biological and Medical Physics, Biomedical Engineering. Springer Science & Business Media, Heidelberg (2012)
6. Mora-Cortes, A., Manyakov, N.V., Chumerin, N., Van Hulle, M.M.: Language model applications to spelling with brain-computer interfaces. Sensors **14**(4), 5967–5993 (2014)
7. Shende, P.M., Jabade, V.S.: Literature review of brain computer interface (BCI) using electroencephalogram signal. In: Book Literature Review of Brain Computer Interface (BCI) using Electroencephalogram Signal, pp. 1–5 (2015)
8. Farwell, L.A., Donchin, E.: Talking off the top of your head: toward a mental prosthesis utilizing event-related brain potentials. Electroencephalogr. Clin. Neurophysiol. **70**(6), 510–523 (1988)
9. Kindermans, P.J., Tangermann, M., Muller, K.R., Schrauwen, B.: Integrating dynamic stopping, transfer learning and language models in an adaptive zero-training ERP speller. J. Neural Eng. **11**(3), 9 (2014)
10. Höhne, J., Schreuder, M., Blankertz, B., Tangermann, M.: A novel 9-class auditory ERP paradigm driving a predictive text entry system. Front. Neurosci. **5**, 1–10 (2011)
11. Kaufmann, T., Holz, E.M., Kübler, A.: Comparison of tactile, auditory, and visual modality for brain-computer interface use: a case study with a patient in the locked-in state. Front. Neurosci. **7**, 1–12 (2013)
12. Schreuder, M., Rost, T., Tangermann, M.: Listen, you are writing! Speeding up online spelling with a dynamic auditory BCI. Front. Neurosci. **5**, 112 (2011)
13. Brouwer, A.-M., Van Erp, J.B.: A tactile P300 brain-computer interface. Front. Neurosci. **4**, 19 (2010)
14. Allison, B.Z., Pineda, J.: ERPs evoked by different matrix sizes: implications for a brain computer interface (BCI) system. IEEE Trans. Neural Syst. Rehabil. Eng. **11**(2), 110–113 (2003)
15. Townsend, G., LaPallo, B., Boulay, C., Krusienski, D., Frye, G., Hauser, C., Schwartz, N., Vaughan, T., Wolpaw, J., Sellers, E.: A novel P300-based brain–computer interface stimulus presentation paradigm: moving beyond rows and columns. Clin. Neurophysiol. **121**(7), 1109–1120 (2010)
16. Sellers, E.W., Krusienski, D.J., McFarland, D.J., Vaughan, T.M., Wolpaw, J.R.: A P300 event-related potential brain–computer interface (BCI): the effects of matrix size and inter stimulus interval on performance. Biol. Psychol. **73**(3), 242–252 (2006)
17. Kaufmann, T., Schulz, S., Grünzinger, C., Kübler, A.: Flashing characters with famous faces improves ERP-based brain–computer interface performance. J. Neural Eng. **8**(5), 056016 (2011)
18. Tangermann, M., Schreuder, M., Dähne, S., Höhne, J., Regler, S., Ramsay, A., Quek, M., Williamson, J., Murray-Smith, R.: Optimized stimulation events for a visual ERP BCI. Int. J. Bioelectromagn. **13**(3), 119–120 (2011)
19. Akram, F., Han, H.S., Kim, T.S.: A P300-based brain computer interface system for words typing. Comput. Biol. Med. **45**, 118–125 (2014)

20. Speier, W., Arnold, C., Lu, J., Taira, R.K., Pouratian, N.: Natural language processing with dynamic classification improves P300 speller accuracy and bit rate. J. Neural Eng. **9**(1), 016004 (2012)

21. Ryan, D.B., Frye, G., Townsend, G., Berry, D., Mesa-G, S., Gates, N.A., Sellers, E.W.: Predictive spelling with a P300-based brain–computer interface: increasing the rate of communication. Int. J. Hum. Comput. Interact. **27**(1), 69–84 (2010)

22. Höhne, J., Schreuder, M., Blankertz, B., Tangermann, M.: Two-dimensional auditory p 300 speller with predictive text system, pp. 4185–4188. IEEE (2010)

23. Li, Y., Guan, C., Li, H., Chin, Z.: A self-training semi-supervised SVM algorithm and its application in an EEG-based brain computer interface speller system. Pattern Recogn. Lett. **29**(9), 1285–1294 (2008)

24. Kindermans P J, Verstraeten D, Buteneers P, et al: How do you like your P300 speller: adaptive, accurate and simple? In: 5th International Brain-Computer Interface Conference, BCI 2011 (2011)

25. Kindermans, P.-J., Verschore, H., Schrauwen, B.: A unified probabilistic approach to improve spelling in an event-related potential-based brain-computer interface. IEEE Trans. Biomed. Eng. **60**(10), 2696–2705 (2013)

26. Verschore, H., Kindermans, P.-J., Verstraeten, D., Schrauwen, B.: Dynamic stopping improves the speed and accuracy of a P300 speller. In: Villa, A.E., Duch, W., Érdi, P., Masulli, F., Palm, G. (eds.) ICANN 2012, Part I. LNCS, vol. 7552, pp. 661–668. Springer, Heidelberg (2012)

27. Schreuder, M., Höhne, J., Blankertz, B., Haufe, S., Dickhaus, T., Tangermann, M.: Optimizing event-related potential based brain–computer interfaces: a systematic evaluation of dynamic stopping methods. J. Neural Eng. **10**(3), 036025 (2013)

28. Hong, B., Guo, F., Liu, T., Gao, X., Gao, S.: N200-speller using motion-onset visual response. Clin. Neurophysiol. **120**(9), 1658–1666 (2009)

29. Zhang, J.X.: Centro-parietal N200: an event-related potential component specific to Chinese visual word recognition. Chin. Sci. Bull. **57**(13), 1516–1532 (2012)

30. Zhu, D., et al.: A survey of stimulation methods used in SSVEP-based BCIs. In: Computational Intelligence and Neuroscience, p. 1 (2010)

31. Parini, S., Maggi, L., Turconi, A.C., Andreoni, G.: A robust and self-paced BCI system based on a four class SSVEP paradigm: algorithms and protocols for a high-transfer-rate direct brain communication. Comput. Intell. Neurosci. **2009**, 1–11 (2009)

32. Colwell, K.A., Ryan, D.B., Throckmorton, C.S., Sellers, E.W., Collins, L.M.: Channel selection methods for the P300 Speller. J. Neurosci. Methods **232**, 6–15 (2014)

33. Yin, E.W., Zhou, Z.T., Jiang, J., Yu, Y., Hu, D.W.: A dynamically optimized SSVEP brain-computer interface (BCI) speller. IEEE Trans. Biomed. Eng. **62**(6), 1447–1456 (2015)

34. Xia, B., Hong, Y., Zhang, Q., et al.: Control 2-dimensional movement using a three-class motor imagery based Brain-computer Interface. In: International Conference of the IEEE Engineering in Medicine & Biology Society, pp. 1823–1826 (2012)

35. Liu, Q., Chen, K., Ai, Q., Xie, S.Q.: Review: recent development of signal processing algorithms for SSVEP-based brain computer interfaces. J. Med. Biol. Eng. **34**, 299–309 (2013)

36. D'albis, T., Blatt, R., Tedesco, R., Sbattella, L., Matteucci, M.: A predictive speller controlled by a brain-computer interface based on motor imagery. ACM Trans. Comput. Hum. Inter. (TOCHI) **19**(3), 20 (2012)

37. Blankertz, B., Dornhege, G., Krauledat, M., Schröder, M., Williamson, J., Murray-Smith, R., Müller, K.-R.: The Berlin Brain-computer Interface presents the novel mental typewriter Hex-o-Spell (2006)

38. Pfurtscheller, G., Brunner, C., Schlögl, A., Da Silva, F.L.: Mu rhythm (de) synchronization and EEG single-trial classification of different motor imagery tasks. Neuroimage **31**(1), 153–159 (2006)
39. Ramoser, H., Muller-Gerking, J., Pfurtscheller, G.: Optimal spatial filtering of single trial EEG during imagined hand movement. IEEE Trans. Rehabil. Eng. **8**(4), 441–446 (2000)
40. Mohanchandra, K., Saha, S., Mohanchandra, K., Saha, S.: Optimal Channel Selection for Robust EEG Single-trial Analysis. Aasri Procedia **9**(9), 64–71 (2014)
41. Wu, B., Su, Y., Zhang, J.-H., Li, X., Zhang, J., Cheng, W., Zheng, X.: A virtual Chinese keyboard BCI system based on P300 potentials. Acta Electron. Sin. **37**(8), 1733–1738 (2009). 1745 (in Chinese)
42. Zidian, X.: 10th. Beijing, China: Shang wu yin shu guan (2004). (in Chinese)
43. Minett, J.W., Zheng, H.Y., Fong, M.C.M., Zhou, L., Peng, G., Wang, W.S.Y.: A Chinese text input brain-computer interface based on the P300 speller. Int. J. Hum. Comput. Interact. **28**(7), 472–483 (2012)
44. Minett, J.W., Peng, G., Zhou, L., Zheng, H.-Y., Wang, W.S.: An assistive communication brain-computer interface for Chinese text input. In: 4th International Conference on Bioinformatics and Biomedical Engineering, pp. 1–4 (2010)
45. Jin, J., Allison, B.Z., Brunner, C., Wang, B., Wang, X., Zhang, J., Neuper, C., Pfurtscheller, G.: P300 Chinese input system based on Bayesian LDA. Biomedizinische Technik/Biomed. Eng. **55**(1), 5–18 (2010)
46. Chen, C., Yang, J., Huang, Y., Li, J., Xia, B.: A cursor control based Chinese-english BCI speller. In: Lee, M., Hirose, A., Hou, Z.-G., Kil, R.M. (eds.) ICONIP 2013. LNCS, vol. 8226, pp. 403–410. Springer, Heidelberg (2013)
47. Huang, T.W., Tai, Y.H., Tian, Y.J., Sun, K.T.: The fastest BCI for writing Chinese characters using brain waves. In: Fourth Global Congress on Intelligent Systems, pp. 346–349 (2013)
48. Xu, X., Fang, H.-J.: A P300-based BCI System for online Chinese input. J. Huaqiao Univ. (Nat. Sci.) **36**(3), 269–274 (2015). (in Chinese)
49. Koun-Tem, S., Tzu-Wei, H., Min-Chi, C.: Design of Chinese spelling system based on ERP. In: 2011 IEEE 11th International Conference on Bioinformatics and Bioengineering, BIBE 2011, pp. 310–313 (2011)
50. Zhao, J.: Steady-state Visual Evoked Potential: the Attentional Mechanism and the Application in Brain Computer Interface, Zhejiang University (2012). (in Chinese)
51. Xu, M., Qi, H., Wan, B., Yin, T., Liu, Z., Ming, D.: A hybrid BCI speller paradigm combining P300 potential and the SSVEP blocking feature. J. Neural Eng. **10**(2), 026001 (2013)
52. Yin, E., Zhou, Z., Jiang, J., Chen, F., Liu, Y., Hu, D.: A novel hybrid BCI speller based on the incorporation of SSVEP into the P300 paradigm. J. Neural Eng. **10**(2), 026012 (2013)
53. Amiri, S., Rabbi, A., Azinfar, L., Fazel-Rezai, R.: A review of P300, SSVEP, and hybrid P300/SSVEP brain-computer interface systems, Brain-Computer Interface Systems—Recent Progress and Future Prospects, pp. 195–213 (2013)
54. Hader, S., Pinegger, A., Kathner, I., Wriessnegger, S.C., Faller, J., Antunes, J.B.P., Muller-Putz, G.R., Kubler, A.: Brain-controlled applications using dynamic P300 speller matrices. Artif. Intell. Med. **63**(1), 7–17 (2015)
55. Bai, L.J., Yu, T.Y., Li, Y.Q.: A brain computer interface-based explorer. J. Neurosci. Methods **244**, 2–7 (2015)

A Wearable Pressure Sensor Based on the Array of Polymeric Piezoelectric Fiber with Metal Core

Ran Chen, Weiting Liu[(✉)], Xiaodong Ruan, and Xin Fu

The State Key Laboratory of Fluid Power and Mechatronic Systems, Zhejiang University,
Zheda Rd. 38, Hangzhou, China
liuwt@zju.edu.cn

Abstract. A wearable pressure sensor is developed based on the array of polymeric piezoelectric fiber with metal core covered by beta phase poly(vinylidene fluoride) layer. Each fiber consists of a polyester-imide enameled copper core, a piezoelectric layer and sputtered gold electrode. Fiber array is fixed on an insulating substrate and well packaged to ensure stable work. This pressure sensor is tested to declare high sensitivity and fast response. Meanwhile, flexibility of this structure indicates its potentiality to be wearable.

Keywords: Wearable sensor · Fiber array · Piezoelectricity · Metal core

1 Introduction

With the development of medical treatment, amount of the olds has increased constantly and brought many requests on health care and management at home [1]. As we know, many diseases are difficult to be detected without a continuous monitoring of physiological signal, such as heartbeat, pulse pressure and so on. Therefore, more and more attentions have been concentrated on the research of unconscious and long-term monitoring of body abnormalities, and development of a wearable device has been considered as an effective solution.

Poly(vinylidene fluoride) (PVDF), one of piezoelectric materials, have attracted attentions as its excellent ferroelectricity and high strain level [2–4]. Compared to the ceramic piezoelectric materials, one advantage of PVDF is its flexibility [5], which makes them suitable for the wearable applications [6].

To date, several researches of wearable sensor based on PVDF have been carried out, mostly based on PVDF film. For example, Wang et al. [7] have developed a PVDF sensor system for unstrained cardiorespiratory monitoring during sleeping. Jiang et al. [8] have developed a PVDF based cardiorespiratory sensor attached on chest wall to recording the actions of heartbeats and respirations. Chiu et al. [9] have developed a patch type PVDF based sensor and further simplified the system.

On the other hand, structure with micro fiber is considered more sensitive than micro film. However, PVDF micro fiber requires a couple of electrodes on the surface, which is very complicated in fabrication. Thus rather than full section fibers, piezoelectric fibers with metal core are considered to be more suitable for integration: the array can

© Springer International Publishing Switzerland 2016
N. Kubota et al. (Eds.): ICIRA 2016, Part II, LNAI 9835, pp. 118–124, 2016.
DOI: 10.1007/978-3-319-43518-3_12

be conductive or not, fibers can be applied on more complex shape, all fibers can be used independently to keep the system work properly even if some fibers fail.

In our previous study [10], we developed a polymeric piezoelectric fiber with metal core. We used PVDF as piezoelectric material and polyester-imide enameled wire as substrate, enamel layer of which can also serve as essential insulating layer in electro-wetting method to improve the adhesion of PVDF. All the morphological and piezo-electric properties have demonstrated that this polymeric piezoelectric fibers can perform stable, sensitive and fast response along with vibration. And therefore, we use this polymeric piezoelectric fiber as sensing element to detect pressure in a wearable application.

In the paper, we develop a pressure sensor based on the array of polymeric piezo-electric fiber with metal core. This sensor is tested in a pulse detecting application to declare high sensitivity and fast response. On the other hand, with an elastic nylon cloth as substrate, this structure shows good flexibility, which also indicates its potentiality to be wearable.

2   Design and Fabrication

Design of sensor is shown in Fig. 1. Piezoelectric fiber is fabricated in our previous study, with an electrowetting aided dry spinning method. Structure of this piezoelectric fiber is shown in Fig. 1(a). Each fiber consists of a 100 μm copper core enameled with 6 μm polyester-imide, a 10 μm piezoelectric poly(vinylidene fluoride) (PVDF) layer and sputtered 100 nm gold layer. This copper core enhances the strength of fiber as well as introduces more convenient solutions to be integrated. When under deformation, electric charges can be detected on the surface of gold layer and copper core.

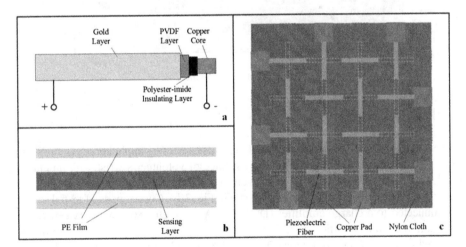

Fig. 1. Design of sensor. (a) Schematic of piezoelectric fiber; (b) Schematic of sensor; (c) Schematic of sensing layer, dash lines stand for structures on the back side.

As shown in Fig. 1(b), the sensor is designed to be 3-layer structure: a 300 μm sensing layer as well as two 30 μm polyethylene (PE) layers covered to protect the piezoelectric fibers and circuits. This sensing layer adopts a fiber stitched structure, shown in Fig. 1(c), which has been proven to be stable by Shimojo et al. [11] In our experiment, we select insulating nylon cloth to be substrate as its good elasticity and flexibility. In the horizontal direction, a row of piezoelectric fibers are stitched into the front and back surfaces of the nylon cloth with back and forth alternated. And then a column of piezoelectric fibers are stitched in the vertical direction as well to form a 4 × 4 array. Finally copper pads are attached on the end of fibers to connect gold layers and copper cores with output wires.

Photo of fabricated prototype is shown in Fig. 2. Sixteen wires are connected to detect output electric charges. Due to its thin and flexible structure, this sensor can be easily attached onto plane surface as well as more complex surfaces to detect pressure, which indicates its potentiality to be wearable.

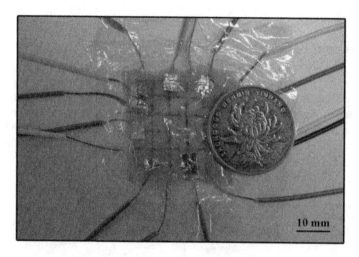

Fig. 2. Prototype of sensor.

3   Testing Setup

As a wearable sensor, this sensor is tested in a pulse detecting application. As shown in Fig. 3(a), the sensor is impacted onto the surface of volunteer's wrist with an elastic bandage. Length of bandage is adjusted to ensure the sensing area being tightly attached onto the surface. The testing system is shown in Fig. 3(b). Copper pads of tested sensor are connected to a charge amplifier (DONGHUATEST DH5862) with sixteen wires. The gain of charge amplifier is set to be 10 mv pC^{-1}. Output of the charge amplifier is acquired through a data acquisition (DAQ, NI USB-6343), after passing through a low-pass filter (with 25 Hz cut-off frequency) to remove noise at high frequency, signal is recorded in PC.

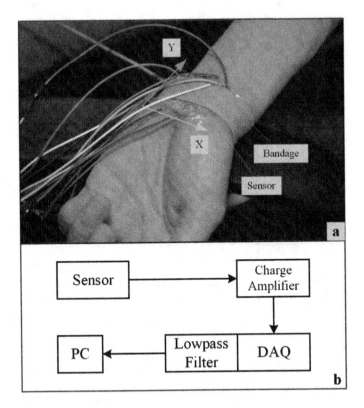

Fig. 3. Testing setup. (a) Photo of sensor setup; (b) Schematic of testing system.

During the testing, the volunteer is asked to keep calm and motionless. Testing lasts for about 15 s and is carried out several times. The most credible data according to the situation at testing moment is selected as our final result.

4   Result and Discussion

Testing result is shown in Fig. 4. Response along axis X (with piezoelectric fiber parallel to the hand) and response along axis Y (with piezoelectric fiber vertical to the hand) are plotted independently, so as to display the distribution of pressure. Several points can be raised from this result.

Firstly, it can be clearly seen that this sensor performs fast and sensitive response to follow the pulse of human. However, significant noise can be also observed, indicating the requirement of an improved packaging in further work.

And then, by comparing the response along axis X and axis Y, it should be noticed that the response along axis X reduces much faster than that along axis Y. This result supports the famous fact that the pulse pressure is almost concentrated on the surface of aorta. And therefore, rather than detecting pressure at a single point, this arrayed sensor

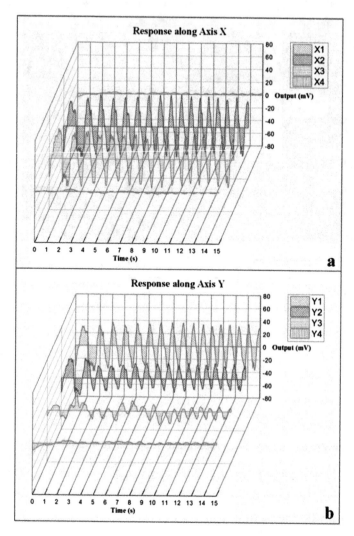

Fig. 4. Testing result. The gain of charge amplifier is set to be 10 mv pC^{-1}.

which can detect pressure on a wider area shows more compatible to be used in health monitoring applications.

Finally, although it's not shown in the testing result, we find that response is affected a lot by the motion of volunteer. Before being used as a wearable sensor, this structure needs more researches, especially on the algorithmic design, being carried out to reduce the influence of human motion.

5   Conclusion

In this paper, we develop a pressure sensor based on the array of polymeric piezoelectric fiber with metal core. This sensor is tested in a pulse detecting application to declare high sensitivity and fast response. On the other hand, with an elastic nylon cloth as substrate, this structure shows good flexibility, which also indicates its potentiality to be wearable. However, as noise from motion and environment still exist in the response of prototype, more researches on packaging and algorithmic should be carried out in our further work.

Acknowledgements. This work was supported in part by the National Basic Research Program (973) of China (2011CB013303), the National Natural Science Foundation of China (51521064) and the Applied Research Project of Public Welfare Technology of Zhejiang Province (2015C31109).

References

1. Choi, S., Jiang, Z.: A novel wearable sensor device with conductive fabric and PVDF film for monitoring cardiorespiratory signals. Sens. Actuators, A **128**, 317–326 (2006). doi: 10.1016/j.sna.2006.02.012
2. Chu, B., Zhou, X., Ren, K., Neese, B., Lin, M., Wang, Q., Bauer, F., Zhang, Q.M.: A dielectric polymer with high electric energy density and fast discharge speed. Science **313**, 334–336 (2006). doi:10.1126/science.1127798
3. Tajitsu, Y., Chiba, A., Furukawa, T., Date, M., Fukada, E.: Crystalline phase transition in the copolymer of vinylidenefluoride and trifluoroethylene. Appl. Phys. Lett. **36**, 286–288 (1980). doi:10.1063/1.91456
4. Zhang, Q.M., Bharti, V., Zhao, X.: Giant electrostriction and relaxor ferroelectric behavior in electron-irradiated poly(vinylidene fluoride-trifluoroethylene) copolymer. Science **280**, 2101–2104 (1998). doi:10.1126/science.280.5372.2101
5. Rathod, V.T., Mahapatra, D.R., Jain, A., Gayathri, A.: Characterization of a large-area PVDF thin film for electro-mechanical and ultrasonic sensing applications. Sens. Actuators, A **163**, 164–171 (2010). doi:10.1016/j.sna.2010.08.017
6. Nilsson, E., Lund, A., Jonasson, C., Johansson, C., Hagström, B.: Poling and characterization of piezoelectric polymer fibers for use in textile sensors. Sens. Actuators, A **201**, 477–486 (2013). doi:10.1016/j.sna.2013.08.011
7. Wang, F., Tanaka, M., Chonan, S.: Development of a PVDF piezopolymer sensor for unconstrained in-sleep cardiorespiratory monitoring. J. Intell. Mater. Syst. Struct. **14**, 185–190 (2003). doi:10.1177/104538903033639
8. Jiang, Y., Hamada, H., Shiono, S., Kanda, K., Fujita, T., Higuchi, K., Maenaka, K.: A PVDF-based flexible cardiorespiratory sensor with independently optimized sensitivity to heartbeat and respiration. Procedia Eng. **5**, 1466–1469 (2010). doi:10.1016/j.proeng.2010.09.393
9. Chiu, Y.-Y., Lin, W.-Y., Wang, H.-Y., Huang, S.-B., Wu, M.-H.: Development of a piezoelectric polyvinylidene fluoride (PVDF) polymer-based sensor patch for simultaneous heartbeat and respiration monitoring. Sens. Actuators, A **189**, 328–334 (2013). doi:10.1016/j.sna.2012.10.021

10. Liu, W., Chen, R., Ruan, X., Fu, X.: Polymeric piezoelectric fiber with metal core produced by electrowetting aided dry spinning method weiting. J. Appl. Polym. Sci. (2016, accepted)
11. Shimojo, M., Namiki, A., Ishikawa, M., Makino, R., Mabuchi, K.: A tactile sensor sheet using pressure conductive rubber with electrical-wires stitched method. IEEE Sens. J. 4, 589–596 (2004). doi:10.1109/JSEN.2004.833152

Practical-Use Oriented Design for Wearable Robot Arm

Akimichi Kojima[✉], Hirotake Yamazoe, and Joo-Ho Lee

Ritsumeikan University, 1-1-1, Nojihigashi, Kusatsu-shi, Shiga, Japan
is0165hp@ed.ritsumei.ac.jp, yamazoe@fc.ritsumei.ac.jp,
leejooho@is.ritsumei.ac.jp
http://www.aislab.org/

Abstract. This paper proposes a wearable robot arm designed for practical use. To reduce the weight of the arm, a passive robotics concept is adopted. By replacing actuators with passive actuators the proposed wearable robot arm is able to reduce its weight. Since the coarse movement of the arm is controlled manually by a user's hand, the user can be safe from unintentional movement of the arm. A prototype which satiates the proposed concept was developed and experiments were performed to verify its usability.

Keywords: Wearable robot arm · Passive robotics · User interface

1   Introduction

In our daily works, we often require an extra hand to hold an object temporarily while both hands are busy. If human have an extra hand, it will improve the work efficiency. Thus research has been carried out to develop a wearable robot arm which fits to a worker and supports his/her tasks. In [1,2], they aim to expand workers ability so that they can cope with the work, which cannot be done by a worker, with two robot arms attached to the worker. It improves work efficiency and it is possible to carry out more difficult tasks alone with the arms. However, the proposed wearable robot arms in these researches, the arm in [1] is about 5 kg and the arm in [2] is about 15 kg. They may become a burden to the people. We cannot deny the risk of accidents caused by people's wrong operation either. It is necessary to reduce the burden on the person by reducing the weight of the robot arm in order to improve the working efficiency and practicality of wearable robotic arm. Figure 1 shows a robot arm on a human, called Assist Oriented Arm (AOA), assisting the task. AOA provides a long period of use since it is lighter than existing wearable arms and working efficiency can be improved additionally. AOA is able to eliminate the possible risk which is caused by wrong operation and the work can be carried out safely too.

We explain details of the existing robot arm's problems in the second section. In the third section, a solution for the problem is mentioned and the fourth section describes experiment, and fifth section concludes this paper.

© Springer International Publishing Switzerland 2016
N. Kubota et al. (Eds.): ICIRA 2016, Part II, LNAI 9835, pp. 125–134, 2016.
DOI: 10.1007/978-3-319-43518-3_13

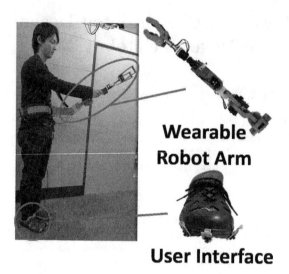

Fig. 1. The proposed wearable robot arm is worn by a man

2   Problem Statement

As described above, there are three problems in the conventional robot arms, which are weight, safety and energy problems.

The first problem is a weight problem. Existing robot arms almost control joint angles by servo motor actuators. However, each joint actuator is not the same one always. Usually in serial link manipulator, if an actuator is close to the base of the manipulator, it requires bigger power than the one which is far from the base. Usually the bigger power actuator weighs more.

The second one is safety problem. This problem is about the possibility of an accident or an injury by wrong operation of person. Usually a robot arm is controlled by a user interface which is joy stick controller or something. If the robot arm moves unexpectedly, it may cause a problem. In the case of the wearable robot arm, the robot arm is closer than usual use, it may cause a big accident.

The third one is energy problem. A joint of conventional robot arms are moved by torque from an actuator. However, the actuator requires power at all times even though the joints are not moving. Therefore a wearable robot should carry heavy batteries to supply energy and this leads to total weight up.

3   Hybrid Actuation System

Passive robotics have been suggested by Goswami, and it controls a movement in passive [3]. In addition Peshkin developed object transportation system based on these concept called Cobot [4]. Concept of passive robotics is that it controls the joints without active. A passive walking robot is a good example to understand

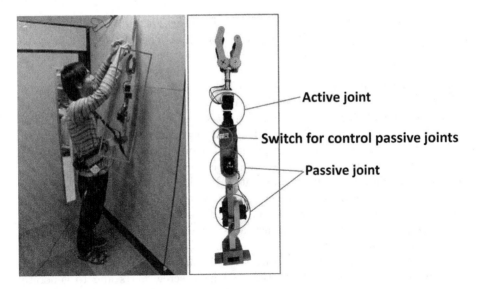

Fig. 2. Hybrid Actuation System

its concept [7]. The robot walks by converting the potential energy into kinetic energy. This energy conversion enables the robot to walk without actuators. However, this concept has not been used in robot arm. Since a robot arm cannot use the potential energy, force from a user is used instead of potential energy. It has two advantages, when this mechanism is used in the robot arm. The first advantages is safety. Since the arm is controlled by direct human force, the arm become safe. There is no risk of wrong operation due to the complex interface. The other is that its weight become light. Most serial link robot's joints have complex structures, and the actuators are heavy. However, the propose mechanism consists of simple structure. Therefore, it can have lighter weight than the conventional serial linked robot arm structure. Thus, it becomes simple and lightweight wearable robot arm. We employ passive robotics as a solution for the problems.

It can reduce the weight, and it can guarantee the safe operation of the robot. However, only with passive joints, it cannot be a robot arm. We have developed Hybrid Actuation System (HAS) for solution of these problems (Fig. 2). This system has two types of actuators which are Passive joints and Active joints. The active joints are same as conventional robot arm has, and an joint angle is only controlled by an actuator. The passive joints are not controlled by the actuators of themselves. They can only fix the joint angle physically. It is not necessary to consider actuator's torque since the joint is not moved by the actuator. Thus, a light and safe robot arm can be realized and it is possible to reduce the weight of the actuator which has ability to move small parts in brake system, and it become a simple wearable robot arm. In addition, a passive actuator does not generate different movements from intentions of the user. Since, angle control of the passive joint depend on a direct operation of user. Therefore, it is possible to reduce the accident risk by a user's wrong operation. The passive joint resolves both problems of increasing the weight of the robot arm and the risk by wrong operation.

In this research, passive joint uses a light and low torque actuator, which is lighter than actuator of conventional robot arm. The most important parts in passive joint of this system is joint brake parts. It is built up with a small actuator and male screw parts and female screw parts (Fig. 3). Passive joint achieves lock and unlock by screw parts and it only fixes the angle physically. Male screw parts are connected to the actuator. When an actuator rotates, male screw parts rotate with the actuator. In addition, male screw parts have semispherical bumps of 2 mm diameter, and body parts have symmetrical dents. The brake system is locked when the bumps enter into the dents. The brake of passive joint is controlled by an actuator. The other important part is guide part. If the actuator rotates without the guide parts, it cannot fix the angle, since female screw parts rotate along with the actuator's rotation. The guide parts are required to prevent rotation of female screw parts and to generate translation of female screw parts. Therefore the actuator does not require high torque. The brake is controlled by a push switch on the robot arm. The brake of the joint is unlocked by the switch, and user can move the joint while the user is pressing the push switch. If the user release the switch, the joint is locked by the actuator.

One of the features of this system is the passive joint, which is unlike the conventional joint mechanism. Usually, when a robot arm is manipulating something, the close joints from the tip of the robot are used more than the close joint from the base. For example, a people attaches a plate to a wall with using a hammer and nails. A hand supports nail and a hammer hits the nail by the other hand. In this case thosen ails and a hammer are handled more by wrists rather than shoulders. Therefore, when a use performs the task with the wearable robot arm, it is possible to perform the task only with the movement of wrist of the robot arm, and base joints are just fixed at a certain angle.

4   Proposed Wearable Robot Arm

4.1   Configuration of Robot Arm

In this section, we will describe a prototype system of wearable robot arm developed based on the concepts describe above. Figure 2 shows the system configuration of joint brake system and Fig. 4 shows the dimensions of AOA. The prototype is consisted of two passive joints, two active actuator (two DOFs) and a gripper part. The passive actuators are used at base joints as shown in Fig. 2. The weight of this prototype is about 800 g. Compared to existing wearable robot arms [1,2], we could achieve a lighter wearable robot arm. This wearable robot arm is attached to the abdomen using a harness. As actuators in both active and passive joints, we adopted ROBOTIS Dynamixel AX-12 [5].

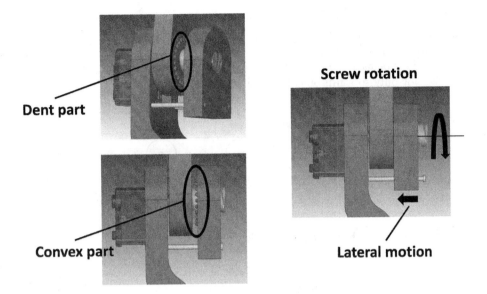

Fig. 3. Passive joint of brake system

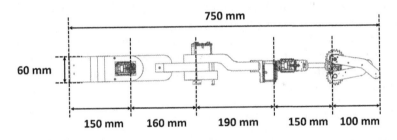

Fig. 4. The dimensions of the AOA

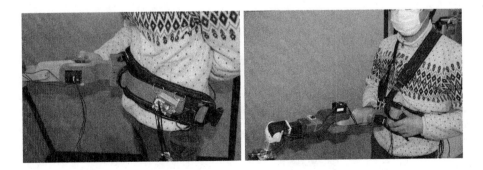

Fig. 5. How the robot arm is attached to the user

4.2 How to Attach

A user can easily attach this robot arm. Left of Fig. 5 shows the first step. The user wears the harness in hip. Subsequently, the user wears another belt from shoulder. This belt fixes the harness. To decrease concentration of load, the belts are fixed not only to a hip but also to shoulders.

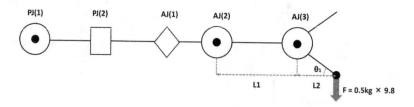

Fig. 6. Simplified view of an active part

4.3 Static Torque Calculation

Next, we calculate the maximum mass of objects that our robot arm can hold. To do this, we calculate the maximum torque at each joint [6]. Figure 6 shows a simplified view of AOA for the torque calculation. The weight of all active parts is 0.3 kg. However, since the AJ(1) rotates in the roll axis, we do not need to consider the torque of the AJ(1). Thus, we exclude AJ(1) from the calculation and the weight of the remaining parts is 0.24 kg. The maximum torques can be calculated as follows:

$$AJ(2) = 0.09 \text{ kg}, \ AJ(3) = 0.15 \text{ kg}, \ L_1 = 0.11 \text{ m}, \ L_2 = 0.07 \text{ m}$$

Calculation of the maximum torque's formula is

$$max.\tau AJ(2) = -F(L_1 + L_2) - AJ(3)gL_1 \tag{1}$$

$$max.\tau AJ(3) = -FL_2 \tag{2}$$

$$max.\tau AJ(2) = 0.72Nm \qquad max.\tau AJ(3) = 0.34Nm$$

The maximum holding torque of Dynamixel AX-12 is 1.5 Nm [5], and we assume the maximum mass of objects as 250 g. From the above calculation, we confirmed that AX-12 have enough power to hold the target objects, even though we consider the safety factor (2.0).

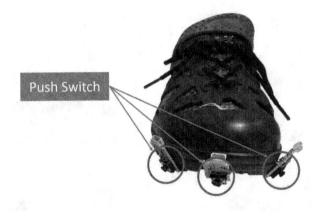

Fig. 7. Operation interface

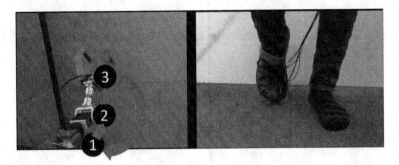

Fig. 8. How to change the target actuators

4.4 User Interface

As a control interface for AOA, various types of interfaces such as voice recognition or using controller devices can be considered. However, voice recognition cannot be used in noisy environment. In addition the proposed arm is assumed to be used when our both hands are busy. Therefore controller device cannot be used either. Thus, we developed shoes-type interface (Fig. 7). This interface consists of three switches and one micro controller (Robotis OpenCM). In the current prototype, the interface and the robot arm are connected by wires. The left and right switches make ccw and cw rotations of an actuator respectively. The center switch shifts the controlled actuator among the active actuators (Fig. 8). If a user pushes the center switch, it change the target actuator from 1 to 2 or 2 to 3 or 3 to 1.

5 Experiment

We conducted the following experiment to evaluate the proposed wearable robot arm. Subjects wear the proposed robot arm, and are instructed to do a defined

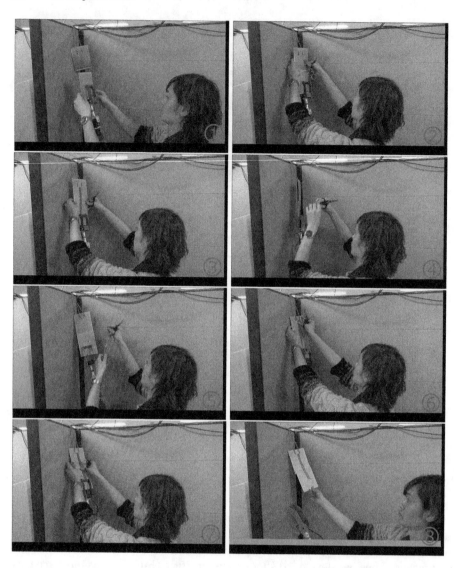

Fig. 9. 1 to 8 show the states of the task. 1 shows that the arm is grabbing a plate. 2 shows that the arm is fixing the plate with another plate. 3 shows taping. 4 shows rotating the plate by the arm. 5 to 7 show taping procedure of opposite surface, and 8 is finish.

task (Fig. 9). The task is to tape a board that is hanging from wall, and is difficult to accomplish alone without the proposed robot arm. After doing task with and without the robot arm, the subjects answered the questionnaire with the five-point scale to evaluate the following five items:

Q1: weight of the robot arm
Q2: length of the robot arm

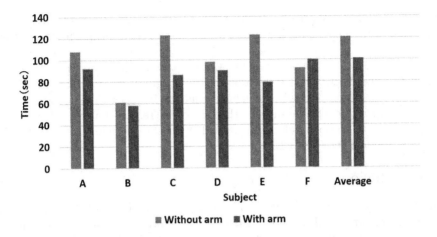

Fig. 10. Average and comparison of working time

Q3: easiness of controlling the arm
Q4: safety of the robot arm
Q5: comparison of with and without the arm

The subjects are six male college students. Figure 10 shows the time required to accomplish the task with and without the arm. Figure 11 is the result of the questionnaire after the task completion. In Q1-Q4 1 means negative and 5 means positive, and in Q5, 1 means high work efficiency without the robot arm and 5 means high work efficiency with the robot arm. Figure 10 shows that the average

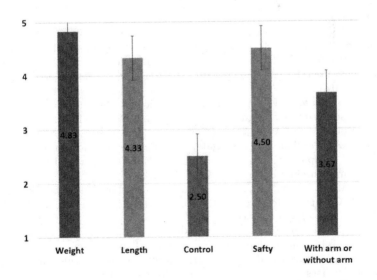

Fig. 11. The average of each question item

time with the arm is shorter than the one without the arm. In this experiment, we assumed that a user start a work with wearing the robot arm. Therefore, it is not included wearing time of robot arm and user interface in Fig. 10. In Fig. 11, the averaged results of evaluation of Q1, Q2, Q4 and Q5 are higher than 3. These results show the effectiveness of the proposed robot arm. However, the result of Q3 is lower than 3. From this result, it can be say that the interface to control the robot arm should be improved. We suppose that one of the reasons for this is lacking of feedback of the operation mode. Thus, we will investigate methods for the feedback such as visual feedback by LEDs.

6   Conclusion

This paper proposed a wearable robot arm to support a user who is doing tasks that is difficult to be done alone. In order to use the proposed robot arm in everyday life, it was designed based on the passive robotics, considering its weight and safety. The robot arm is consisted of two types of actuators: active joints and passive joints. The passive joints have a feature that they do not require power to keep fixing joints. We developed the shoes-type user interface to control the robot arm by foot. By using this interface, we can control the robot arm without using hands. We conducted experiment to evaluate the effectiveness of the proposed arm and the results showed the advantages of the robot arm. However, the results also showed the difficulties to control the arm. Future works include improvements of interface and we have to improve degree of freedom of the robot arm to perform a lot of task. Operation interface by foot switch is simple and easy to understand. However it is difficult operate the robot arm and it reduced the efficiency. Thus the interface should be improved.

References

1. Bonilla, B.L., Asada, H.H.: A robot on the shoulder: coordinated human-wearable robot control using coloured petri nets and partial least squares predictions. In: 2014 IEEE International Conference on Robotics and Automation (ICRA), pp. 119–125 (2014)
2. Parietti, F., Asada, H.H.: Supernumerary robotic limbs for aircraft fuselage assembly: body stabilization and guidance by bracing. 2014 IEEE International Conference on Robotics and Automation (ICRA), pp. 1176–1183 (2014)
3. Goswami, A., Peshkin, M.A., Colgate, J.E.: Passive robotics: an exploration of mechanical computation. In: Proceedings of IEEE International Conference on Robotics and Automation, pp. 279–284 (1990)
4. Peshkin, M.A., Colgate, J.E., Moore, C.A., Wannasuphoprasit, W., Gillespie, R.B., Akella, P.: Cobot architecture. IEEE Trans. Robot. Autom. **17**(4), 377–390 (2001)
5. ROBOTIS. http://jp.robotis.com/index/
6. Piltan, F., Yarmahmoudi, M.H., Shamsodini, M., Mazlomian, E., Hosainpour, A.: PUMA-560 robot manipulator position computed torque control methods using MATLAB/SIMULINK and their integration into graduate nonlinear control and MATLAB courses. Int. J. Robot. Autom. **3**(3), 167–191 (2012)
7. Collins, S.H., Wisse, M., Ruina, A.: A three-dimensional passive-dynamic walking robot with two legs and knees. Int. J. Robot. Res. **20**(7), 607–615 (2001)

Robot Fault Diagnosis Based on Wavelet Packet Decomposition and Hidden Markov Model

You Wu[1], Zhuang Fu[1(✉)], Shuwei Liu[1], Jian Fei[2(✉)], Zhen Yang[1], and Hui Zheng[1]

[1] State Key Lab of Mechanical System and Vibration, Shanghai 200240, China
zhfu@sjtu.edu.cn
[2] Ruijin Hospital, Shanghai Jiao Tong University, Shanghai 200240, China
feijian@hotmail.com

Abstract. Fault diagnosis has great significance in industrial robots. The Selective Compliance Assembly Robot Arm (SCARA) is a widely used robot in the industry. In this paper, SCARA robot is taken as an example to do fault diagnosis. The electromechanical actuator model of SCARA was built to simulate typical faults, laying the foundation for the diagnose work. Then based on Wavelet Packet Decomposition and Hidden Markov Model (HMM), a new fault diagnosis method is proposed. A maximum likelihood estimator is derived to evaluate the fault. Finally, experiment is done to verify the accuracy of the fault diagnosis method.

Keywords: Fault diagnosis · Wavelet packet decomposition · HMM

1 Introduction

From 1970s, industrial robots are widely used in the industry. The application of industrial robots has expanded from the automotive industry to the electronic manufacturing, food, drug and plastic industries. A robot is a complex system composed of mechanical system, electronic system, hydraulic system, pneumatic system and other components. The performance of robot modules may deteriorate during runtime. Furthermore, some parts may even go out of action [1]. If the failure cannot be timely detected and processed in time, the faults will lead to the product quality decreasing and serious economic losses [2]. In order to complete the complex process, industrial robots usually cooperate in the production line [3]. If a robot fails, the entire production line may fall into a standstill, resulting in losses and disasters. So it is very important and necessary to study the fault diagnosis of industrial robots.

Fault diagnosis is the floorboard of fault separation, fault detection and fault identification. It is generally believed that analytical redundancy is the earliest fault diagnosis technique. Analytical redundancy method is to find the relationship between the measurable variables.

In 1971, Mehra and Peschon proposed a method to obtain the system fault information by comparing the output of the observer [4]. In 1983, Gini published a paper on

© Springer International Publishing Switzerland 2016
N. Kubota et al. (Eds.): ICIRA 2016, Part II, LNAI 9835, pp. 135–143, 2016.
DOI: 10.1007/978-3-319-43518-3_14

fault diagnosis and repair of the robot's perception, marking the beginning of the application of fault diagnosis technology to the field of robotics [5]. Later, Daniele et al. proposed a model based fault diagnosis scheme. In the scheme the SOSM algorithm is used to design the input rule [6]. In 2007, Anand discussed a fault diagnosis method based on fuzzy neural network for robot arm. The fuzzy inference system is used to identify and isolate the fault [7]. Saeed used simplified Euler Lagrange equation to reduce the complexity of the fault diagnosis method. The method can also deal with influence of gravity [8]. Gspandl et al. studied the robot fault diagnosis based on historical information of autonomous [9]. Li proposed a robot intelligent diagnosis system based on knowledge decision. The system combines the intelligent decision technology and robot fault diagnosis knowledge. The intelligent diagnosis system can improve the efficiency of intelligent fault diagnosis [10].

Our fault diagnosis method differs from earlier work. Hidden Markov Model and Wavelet Packet Decomposition are used to obtain monitoring conditions. Finally, the validity and correctness of the method are verified by experiments and tests.

2 Electromechanical Actuator Model

In the design of fault diagnosis system, the modular structure of the robot system can be used to analyze the system. Generally speaking, the robot system can be divided into: computer control system, mechanical execution unit and electric servo system. The electromechanical actuator is composed of the electric servo system and the mechanical execution unit [11]. For SCARA robot, each degree of freedom is equivalent to an electromechanical actuator. It is a rotation pair, which is the general structure of SCARA robot. The sliding pair is slightly different, by using the application of equivalent torque method we can obtain similar model.

In this paper, the SCARA robot has $5°$ of freedom. There are 4 rotation pairs and 1 sliding pair. The driving modules are hybrid stepping motors. Taking the two-phase hybrid stepping motor as an example, we establish the mathematical model of the electromechanical actuator. The model is shown as follows:

$$\begin{cases} \frac{d\theta}{dt} = \omega \\ \frac{d\theta}{dt} = [-K_m i_a \sin(N_r\theta) + K_m i_b \cos(N_r\theta) - B\omega - T_l]/J \\ \frac{di_a}{dt} = [u_a - Ri_a + K_m\omega \sin(N_r\theta)]/L \\ \frac{di_b}{dt} = [u_b - Ri_b + K_m\omega \sin(N_r\theta)]/L \end{cases} \tag{1}$$

where R is the motor winding resistance, L is winding inductance, K_m is the back electromotive force coefficient, ω is the motor speed, N_r is the number of rotor teeth, J is the inertia of mechanical components, B is viscous friction coefficient, T_l is load torque. According to the mathematical model above, the simulation model is established in Simulink, which is shown in Fig. 1. The step angle of two-phase stepper motor $1.8°$. Without segmentation the motor will rotate $1.8°$ when receives a pulse. Through the model of single axis we can do qualitative analyses of the fault type.

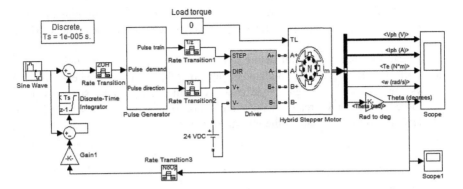

Fig. 1. Single axis simulation model

3 Fault Analysis

We will analyze some typical failure modes according to the experiment or simulation in this section.

- Connection failure
  Connection failure is a kind of common fault. The sign of connection failure is that the motion error increases drastically. Motion error is the difference between the target position and the return value of the encoder. For the SCARA robot, the motion of each axis is simulated in Simulink. It is supposed that the gear of the fourth axis gets loose. The error of each axis is shown in Fig. 2.

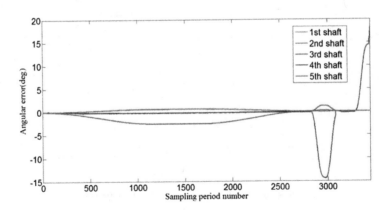

Fig. 2. Error of each axis

- Follow error being out of range
  The follow error refers to the dynamic error between the command position and the actual position of the servo system. Several factors may cause the follow error being out of range. The common situation is that the servo system loses efficacy (load too

heavy or lubrication failure). Motor speed cannot meet the requirements. The tracking error accumulates with time, and finally exceeds the allowable value of error.

We can use the model built before to do the analysis. In the simulation model, the lubrication failure can be simulated by adjusting the friction coefficient. Figure 3 shows the torques on different lubrication conditions. It can be seen that high friction coefficient will greatly increase the motor torque. As can be seen in Fig. 4, lubrication failure will lead to the growth of the following error.

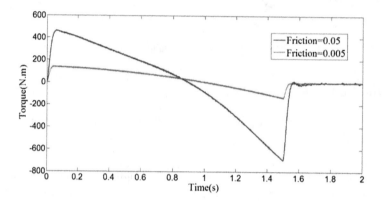

Fig. 3. Torque with different lubrication conditions

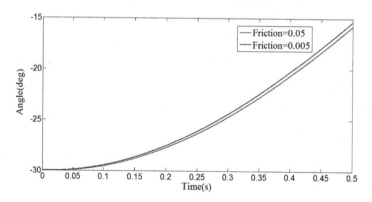

Fig. 4. Angle with different lubrication conditions

In addition to the above cases, the faults can be caused by other factors, such as mechanical connection obstruction, motor coil short circuit and severe vibration.

4 Fault Diagnosis Based on HMM

A complete fault diagnosis system consists fault monitoring subsystem and fault diagnosis subsystem. The fault monitoring subsystem collects state information. Fault diagnosis subsystem classifies the fault information and finds the location of faults. The rule base also consists of monitoring rule base and fault diagnosis rule base.

The information acquisition units in this paper are encoders and gyroscope accelerometers. We take the third axis data for an example to do the evaluation according to motion error.

Wavelet packet decomposition is used to decompose the robot's motion error signals, and the energy of each node is extracted to form a feature vector. Wavelet transformation is using a cluster of orthogonal wavelet function to represent the signal. When the wavelet packet transformation is used to decompose the signal, the signal of each node is decomposed into low frequency part and high frequency part, and the analysis bandwidth of the signal is reduced to half of the bandwidth of the original signal. The normal signal is decomposed into three layers with wavelet set transform. The result of the transform is an eight dimensional vector. The result is shown in Fig. 5. A represents the low frequency component, and D represents the high frequency component.

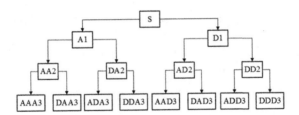

Fig. 5. Wavelet packet decomposition tree

Figure 6 shows the results of the wavelet packet decomposition. We can get eight nodes of wavelet package decomposition. The energy of the 8 nodes of the three layers wavelet tree of the normal signal is extracted. Then the pattern recognition of the fault state can be done by Hidden Markov Model (HMM).

HMM is a statistical Markov model in which the system being modeled is assumed to be a Markov process with unobserved (hidden) states. The energy of the 8 nodes of the three layer wavelet tree constitute the observation vector of HMM. Select the normal state data as the training data, we can train hidden Markov model λ. Model λ is HMM model library. Baum-Welch algorithm is used in the process of training Hidden Markov Model. There is a forward variable and a backward variable in Baum-Welch algorithm. The definition of the forward variable is that it is the probability of arriving an intermediate state, that is, the probability of partial observation sequence before the time t, denoted by $\alpha_t(i)$.

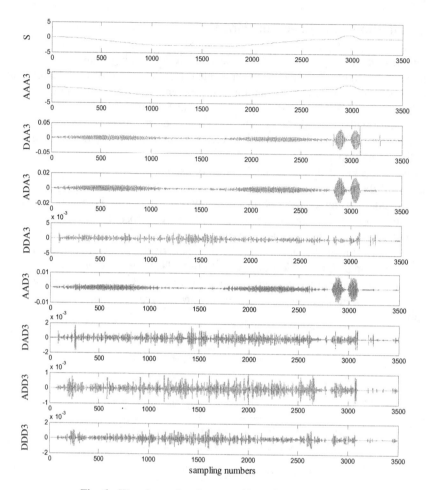

Fig. 6. Wavelet packet decomposition of normal signal

$$\alpha_t(i) = P(o_1 o_2 \ldots o_i, q_i = S_i | \lambda) \tag{2}$$

In the above formula, the state of the t time is S_i. Backward variable is the probability of the ending partial sequence $o_{t+1}, \ldots, o_T$ given starting state i at time t.

$$\beta_t(i) = P(o_{t+1} o_{t+2} \ldots o_T, q_t = S_i | \lambda) \tag{3}$$

We can now calculate the temporary variables, according to Bayes' theorem:

$$\gamma_t(i) = P(q_t = S_i | O, \lambda) \tag{4}$$

which is the probability of being in state i at time t given the observed sequence O and the parameters λ

$$\xi_t(i,j) = P(q_t = S_i, q_{t+1} = S_{i+1}|O, \lambda) \tag{5}$$

which is the probability of being in state i and j at times t and $t + 1$ respectively given the observed sequence O and parameters λ.

Formula (5) can be further expanded by substituting forward variable and backward variable.

$$\xi_t(i,j) = \frac{\alpha_t(i)a_{ij}b_j(O_{t+1})\beta_{t+1}(j)}{P(O|\lambda)} = \frac{\alpha_t(i)a_{ij}b_j(O_{t+1})\beta_{t+1}(j)}{\sum_{i=1}^{N}\sum_{j=1}^{N}\alpha_t(i)a_{ij}b_j(O_{t+1})\beta_{t+1}(j)} \tag{6}$$

Formula (4) can be further expanded by substituting forward variable and backward variable.

$$\gamma_t(i) = \frac{\alpha_t(i)\beta_t(j)}{\sum_{i=1}^{N}\alpha_t(i)\beta_t(j)} \tag{7}$$

After the preparation of the previous work, we can get the Baum-Welch model.

- Parameter initialization of HMM
  A random set of HMM parameters are given as the initial estimates, denoted by $\lambda_0 = (\pi_0, A_0, B_0)$.
- Calculate model probability and update model parameters
  Use Formulas (6) and (7) to calculate $\xi_t(i, j)$ and $\gamma_t(i)$, and then update Hidden Markov parameters, we can get a new HMM model, denoted by $\bar{\lambda}_0 = (\bar{\pi}_0, \bar{A}_0, \bar{B}_0)$.

$$\bar{\pi} = \gamma_1(i) \tag{8}$$

$$\bar{a}_{ij} = \frac{\sum_{t=1}^{T-1}\xi_t(i,j)}{\sum_{t=1}^{T-1}\gamma_t(i)} \tag{9}$$

$$\bar{b}_j(k) = \frac{\sum_{\substack{t=1 \\ s.t.o_t=v_k}}^{T}\gamma_t(j)}{\sum_{t=1}^{T}\gamma_t(j)} \tag{10}$$

- Stop condition
  If $P(O|\bar{\lambda}) - P(O|\lambda)$ is smaller than the set value, then the maximum likelihood estimator $P(O|\lambda)$ has been obtained.
  Input the data to be measured data O into the model λ, we can calculate of the output $P(O|\lambda)$. The model trained on the state of normal data, so $P(O|\lambda)$ reflects the

probability that data O generated on conditions of normal operation. Therefore, $P(O|\lambda)$ can be used to describe the performance of the robot. If $P(O|\lambda)$ is within the normal range, the signal can be considered normal.

5 Experiment and Summary

In order to verify the accuracy of the fault diagnosis method, the third axis HMM model is used to test the algorithm. As it is shown in Fig. 7, the 4 sets of error data are used to test. The values of Test1 and test4 exceed the threshold ± 5. They can be considered as fault data, Test3 and test2 are normal data.

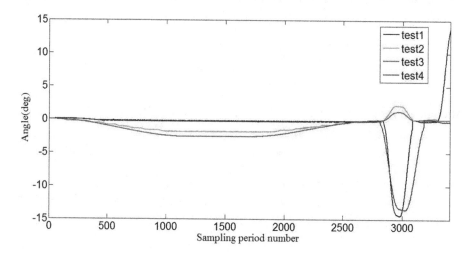

Fig. 7. Error of the second shaft

Using the HMM forward algorithm, The $P(O|\lambda)$ of test data is calculated. The result is processed by logarithm operation:

Test1: $P_1(O|\lambda) = -\text{Inf}$
Test2: $P_2(O|\lambda) = 76.1$
Test3: $P_3(O|\lambda) = 196.1$
Test4: $P_4(O|\lambda) = -794.6$

Test1 and test4 are the fault data, Likelihood probabilities of which are far less than normal data. It is can be proved that method used in this paper is valid and accurate.

In general, it is difficult to diagnose the faults of the robot, which is caused by the complexity of the robot's structure. In this paper, SCARA robot is taken as an example to realize fault diagnose. The electromechanical actuator model of SCARA was built to simulate several typical faults, laying the foundation for the diagnose work. Then based on HMM and wavelet packet decomposition, a new fault diagnosis method is proposed. A maximum likelihood estimator is derived to evaluate the fault.

Finally, experiment is done to verify the accuracy of the fault diagnosis method. Expert system based on monitoring knowledge and diagnosis knowledge is to be studied in the future.

Acknowledgement. This work was partially supported by the National Natural Science Foundation of China (Grant No. U1401240, 61473192) and National Basic Research Program of China (2014CB046302).

References

1. Zhao, B., Skjetne, R., Blanke, M., et al.: Particle filter for fault diagnosis and robust navigation of underwater robot. IEEE Trans. Control Syst. Technol. **22**(6), 2399–2407 (2014)
2. Defoort, M., Veluvolu, K.C., Rath, J.J., et al.: Adaptive sensor and actuator fault estimation for a class of uncertain Lipschitz nonlinear systems. Int. J. Adapt. Control Sig. Proc. **30**(2), 271–283 (2016)
3. Craig, W.S.: Data driven approach to non-stationary EMA fault detection and investigation in-to remaining useful life. Dissertations Theses Gradworks **36**(36), 378–380 (2014)
4. Mehra, R.K., Peschon, J.: An innovations approach to fault detection and diagnosis in dynamic systems. Automatica **7**(5), 637–640 (1971)
5. Gini, G., Gini, M.: Explicit programming languages in industrial robots. J. Manuf. Syst. **2**(1), 53–60 (1983)
6. Daniele, B., Massimiliano, C., Antonella, F., Pierluigi, P.: Fault detection for robot manipulators via second-order sliding modes. IEEE Trans. Ind. Electron. **55**(11), 3954–3963 (2008)
7. Anand, D.M., Selvaraj, T., Kumanan, S., Janarthanan, J.: Fault diagnosis system for a robotmanipulator through neuro fuzzy approach. Int. J. Model. Ident. Control **3**(2), 181–192 (2008)
8. Saeed, M., Mehrzad, N.: Fault diagnosis in robot manipulators in presence of modeling uncertainty and sensor noise. In: Proceedings of the IEEE International Conference on Control Applications, pp. 1750–1755 (2009)
9. Gspandl, S., Pill, I., Reip, M., Steinbauer, G.: Belief management for autonomous robots using history-based diagnosis. In: Mehrotra, K.G., Mohan, C., Oh, J.C., Varshney, P.K., Ali, M. (eds.) Developing Concepts in Applied Intelligence. SCI, vol. 363, pp. 113–118. Springer, Heidelberg (2011)
10. Duan, Z.H., Cai, Z.X.: Particle filters based fault diagnosis for internal sensors of mobile robots. In: 2009 International Conference on Measuring Technology and Mechatronics Automation (ICMTMA), vol. 1, pp. 47–50 (2009)
11. Van, M., Kang, H.J., Suh, Y.S., et al.: A robust fault diagnosis and accommodation scheme for robot manipulators. Int. J. Control Autom. Syst. **11**(2), 377–388 (2013)

Assistive Robotics

Development of Pneumatic Myoelectric Hand with Simple Motion Selection

Kotaro Nishikawa[1,3(✉)], Masayuki Shakutsui[2], Kentaro Hirata[3],
and Masahiro Takaiwa[4]

[1] Technical Department, National Institute of Technology,
Tsuyama College, 624-1, Numa, Tsuyama-Shi, Okayama, Japan
nisikawa@tsuyama-ct.ac.jp
[2] Toyota Motor Corporation, 1, Toyota-Cho, Toyota-Shi, Aichi, Japan
[3] Graduate School of Natural Science and Technology, Okayama University,
3-1-1, Tsushima-Naka, Kita-Ku, Okayama, Japan
[4] Graduate School of Institute of Technology and Science,
Tokushima University, 2-1, Minamijyousanjima-Cho,
Tokushima-Shi, Tokushima, Japan

Abstract. Of the 82 thousand upper-extremity amputees in our country, the majority use prosthetic hands. Recently, research and development of myoelectric hand have come to the forefront. However, due to the usage of multiple electric actuators, the myoelectric hand is heavy, structurally complicated and expensive. Thus, the authors propose a pneumatically-driven prosthetic hand with bellows incorporated in the joint. This grip force exceeds that of general pneumatically-driven prosthetic hands. Also it is lightweight of 240 g, reducing user fatigue. The hand's control performance showed quick response to the reference angle and pressure. Moreover, this prosthetic hand enables grasping, pinching, and typing by simple motion selection although it is not equipped with motion estimation.

Keywords: Pneumatic · Myoelectric hand · Bellows · Motion selection

1 Introduction

Of the 82 thousand upper-extremity amputees [1] in our country, the majority use prosthetic hands. Recently, research and development of myoelectric hand have come to the forefront. However, due to the usage of multiple electric actuators, the myoelectric hand is heavy, structurally complicated and expensive. For these reasons, the adoption rate of myoelectric hand is only 2.3 % [2], hardly become widespread in society. So far, light-weight pneumatic prosthetic hand and robot hand have been researched, and McKibben artificial muscles [3] or pneumatic balloons [4] are adopted to them. These have the low mechanical efficiency and small grip force because the pneumatic actuator indirectly drives fingers with the wires. Thus, the authors propose a pneumatically-driven prosthetic hand with bellows incorporated in the joint.

Also, in order to achieve multiple grasping by the myoelectric hand, there have been many studies [5–9] on the motion estimation of EMG. For example, Most of those

© Springer International Publishing Switzerland 2016
N. Kubota et al. (Eds.): ICIRA 2016, Part II, LNAI 9835, pp. 147–157, 2016.
DOI: 10.1007/978-3-319-43518-3_15

using the neural network cannot estimate the motion correctly. In addition, they requires a lot of myoelectric sensors. It is impractical in terms of complexity of the system and increase in cost. Thus, the authors propose a myoelectric hand without motion estimate. Specifically, it is to toggle the grasp mode set in advance. In addition, there is a myoelectric hand choosing a many grasp mode with a smartphone [10]. However, the smartphone is big, and operation is troublesome. In this paper, the authors report control performance of pneumatic myoelectric hand and grasp experiment using grasp mode switching system without the motion estimation of EMG.

2   Pneumatic Myoelectric Hand

2.1   Mechanism and Drive Principle of Joints

The overview of pneumatic prosthetic hand is shown in Fig. 1. This prosthetic hand has five fingers, and each joint is independent. Therefore, it can achieve multiple grasping such as grasping, pinching, and typing. The each joint has one of two types of bellows made of polyethylene. The correspondence of bellows and joint are shown in Table 1. The table color is bellows color in Fig. 1.

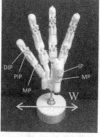

(a) Front view   (b) Side view

Fig. 1.   Pneumatic prosthetic hand

Table 1.   Bellows of joints

Fingers	Joints			
	IP	DIP	PIP	MP
First	Φ20.0 mm	—	—	Φ20.0 mm
Second	—	Φ15.0 mm	Φ15.0 mm	Φ20.0 mm
Third	—	Φ15.0 mm	Φ15.0 mm	Φ20.0 mm
Fourth	—	Φ15.0 mm	Φ15.0 mm	Φ20.0 mm
Fifth	—	Φ15.0 mm	Φ15.0 mm	Φ15.0 mm

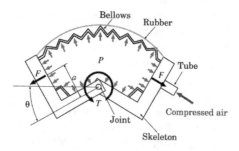

Fig. 2.   Drive principle of joints

The bellows has a compact and large pressure-receiving area, generating a large thrust force relatively. The drive principle of joints is shown in Fig. 2. As compressed air comes into bellows, it expands to longitudinal direction with generating a thrust force to bend the finger. Since the side of bellows are fixed, the joint

Table 2. Specifications

Weight	240 g
Maximum supply pressure	100 kPa
Maximum angle of joint	approx. 90 deg
Size	H 211.0 mm × W 128.1 mm × L 107.4 mm

bends angle θ by generating a driving torque T around the link to fixed side. On the other hand, when the finger is extended, it moves by restoring force of rubber on the other side, which is to open the compressed air to the atmosphere. The prosthetic hand controls driving torque T of joint by pressure P in bellows.

Thus, the prosthetic hand drives a finger by bellows directly. The fingers were designed to first finger 65.0 mm, second finger 80.0 mm, third finger 90.7 mm, forth finger 86.3 mm, and fifth finger 72.1 mm in reference to a statistical data [11]. Specifications of the prosthetic hand are shown in Table 2. General myoelectric hands are over approx. 500 g, but this prosthetic hand made of PLA is lighter than them. It would contribute to reduce users' fatigue. The pneumatic circuit of a finger is shown in Fig. 3. The pneumatic circuit consists of two lines. The lines are DIP · PIP joint line and MP joint line. The bending angle of fingers is detected by angle sensor. The pressure of bellows is detected by pressure sensor. These signals are converted A/D converter with a resolution of 12 bit, and input to the PC. The control signal u from PC are converted D/A converter with a resolution of 12 bit, and input to the microcomputer. The microcomputer carries out PWM control to valve-opening area of three ports solenoid valve so that it regulates a flow of compressed air in bellows. Maximum supply pressure P_S is 100 kPa. The prosthetic hand having seven solenoid valves performs multiple grasping by solenoid valve drive switching. The correlation between grasp and driving solenoid valves are shown in Tables 3, 4 and 5. The yellow part on the table is driving solenoid valve. The prosthetic hand control program is constructed by real-time Linux.

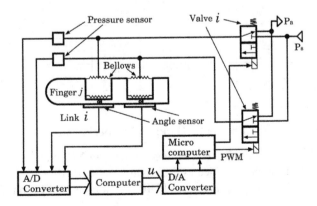

Fig. 3. Pneumatic circuit

2.2   Mechanism of Grasp Mode Switching System

Overview of the grasp mode switching system is shown in Fig. 4. It is possible for the system to toggle grasp mode. The correlation between toggle and grasp mode are shown in Tables 3, 4 and 5. It has four grasp modes: soft grasp [mode1], grasp [mode2], pinch [mode3], and typing [mode4], which are essential in daily life. These modes can be switched by a flow chart shown in Fig. 5. User can choose the optionally grasp mode by the system. Therefore, the conventional motion estimate of EMG is not required for this prosthetic hand. Myoelectric sensor is one-site system, which has only one sensor for control signals. It can reduce the cost.

Table 3. Valves of joints (Soft grasp:[mode1], Grasp:[mode2])

Fingers	Joints			
	IP	DIP	PIP	MP
First	Valve 1	—	—	Valve 2
Second	—	Valve 3	Valve 3	Valve 4
Third	—	Valve 5	Valve 5	Valve 6
Fourth	—	Valve 5	Valve 5	Valve 6
Fifth	—	Valve 5	Valve 5	Valve 7

Table 4. Valves of joints (Pinch:[mode3])

Fingers	Joints			
	IP	DIP	PIP	MP
First	Valve 1	—	—	Valve 2
Second	—	Valve 3	Valve 3	Valve 4
Third	—	Valve 5	Valve 5	Valve 6
Fourth	—	Valve 5	Valve 5	Valve 6
Fifth	—	Valve 5	Valve 5	Valve 7

Table 5. Valves of joints (Typing:[mode4])

Fingers	Joints			
	IP	DIP	PIP	MP
First	Valve 1	—	—	Valve 2
Second	—	Valve 3	Valve 3	Valve 4
Third	—	Valve 5	Valve 5	Valve 6
Fourth	—	Valve 5	Valve 5	Valve 6
Fifth	—	Valve 5	Valve 5	Valve 7

2.3   Control Law of the Prosthetic Hand and Experimental

The finger model is shown in Fig. 6. The finger moving in X-Y plane is two DOF link. Bending angle of MP joint is defined as θ_1, bending angle of DIP · PIP joint is defined as θ_2. Pressure of MP joint is P_1. Pressure of DIP · PIP joint is P_2. PI control law is used

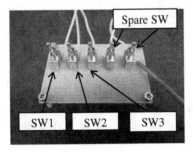

Fig. 4. Grasp mode switching system

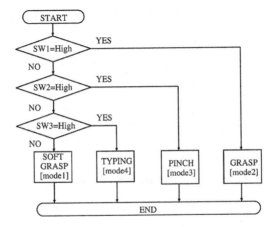

Fig. 5. Flow chart of simple motion selection

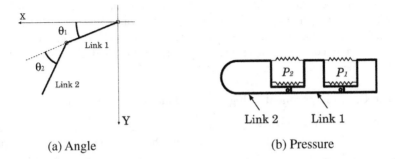

(a) Angle   (b) Pressure

Fig. 6. Model of finger

for angle and pressure control of the prosthetic hand, that is expressed by the following formula.

$$U(s) = \left(K_p + \frac{K_i}{s}\right)E(s) \tag{1}$$

Where $U(s)$ is Laplace transform of control input, K_p is Proportional gain, K_i is Integral gain, $E(s)$ is Laplace transform of deflection (Reference angle − Measured angle), (Reference pressure − Measured pressure).

Response to step input of reference angle and pressure are as follows. Furthermore, the authors conducted experiments which are to grasp certain objects by grasp mode switching system. Sampling frequency is 5.0 ms. Block diagram is shown in Fig. 7.

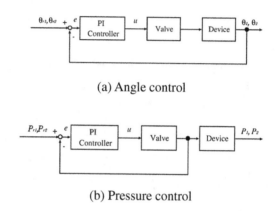

(a) Angle control

(b) Pressure control

Fig. 7. Block diagram

3   Test Results and Considerations

3.1   Response to Step Input

Results of angle response are shown in Fig. 8(a). Both θ_1 and θ_2 are tracked reference angle within 1 s. Results of pressure response are shown in Fig. 8(b). P_1 and P_2 are tracked reference pressure within 1 s. The hand's control performance showed quick response. In angle control, a finger does not return to 0 rad influence of rubber.

3.2   The Grasp Experiment Using Grasp Mode Switching System

The grasp experiment with pressure control by EMG was conducted. The objects are a 500 ml (500 g) PET bottle and an eraser. The soft objects are a rubber ball and a paper cup. A keyboard was used for typing. Grasping, pinching and typing were performed by grasp mode switching system. A test subject is an adult male. The test subject wore the prosthetic hand on his right arm using a L shaped socket as shown in Fig. 9. He also wore dry-type myoelectric sensor and a reference band on same brachioradialis muscle

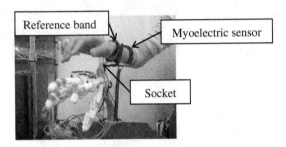

(a) Angle control   (b) Pressure control

Fig. 8. Response to step input

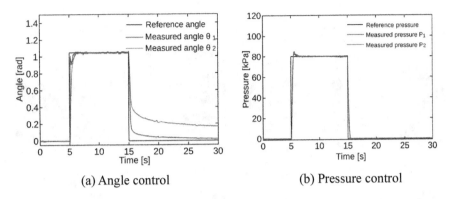

Fig. 9. Installation

of right arm. EMG envelope was used to control the prosthetic hand. The experiment was conducted using the prosthetic hand without slip resistance or surface skin. As for the control system, reference pressure was set to be proportional to EMG envelope as trial and error. In order to keep angle and pressure to be constant with EMG envelope, user needs to generate precisely, which is burden for user. The prosthetic hand cannot grasp because EMG envelope shows ringing and it makes prosthetic hand's finger vibrate. The user needs to watch object not to fall or crush. Thus, authors set prosthetic hand's threshold of EMG envelope to 5 V. Therefore, the prosthetic hand cannot control angle continuously, but prosthetic hand could grasp object stably and burden of user is small. When user typing, the prosthetic hand does not need angle control continuously because user can regulate the finger position by bending their elbow. [Mode1] is used to grasp soft object. The prosthetic hand could grasp a rubber ball and a paper cup with bending fingers by EMG envelope, shown in Figs. 10 and 11(a). Reference pressures were $P_{r1} = 40$ kPa, $P_{r2} = 20$ kPa. Mean pressure was 30 kPa. P_1 and P_2 are tracked reference pressure within 1.5 s and 1.0 s. In addition, an overshoot showed P_1, because bellows using for each joint is different. The prosthetic hand also could grasp soft object. [Mode2] is used to grasp a PET bottle. The prosthetic hand could grasp a PET bottle bending fingers by EMG envelope, shown in Fig. 11(b). Reference pressures were $P_{r1} = 90$ kPa, $P_{r2} = 30$ kPa. Mean pressure was 60 kPa. The prosthetic hand may grasp

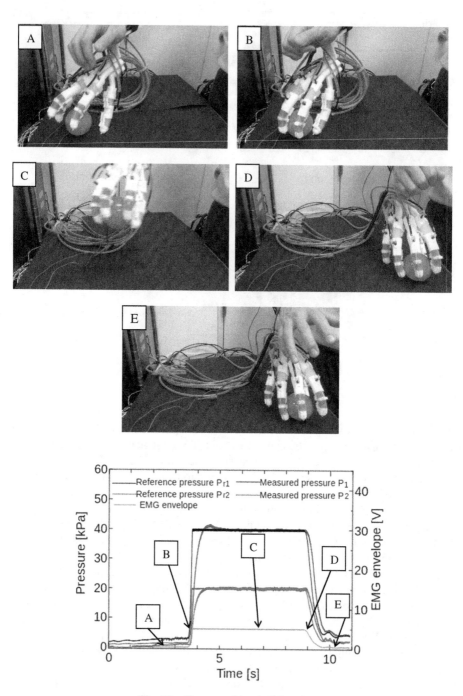

Fig. 10. Grasp a rubber ball [mode1]

(a) Paper cup [mode1]

(b) PET bottle [mode2]

(c) Eraser [mode3]

(d) Typing [mode4]

Fig. 11. All mode of the prosthetic hand

heavier object, because maximum supply pressure is 100 kPa. [Mode3] is used to pinch an eraser. The prosthetic hand could pinch an eraser bending first and second fingers by EMG envelope, shown in Figs. 11(c) and 12. Reference pressure were P_{r1} = 50 kPa, P_{r2} = 30 kPa. Mean pressure was 40 kPa. P_1 and P_2 are tracked reference pressure within 1.2 s and 1.0 s. The prosthetic hand could pinch small objects. [Mode4] is used for typing. The prosthetic hand could type bending second finger only by EMG envelope, shown in Figs. 11(d) and 13. Reference pressure were P_{r1} = 40 kPa, P_{r2} = 20 kPa. Mean pressure was 30 kPa. P_1 and P_2 are tracked reference pressure within 1.1 s and 1.0 s.

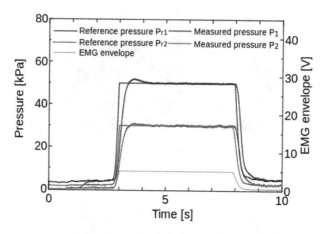

Fig. 12. Pinch an eraser [mode3]

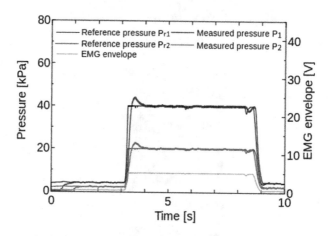

Fig. 13. Typing [mode4]

4   Conclusion

In this paper, authors developed pneumatic myoelectric hand using bellows actuator. The hand's control performance showed quick response to the reference angle and pressure. In the grasp experiment, it could grasp soft objects by grasp mode switching system according to objects. The prosthetic hand could grasp a 500 ml (500 g) PET bottle with large grip force. The prosthetic hand could pinch small objects like erasers. The prosthetic hand could type.

References

1. Cabinet Office: http://www8.cao.go.jp/shougai/whitepaper/h25hakusho/zenbun/index.html
2. Kashimoto, O.: Issues through financial support for the myoelectric upper limb prosthesis under the services and supports for persons with disabilities act. Japan. J. Occup. Med. Traumatol. **61**(5), 305–308 (2013). (in Japanese)
3. Tsujiuchi, N., Koizumi, T., Shirai, S., Nishino, S., Kudawara, T., Shimizu, M.: Development of low-pressure driven pneumatic actuator and its application to multi-finger robot hand. In: Proceedings of the 2006 JSME Conference on Robotics and Mechatronics, vol. 2006, pp. 1A1-C04(1)–1A1-C04(4), Waseda, Japan (2006) (in Japanese)
4. Nagase, J., Saga, N.: Development of a tendon driven robot hand using a pneumatic balloon. Trans. Soc. Instrum. Control Eng. **9**(11), 76–83 (2010). (in Japanese)
5. Harada, A., Ishii, C., Nakakuki, T., Hikita, M.: Recognition of finger operation via surface EMG. In: Proceedings of the 2010 JSME Conference on Robotics and Mechatronics, vol. 2010, pp. 1P1-G04(1)–1P1-G04(4), Asahikawa, Japan (2010) (in Japanese)
6. Tsuji, T., Shigeyoshi, H., Fukuda, O., Kaneko, M.: Bio-mimetic control of an externally powered prosthetic forearm based on EMG signals. Trans. Jpn. Soc. Mech. Eng. Ser. C **66** (648), 2764–2771 (2000)
7. Uchida, M., Ide, H., Yokoyama, S.: Robot of hand and finger motion using EMG. Trans. Inst. Electr. Eng. Jpn. **118**(3), 204–209 (1998). A publication of Fundamentals and Materials Society (in Japanese)
8. Nishikawa, D., Yu, W., Yokoi, H., Kakazu, Y.: On-line learning method for EMG prosthetic hand controlling. Trans. Inst. Electron. Inf. Commun. Eng. **J82-D-II**(9), 1510–1519 (1999). (in Japanese)
9. Tanabe, Y., Kosaka, M.: New motion recognition system using the ratio of Electoromyogram. In: Mechanical Engineering Congress, Japan 2013, pp. J164011-1–J164011-4 (2013) (in Japanese)
10. Touch Bionics Inc. http://www.touchbionics.com/
11. Kouchi, M.: https://www.dh.aist.go.jp/database/hand/index.html

Research of Rehabilitation Device for Hemiplegic Knee Flexion Based on Repetitive Facilitation Exercise

Yong Yu[(⊠)], Mizuki Kodama, Hirokazu Matsuwaki, Koutaro Taniguchi,
Shuji Matsumoto, Hiroko Yamanaka, Isamu Fukuda, Megumi Shimodozono,
and Kazumi Kawahira

Graduate School of Science and Engineering, Kagoshima University,
1-21-40 Korimoto, Kagoshima 890-0065, Japan
yu@mech.kagoshima-u.ac.jp, taniguchi@eng.kagoshima-u.ac.jp
http://grad.eng.kagoshima-u.ac.jp/en/

Abstract. This paper proposes a functional recovery training device for hemiplegic knee flexion based on the Repetitive Facilitation Exercises, which can realize plasticity of the brain by a special procedure. Based on the Repetitive Facilitation Exercises, a muscle-rapidly-extended facilitating stimulus is proposed to induce knee flexion stretch reflex and cause voluntary knee flexion. Then a power assist control is given to help the voluntary movement and a mechanism is devised to realize the knee rehabilitation. And, it was confirmed that to induce the stretch reflex effectively, using this device by the verification experiment of EMG measurement. In addition, the assist effect at the time of RACC was confirmed.

Keywords: Hemiplegic knee rehabilitation · Repetitive Facilitation Exercises · Muscle-rapidly-extended facilitating stimulus · Knee flexion stretch reflex

1 Introduction

Cerebrovascular disease patients is increasing every year, typical sequelae of cerebral vascular disease is a hemiplegia. In recent years, among the motor function recovery rehabilitation hemiplegia, rehabilitation using brain plasticity are be effective. The ability to acquire other function that have acquired once a specific brain ability is called the brain plasticity. There is a Repetitive Facilitation Exercises as an effective a therapeutic method using the brain plasticity [1]. By that training method, paralysis can expect to recover, even if it has been considered not recover.

Expression of brain plasticity in RFE (Repetitive Facilitation Exercises) is dependent on frequency of use, it is necessary to train continuously by therapist for paralyzed patients in hemiplegia recovery. Furthermore, it is impossible for

© Springer International Publishing Switzerland 2016
N. Kubota et al. (Eds.): ICIRA 2016, Part II, LNAI 9835, pp. 158–167, 2016.
DOI: 10.1007/978-3-319-43518-3_16

a human to measure accurately speed and distance of the training joint. Therefore, in order to realize labor saving training, and enhance the training accuracy and effectiveness by optimizing the timing and strength of the various facilitation stimulus intervention during the training, there are demands of automation robots based on RFE theory.

Rehabilitation training device with RFE rehabilitation for hemiplegic upper limbs has been developed, but the lower limbs of rehabilitation training device has not been developed yet. Also, knee joints in particular among lower limbs joints involved heavily on the day-to-day activities, such as walking operation and the standing sitting behavior. In the previous study, rehabilitation robots based on RFE theory for hemiplegic upper limbs have been developed, for example, such as "Finger-Expansion Function Recovery Training Device [2]", "Active Arm Weight-Bearing Unit [3]", and "Pronation-Supination Function Recovery Training Device [4]". In addition, although synergic voluntary movement training robots have been researched as a rehabilitation robot based on RFE theory for hemiplegic lower limbs, for example, such as "Wearable Walk Rehabilitation Device for Hemiplegic Legs with Facilitative Vibration Stimulus and Power Assistance [5]", "Robotic Rehabilitaion System for Recovery of Lower Extremity [6]", and "Exoskeleton Power Assist System HAL [7]". However, separate movement training robots based on RFE theory for hemiplegic lower limbs have never been researched. In addition, 90 % of hemiplegic knee patients can exercise extension, but it is considered flexion is difficult. By the above backgrounds, this research is aimed at the development of knee flexion motor function recovery training device.

2 RFE for Knee Joint

2.1 Motion of Knee Flexion

The movements of knee joints have two types motion which flexion and extension. Especially flexion movement of the knee is caused by biceps femoris and gastrocnemius to work at the same time [8]. Figure 1 shows triceps muscle of calf and hamstrings. Biceps femoris is a two-joint muscle to work in two of the movement of the hip joint extension, and knee joint flexion, gastrocnemius is a two-joint muscle of ankle joint plantar flexion, knee flexion.

2.2 RFE and Stretch Reflex

In RFE, the therapist gives a stimulus to the nerve pathway to operate the paralyzed limb of a patient to induce target motion. This operation is called a facilitation, it will go to strengthen and rebuild the neural circuit between the brain and muscle by repeating to the patient's paralyzed limb. To induce patient's voluntary movement, the therapist gives tapping stimulus or stretch reflex (muscle reflection when you stretch the muscle to rapidly shrink [9]). It is possible to cause the voluntary movement of paralyzed patients, at the moment of increasing the excitement of the motor cortex by the stretch reflex and tapping stimulation of paralyzed limbs by sending an optional command from the brain to the paralyzed limb.

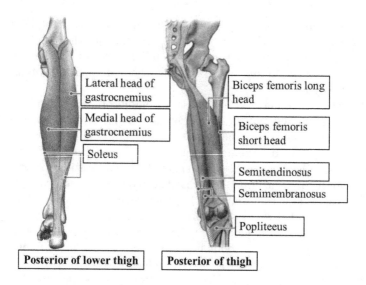

Fig. 1. Triceps muscle of calf and Hamstrings [8]

2.3 RFE Training Procedure of Knee Flexion (Sitting Position)

RFE training procedure of knee flexion of knee joint flexion as follows. It is shown in Fig. 2 training appearance of facilitation repeat therapy of hemiplegic knee flexion.

1. With one hand, grasp the toes, and the other hand grasp the hamstrings, make a passively knee extension at a constant speed.
2. Just before the end of stretching the paralyzed limb moves in quickly the direction of extension to the vicinity of the range of motion, to induce the stretch reflex of knee joint flexion muscles.
3. Perform quickly an inner-back operation of the ankle joint, to maintain the tension of the knee joint flexion muscles. At the same time, given the tapping stimulus to the hamstrings, enhance the excitement of the motor cortex.
4. Voluntary flexion movement is occurred by the stretch reflex and tapping, and patient actively move flexion motion. At that time, the therapist read the motion intention, while providing resistance to the automatic movement in the opposite direction, to assist the movement.

Compared to the case of standing position, the biceps femoris is more stretched by the sitting position, the patient easily cause a stretch reflex. Because, biceps femoris is the main operation muscles of the knee joint. And, it is a two-joint muscle to work as extension of the hip joint and flexion of the knee joint. And, to stretch the biceps femoris by the operation of the paralysis limb of from 1 to 2 operation, the gastrocnemius muscle is stretched by the inner-back operations of 3. By stretching the biceps femoris and the gastrocnemius which a two-joint muscles related in knee flexion, are trained to facilitate the stretch reflex of knee joint flexion.

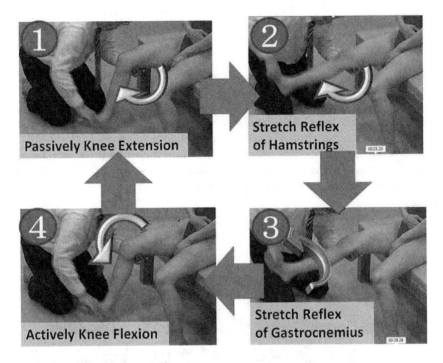

Fig. 2. Appearance of the RFE of knee flexion [1]

3 Mechanisms Required for the Device

3.1 Proposal of Foot Horizontal Driving Mechanism

Inner-back of the ankle joint is a complex exercise that combines the three directions of motion of the ankle of supination-adduction-plantar flexion. It takes a plurality of actuators in order to achieve these movements. So, there are some problems such as "Cost increase", "complex control", and so on. So, we propose a device to achieve the training effect of RFE, and realizes enough training effect only by a single actuator.

The purpose of Inner-back operation of the ankle joint is extending the gastrocnemius as a two-articular muscle of the knee joint flexion muscle, ankle plantar flexion. Therefore, it was thought to be able to reproduce the effects of the ankle joint in-back by to stretch the gastrocnemius muscle in the ankle joint in plantar flexion. Because knee extension and ankle plantar flexion is the same planar motion, device drive mechanism uses the foot horizontal driving mechanism of one axis. This mechanism realize two joints movements of the knee extension and ankle plantar flexion at the same time, in only 1 degree of freedom. In addition, ankle dorsiflexion movement in walking behavior is an important motion in order to walk without dragging the foot. Because it can also cause stretch reflex of ankle dorsiflexion by the movement of foot horizontal driving mechanism, it is possible to carry out the training of ankle dorsiflexion at the time of walking motion.

3.2 Proposal of High Sensitivity Force Sensing Mechanism

Therapist have to read the patient's motion intent in performing active flexion motion. Because the force of the hemiplegic limbs are weak, it is required that the sensing mechanism which can detect small forces in order to read the motion intent. Therefore, in order to get the patient's motion intent, it was adopted the sensing system using the strain expansion mechanism [10]. Figure 3 shows designed device and force sensing mechanism.

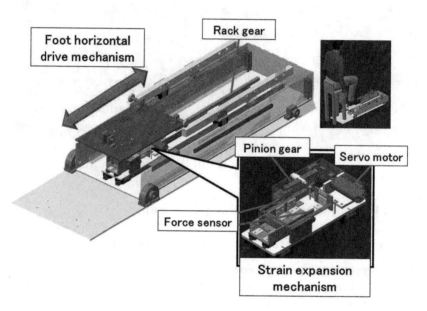

Fig. 3. Designed device and force sensing mechanism

4 Control of Hemiplegic Knee Motor Function Recovery Training Device

4.1 Proposal of Knee Flexion RFE Using Device

The training method of knee flexion using the device will be described below. A schematic diagram of a training method shown in Fig. 4.

1. Move the paralyzed limb by foot horizontal driving mechanism at a constant speed passively in the direction of extension.

2. Induce the stretch reflex of the knee joint and the ankle joint at the same time, by accelerating the drive mechanism in the vicinity of the limit of the motion range.
3. Voluntary movement of the patient is induced by the stretch reflex, the patient start active knee flexion. When the active exercise of the patient, provide reverse resistance to the direction of patient's movement, and assist by sensing the motion intent.

RFE of knee flexion will be performed using the device by the repetition of 3 operations above-mentioned.

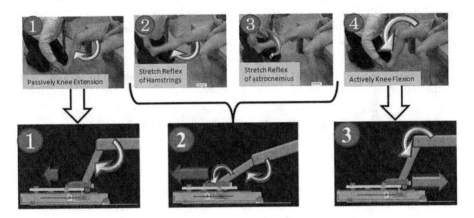

Fig. 4. Hemiplegic knee flexion RFE using the device

4.2   Proposal of Resistance-Accompanying Cooperation Control

The case of patient's active exercise, the therapist give a small resistance in a direction opposite to patient's movement in order to maintain muscle tone, and assist the motion with sensing the intent of patient's motion. In order to realize this operation, Resistance-Accompanying Cooperation Control (RACC) adopted at the time of the patient's active motion. The RACC is a control method to assist the movement by sensing the force of the paralyzed limb, while gives the device driving unit so as to have a constant impedance characteristic with respect to the force information resistance to paralysis limb [2]. There is a need to generate a target speed from the force information at the time of knee flexion. The impedance equation for the force of the paralyzed limb F is expressed by the following equation.

$$F = M\ddot{p} + C\dot{p} + K\Delta p \tag{1}$$

p represents the current position of the device drive unit, M represents the virtual mass, C is a virtual viscosity, K represents a virtual elasticity. Discretize the Eq. (1), and set $K = 0$, the following equation is obtained.

$$\dot{p}_{n+1} = \left\{ F_n - \left(\frac{C}{2} - \frac{M}{\Delta t} \right) \dot{p}_n \right\} \Big/ \left(\frac{C}{2} + \frac{M}{\Delta t} \right) \qquad (2)$$

Δt is the sampling time. The impedance characteristics can change by adjusting two coefficients M, C.

4.3   Tuning Control for the Stretch Reflex Time Lag

Operations 1 and 2 are controlled by speed control, because training motions are passive exercises. However, the operation 3 is controlled by RACC, because the training motion is active exercise. Between the operation 2 and the operation 3, the motion of the knee change from passive exercise to active exercise. Then, a time lag occurs until inducing motion by stretch reflex of the patient. Therefore, the waiting time was set between the finish time of passive exercise and the start time of active exercise. Consequently, the operation of device motion well tuned to the active exercise and rhythm of the patient, it is possible to improve the training effect.

5   Verification of the Realization of RFE by the Prototype Device

The device based on designed mechanism and control method mentioned above was prototyped. Figure 5 shows the prototyped device. In order to realize the RFE, it is necessary to realize the power assistance of active exercise, and to induce of the stretch reflex of the semitendinosus. A verification experiment to confirm the performance of the device was conducted. The relationship between velocity and force during the training using the device is shown in Fig. 6.

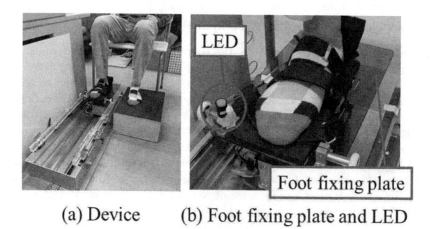

(a) Device   (b) Foot fixing plate and LED

Fig. 5. Hemiplegic knee motor function recovery training device

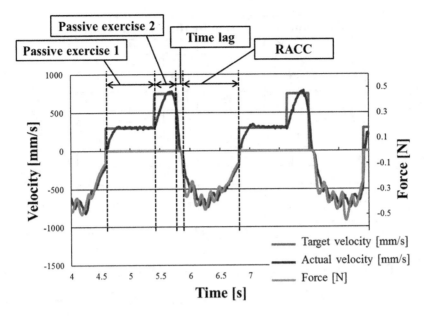

Fig. 6. Relationship between the velocity and force at the time of the training by the device

5.1   Verification of Inducing Stretch Reflex by the Device

The verifying test was conducted whether stretch reflex induced to paralysis limb during training. Paste the surface electrode to semitendinosus which is a knee flexion muscles of the subject, the surface muscle potential at the time of training by the device was measured. Verification was carried out to able-bodied subjects. Figure 7 shows a relationship between the velocity and EMG during training. The results indicate that EMG was increasing rapidly, immediately after stretching of the muscle by the acceleration from the first speed to the second speed. Consequently, it was confirmed that the stretch reflex of the knee flexion muscles induced by the training using the device.

5.2   Verification of the Assist Effect by the RACC

The assist effect by RACC was verified. Paste the surface electrode to semitendinosus of able-bodied subject, and verify by comparing EMG.

Firstly, in order to measure EMG needed for the normal knee joint flexion,

1. Attach to the device to foot, perform the flexion and extension in a state of not power on.

By changing the values M, C of RACC formula, it can be adjusted assist effect. Therefore, under the following conditions, EMG at the time of knee flexion-extension was measured.

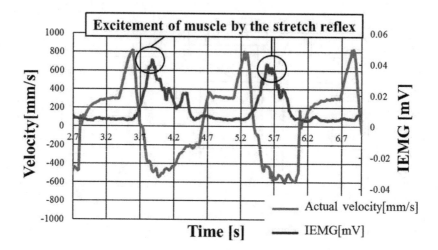

Fig. 7. Velocity and EMG at the time of training using the device

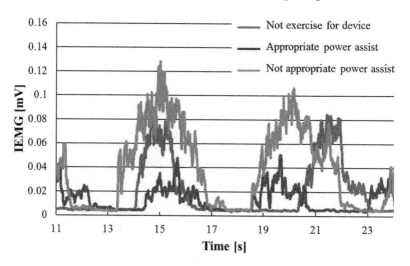

Fig. 8. Evaluation of the power assist effect

2. Appropriate virtual mass such that the device is moved by a small force, RACC at the time of the virtual viscosity.
3. Virtual mass the assist becomes less effective as opposed to 2, RACC at the time of the virtual viscosity.

Experiments were carried out under the conditions of the above. The results are shown in Fig. 8. The maximum EMG value of normal flexion output by the condition 1 was 0.08 [mV]. In the case of not appropriate assist effect, the maximum EMG value of flexion was 0.12 [mV]. Consequently, under the conditions of 3 was poor assist effect, because it takes a big load on the movement compared to the flexion normal, it can be seen that large force need for flexion.

The maximum EMG value of flexion in suitable assist effect is 0.04 [mV], that is one half by compared with the normal one. As a result, in the optimal assist condition, it was confirmed that the device can operate with half of the force required to the original flexion movement. Therefore, for hemiplegic patients who do not put out only a weak force, by using the RACC, it is possible to train while assisting the active exercise of the patient.

6 Conclusion

The device it is possible to reproduce the RFE which is an effective hemiplegia recovery training method has been developed. And, it was confirmed that to induce the stretch reflex effectively, using this device by the verification experiment of EMG measurement. In addition, the assist effect at the time of RACC was confirmed.

As future works, developing a software required for the clinical trials, the realization of hemiplegic knee motor function recovery training device by going through the clinical trials are expected.

References

1. Kawahira, K.: Exercise therapy for hemiplegia recovery - Theory of Kawahira method and the neural pathways reinforcing Repetitive Facilitation Exercises: IGAKU-SHOIN Ltd., Japan (2006)
2. Yong, Y., Iwashita, H., Kawahira, K., Hayashi, R.: Development of functional recovery training device for hemiplegic fingers with finger-expansion facilitation exercise by stretch reflex. Trans. Soc. Instrum. Control Eng. **48**(7), 413–422 (2012)
3. Yu, Y., Nagai, M., Matsuda, J., Kawahira, K., Hayashi, R.: Development of an active arm weight-bearing unit and its application in hemiplegic arm rehabilitation. In: Proceedings of SICE Annual Conference, pp. 2082–2087 (2014)
4. Yu, Y., Nakanishi, Y., Kawahira, K., Shimodozono, M., Hayashi, R.: Production of effective stretch reflex by a pronation and supination function recovery training device for hemiplegic forearms. In: Proceedings of 2015 IEEE International Conference on Robotics and Biomimetics, pp. 150–157 (2015)
5. Yong, Y., Toyama, T., Shimodozono, M., Kawahira, K., Hayashi, R.: Development of wearable walk rehabilitation device for hemiplegic legs with facilitative vibration stimulus and power assistance. J. Robot. Soc. Jpn. **33**(7), 497–504 (2015)
6. Hirata, R., Nakamoto, Z., Hachisuka, K., Wada, F., Makino, K.: Robotic rehabilitation system for recovery of lower extremity. J. Soc. Biomech. **30**(4), 205–210 (2006)
7. Sankai, T., Kawamura, Y., Okamura, J., Woong, L.S.: 915 study on hybrid power assist system HAL-1 for walking aid using EMG. In: Ibaraki District Conference, pp. 269–270 (2000)
8. Takei, H.: Dictionary to understand the mechanism of muscle and joint. Seitosha Co. Ltd., Japan (2013)
9. Sakai, K.: The normal structure and function of the human body. Jpn. Med. J., Japan (2012)
10. Yong, Y., Ishitsuka, T., Tsujio, S.: Torque sensing of finger joint using strain-deformation expansion mechanism. J. Robot. Soc. Jpn. **20**(5), 526–538 (2002)

Development of an Add-on Driving Unit for Attendant Propelled Wheelchairs with Sensorless Power Assistance

Masashi Shibayama, Chi Zhu$^{(\boxtimes)}$, and Wang Shui

Department of Environmental and Life Engineering, Graduate School of Engineering,
Maebashi Institute of Technology, Kamisadori 460-1,
Maebashi, Gunma 371-0816, Japan
{m-sibayama_g,zhu}@maebashi-it.ac.jp

Abstract. In this study, we are developing a power add-on unit (PAU) for attendant propelled wheelchairs with sensorless speed control and power assistance to help an attendant who are pushing a conventional commercialized manual wheelchair in which a patient is sitting in the wheelchair. This development is based on the following three concept: to make our PAU (1) easy to be attached to and detached from the wheelchair; (2) able to be used to the most of the conventional commercialized wheelchairs; (3) low cost. Further, in order to make the attendant more easily and comfortably propel the wheelchair, a sensorless power assistance approach is proposed. The experimental results verify the effectiveness of the proposed approaches.

Keywords: Wheelchair · Sensorless power assistance · Disturbance observer

1 Introduction

With the rapid progress of aging society, more and more people whose walking is difficult or impossible due to aging, illness, injury, or disability, need assistance for mobility. Most commonly, the alternative mode of mobility chosen is a manual wheelchair. A manual wheelchair is the preferred prescription if the user possesses the capabilities to operate one. It encourages the operator to remain physically active because it takes a tremendous effort to propel the large drive wheels. However, a manual wheelchair is a very inefficient form of transportation [1]. Moreover, use of a manual wheelchair requires significant upper extremity strength, especially to traverse up or down inclines and to propel for long distances. Recently, development of various assistive technologies to support the elderly, the disabled or the attendants (caregivers) has attracting researcher's attention. Now, many walking support systems and electrical powered wheelchairs (EPW) are developed to reduce the burden of the user (the elderly, the disabled or the attendant) [2–10]. But, people who use a powered wheelchair encounter two problems. The first one is a powered wheelchair is usually much

© Springer International Publishing Switzerland 2016
N. Kubota et al. (Eds.): ICIRA 2016, Part II, LNAI 9835, pp. 168–178, 2016.
DOI: 10.1007/978-3-319-43518-3_17

more expensive than a manual wheelchair. The second one is many powered wheelchairs are usually large, heavy, bulky and generally cannot be folded or lifted manually for transportation. These two problems significantly inhibit the wide use of the powered wheelchairs. The power add-on unit (PAU, [11]) is a good solution that allows the convenience of a powered wheelchair with low cost and the ease of transportation provided by a manual wheelchair. Though PAUs can greatly enable the users of the manual wheelchairs much easier to propel the wheelchairs, there are seldom academic papers on them. Contrarily, in market there have existed a number of commercialized PAUs [12].

In this paper, we focus on two aspects. One is how to design the mechanical structure of the PAU so that the brackets or adapter plates for PAU docking no longer need to be mounted on the manual wheelchair in advance. Therefore, any untrained person can simply attach the PAU to and detach it from the wheelchair just by bare hands and any tools such as wrench are not needed. The other is how to develop a control that makes the powered wheelchair with PAU yield to the attendant's walking intent, not conversely. This not only guarantees the safety of the attendant and the seat occupant but also makes the attendant much easier and more comfortable to propel the wheelchair.

2 Structure of the Novel Power Add-on Unit

2.1 Design Concept of the Novel Power Add-on Unit

Based on the investigation of the commercialized PAUs for attendant propelled wheelchairs (APWs) on market, we determine the concept of our development of a novel PAU for APW as follows:

1. The novel PAU should be easy to be attached to and detach from the wheelchair. Concretely, the mechanical structure of the PAU should be as simple as possible and the brackets or adapter plates for PAU attachment do not need to be mounted on the manual wheelchair in any time. Thus, anyone can simply attach the PAU to and detach it from the wheelchair just by his/her hand and any tools such as wrench no longer need.
2. The novel PAU should be attachable to almost all of the manual wheelchairs on the market.
3. The PAU should be cheaper and more compact than most of the PAUs on the market.
4. The novel PAU should completely yield to the attendant's walking ability, or saying, walking speed, not conversely that the attendant has to yield to the powered wheelchair's running speed. Note that the latter is the main drawback of the most speed controlled PAUs. This function can be implemented with the admittance control applied in our previous research [9], in which, the target speed is not directly pre-selected, but is generated by the exerted force of the attendant via a mechanical admittance model. With our control approach, the novel PAU not only can guarantee the safety of the attendant and the seat occupant but also can make the attendant much easier and more comfortable to maneuver the wheelchair.

Fig. 1. A prototype of the novel PAU for attendant propelled wheelchair

Fig. 2. The tipping lever exists in almost all of manual wheelchairs

2.2 Structure of the Novel Power Add-on Unit

Based on the above development concept, the prototype of our novel PAU in development is shown in Fig. 1. The structure of the PAU is seemingly the same as the conventional ones on market, which usually consist of a handle control panel, a control unit, a battery pack, and a drive unit. But in our new PAU, there are some significant differences from the conventional ones.

1. Easy attachment and detachment: To satisfy our first design concept, we consider what is the part existed in almost all of the manual wheelchairs and the place of the part should be easily accessed. We find that this part is tipping lever as shown in Fig. 2. It exists in almost all of manual wheelchairs as well as it is at the rear of a manual wheelchair, the most convenient position to access by attendant. For this reason, we determine our PAU should be attached to the tipping lever. As shown in Fig. 3, two clamps are used to fasten the PAU to the two tipping levers on the two sides of the frame of the wheelchair, respectively. In this way, the PAU can be very easily attached to and removed from the manual wheelchair only with bare hands no need the help of any tools by any untrained person including the attendant who

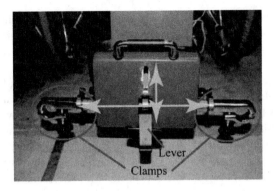

Fig. 3. The adjustable mechanism to fit for various manual wheelchairs

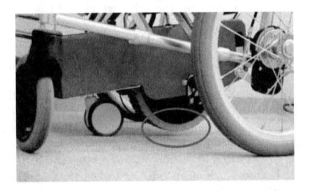

Fig. 4. The drive wheel of PAU is in lifted state

pushes the wheelchair. In addition, since the tipping levers are at the rear of the wheelchair, the PAU is at the rear of the wheelchair, too.

2. Adaptability to various wheelchairs: To satisfy the second development concept, the PAU should possess adaptability to fit for the different manual wheelchairs. As shown in Fig. 3, the two clamps at two ends of the bar are slidable along the bar and the height of the bar itself are adjustable by a lever. In this way, our PAU can be attached to almost all of commercialized manual wheelchairs.

3. Conversion between drive mode and free mode: It is worth mentioning that a manual wheelchair is not always necessary to be powered, for instance, in indoor environment. Contrarily, in outdoor, especially when running on up/down slope, there are great benefits for a manual wheelchair to be powered by a PAU. This means that a PAU mounted wheelchair has two motion modes. One is drive mode, in which the PAU is engaged in drive to support the attendant. The other is free mode, in which the PAU is not activated and the wheelchair is completely propelled by an attendant like a normal manual wheelchair.

Consequently, there is a problem, that is, how to switch these two modes. A mechanical lever is used to manually adjust and lock the height of the drive wheel so that the drive mode and free mode can be easily switched while the entire structure of our PAU is kept very simple. When the assistance is no longer needed, for instance, in indoor environment, the drive wheel can be lifted so that the powered wheelchair can be maneuvered as a manual wheelchair without extra resistance caused by the PAU. This is free mode. Figure 4 shows the PAU is in lifted state in which the drive wheel of the PAU does not contact the ground and the wheelchair is operated as a normal manual wheelchair.

Meanwhile in our PAU, we use shield lead acid battery, that is cheap and heavy. As shown in Fig. 1, the battery pack is at the front of the PAU. This is to increase the exerted normal force of the drive wheel (in-wheel motor) to the ground so that the grip performance and traction force of the PAU can be enhanced and the slippage between the drive wheel and the ground could be greatly prevented.

3   Speed Observer and Sensorless Speed Control

The ultimate purpose of a PAU is to help an attendant to push the wheelchair. Therefore, its most basic function is to control the speed that is usually pre-selected via the handle control panel or something like that to prevent the speed variance mainly caused by load or environment condition. In order to realize a speed control, the actual speed of the motor is necessary and it is conventionally measured by speed sensor such as encoder or tachogenerator. However, in our PAU, we use an in-wheel motor that is in fact a hub motor incorporated with a DC motor, a gear, and a brake, but there are no any speed sensors in the motor. In this study, we employ a speed observer to estimate the motor's speed and then control this estimated speed with a simple PID control for PAU.

A DC motor model is shown in Fig. 5(a), in which, a current I flows through the armature according to the terminal voltage U, the motor's inductance L, resistance R, and the back emf voltage U_e. Its equation is expressed as

$$U = L \cdot \frac{dI}{dt} + R \cdot I + U_e \tag{1}$$

The back emf voltage U_e is proportional to the motor's speed ω_m,

$$U_e = k_e \cdot \omega_m \tag{2}$$

where k_e is the back emf constant. For a miniature motor, its inductance L is usually small and it can be neglected. Therefore, the speed of output axis with a reduction gear ratio gr can be expressed as

$$\omega = \frac{U - R \cdot I}{gr \cdot k_e} \tag{3}$$

Since R, k_e, and gr are constants, the output speed ω as shown in Fig. 5 can be easily estimated with the above equation from the motor's terminal voltage U

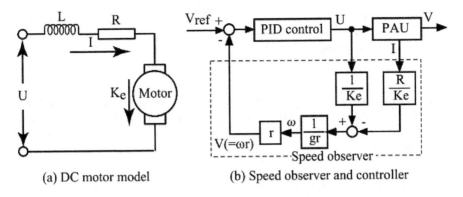

(a) DC motor model   (b) Speed observer and controller

Fig. 5. (a) DC motor model; (b) Speed observer and controller

and current I. This is in fact a speed observer that estimates the speed of the PAU without speed sensor. With this estimated speed, a simple PID controller is used to control the speed of the PAU and wheelchair. The developed speed observer and controller are shown in Fig. 5(b), in which, r is the radius of PAU's drive wheel.

4   Sensorless Power Assistance

4.1   Principle of Power Assistance

As aforementioned in Subsect. 2.1 with regard to the conventional PAU, the attendant has to yield to the powered wheelchair's pre-selected speed regardless of his/her walking ability. Contrarily, our PAU can completely yield to the attendant's walking ability, or saying, walking speed. This function is implemented with the admittance control. Here, we briefly describe the principle of admittance control for power assistance.

Transfer function of admittance of a mechanical system is defined as

$$G(s) = \frac{V(s)}{F(s)} = \frac{1}{M_s + D} \tag{4}$$

where, V is the output speed, F is the input applied force; M and D are respectively the virtual mass and virtual damping of the defined system, and these two parameters can be arbitrarily determined as needed. The time response $V_{ref}(t)$ for a step input F of the above transfer function is:

$$V_{ref}(t) = \frac{F}{D}(1 - e^{-t/\tau}) \tag{5}$$

where, τ is the time constant defined by $\tau = M/D$. The $V_{ref}(t)$ is the speed command of our PAU. Further, with the estimated speed of PAU as shown in Eq. (3) that is assumed as the actual speed of PAU. A simple PID speed controller is used to realize the power assistance. In steady state the necessary pushing force F_s, or saying, the burden that the attendant feels reacted from the powered wheelchair should be

$$F_s = V_s \cdot D \tag{6}$$

Thus, by simply adjusting the virtual damping coefficient D, the attendant will have different burden feeling. By the way, with different virtual mass coefficient M (therefore τ is changed), the different dynamic corresponds of the system will be obtained.

4.2   External Force Estimation

In the above admittance control, the force the attendant applies to the handle of the wheelchair is usually measured by an expensive multi-axis force/torque sensor. To make our PAU affordable for most of users, we do not use such expensive sensor. In stead, we develop a disturbance observer to estimate the applied force of the attendant. As shown in Fig. 6, F_{xs} is the propelling force the attendant applies to the handle of the wheelchair; F_{rs} represents the rolling friction between the drive wheel and the ground; F_{ts} is the traction force of the drive wheel of PAU; m is the total mass including the wheelchair, the PAU, and the seat occupant. On a slope with a angle θ about to the ground,

$$F_{rs} = c \cdot mg \cdot cos\theta \tag{7}$$

where, c is the coefficient of rolling friction. F_{ts} is caused from the motor's electromagnetic torque and it drives the drive wheel of the PAU by the following equation:

$$F_{ts} = \frac{gr}{r}(K_\tau I - J\frac{d\omega_m}{dt} - D_m\omega_m) - mg \cdot sin\theta \tag{8}$$

where, K_τ, J, and D_m are the motor's torque coefficient, motor's initial moment and damping, respectively. On the other hand, as shown in Fig. 6, the resultant force F along the slope is

$$F = F_{xs} + F_{ts} - F_{rs} = mr\frac{d\omega}{dt} + D_W r\omega \tag{9}$$

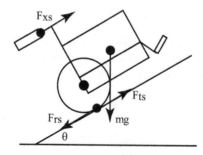

Fig. 6. Force model of a powered wheelchair on a slope

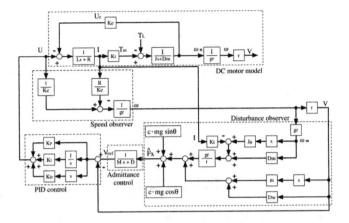

Fig. 7. The whole system model

where, D_W is assumed to be the damping of the powered wheelchair. In this way, the applied force F_{xs} of the attendant is estimated by the following formulation:

$$F_{xs} = F + F_{rs} - F_{ts} \tag{10}$$

This equation is a disturbance observer for sensorlessly estimating the applied force of the attendant. The whole system model including speed observer, disturbance observer, admittance control, and speed control is shown in Fig. 7.

4.3   Experimental Results

First we verify the estimation method for the applied force of the attendant by the disturbance observer. The estimated force and the measured force by 6-axis force/torque sensor are shown in Fig. 8. We can find that, the estimated force approximately coincides with the measured force, though the error is comparatively big at the start and stop phases. Further, the developed admittance based

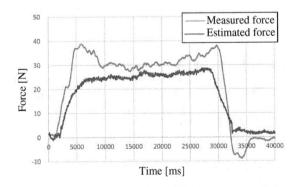

Fig. 8. Experiments result: the measure force and the estimated force

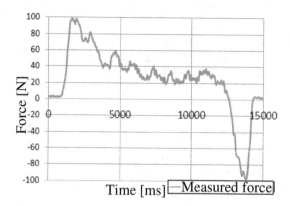

Fig. 9. Experimental results: pushing force without power assistance

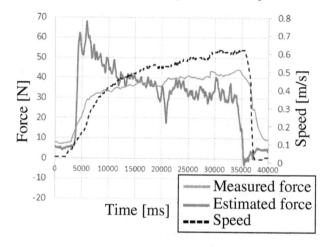

Fig. 10. Experimental results: with power assistance (as $D = 50\,[\text{N·s/m}]$)

interaction control is tested. The sampling and control period is 1 [kHz]. The attendant is holding the robot handle to move the robot forward while a 90 [kg] person sitting in the seat.

Experiment Without Power Assistance. We turn off the PAU and do the experiment without power assistance. In this case, The walking speed was measured separately. The result is shown in Fig. 9. We can find:

1. To start to move the robot, a big pushing force as large as 100 [N] is needed.
2. In steady state, the pushing force is about 30 [N].
3. To stop the robot, a big negative (pulling) force is needed.
4. The user's walking speed is about 0.6 [m/s].

Experiments with Power Assistance as $D = 50\,[\text{N·s/m}]$. In this case, we try to keep the attendant's burden with power assistance to about 30 [N].

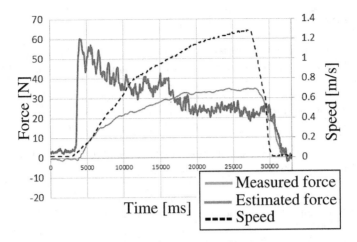

Fig. 11. Experimental results of: with power assistance (as $D = 25$ [N·s/m])

Since the user's walking speed is about 0.6 [m/s], according to Eq. (6) we set $D = F/V = 50$ [N·s/m]. Figure 10 show the experimental results. The results show that the pushing force has no change when the steady walking speed is about 0.6 [m/s] as we planned, and no pulling (negative) force is needed to stop the wheelchair, which can be directly interpreted. The purpose of power assistance for the caregiver is achieved.

Experiments with Power Assistance as $D = 25$ [N· s/m]. The result of power assistance experiment with $D = 25$ [N·s/m] is shown in Fig. 11. The result shows that the measured pushing force is about 25 [N] and the estimated force is about 30 [N] when the walking speed is about 1.2 [m/s]. Same as the above, these results demonstrate that the pushing force (the user's burden) is well estimated and controlled by the admittance based controller. How to optimally select the parameters is one of our future tasks.

5   Conclusions

In this study, we developed a novel power add-on unit (PAU) for attendant propelled wheelchair with sensorless power assistance to help an attendant who are pushing an conventional commercialized wheelchair in which a patient is sitting in the wheelchair. The most two significant characters of this PAU are no brackets and plates needed to be installed on the wheelchair and easiness to be attached to and removed from the wheelchair. To enhancement its performance and in order to make the attendant more easily and comfortably propell the wheelchair, a sensorless power assistance approach is proposed and experimentally verified.

References

1. Beekman, C., Leslie, M., Marion, S.: Energy cost of propulsion in standard and ultralight wheelchairs in people with spinal cord injuries. Phys. Ther. **79**(2), 146–158 (1990)
2. Abel, E., Frank, T., Boath, G., Lunan, N.: An evaluation of different designs of providing powered propulsion for attendant propelled wheelchair. In: Annual International Conference of the IEEE Engineering in Medicine and Biology Society, pp. 1863–1864 (1991)
3. Cooper, R.A.: Intelligent control of power wheelchairs. IEEE Eng. Med. Biol. Mag. **14**(4), 423–431 (1995)
4. Oh, S., Hata, N., Hori, Y.: Proposal of human-friendly motion control: control design for power assistance tools and its application to wheelchair. In: Proceedings of the 30th Annual Conference of the IEEE Industrial Electronics Society, pp. 436–441 (2004)
5. Ding, D., Cooper, R.: Electric powered wheelchairs. IEEE Control Syst. **25**(2), 22–34 (2005)
6. Miyata, J., Kaida, Y., Murakami, T.: Coordinate-based power-assist control of electric wheelchair for a caregiver. IEEE Trans. Industr. Electron. **55**(6), 2517–2524 (2008)
7. Zhu, C., Oda, M., Yoshioka, M., Nishikawa, T., Shimazu, S., Luo, X.: Admittance control based walking support and power assistance of an omnidirectional wheelchair typed robot. In: Proceedings of the 2010 IEEE/RSJ IROS, pp. 381–386 (2010)
8. Suzuki, T., Zhu, C., Yan, Y., Yu, H., Duan, F.: Power assistance on slope of an omnidirectional hybrid walker and wheelchair. In: Proceedings of the 2012 IEEE ROBIO, pp. 974–979 (2012)
9. Clark, L.L.: Design and testing of a quick-connect wheelchair power addon unit, Doctor Dissertation of the Department of Industrial and Systems Engineering, The Virginia Polytechnic Institute and State University (1997)
10. Duke, S.: Motor Attachment for Wheelchairs, United States Patent Number 2,495,573 (1950)
11. Yamaha Motor. http://www.yamaha-motor.co.jp/wheelchair/
12. http://www.usatechguide.org/reviews.php?vmode=1&catid=144

Modeling Dynamic of the Human-Wheelchair System Applied to NMPC

Gabriela M. Andaluz[3], Víctor H. Andaluz[1,2(✉)], Héctor C. Terán[1],
Oscar Arteaga[1], Fernando A. Chicaiza[1], José Varela[1],
Jessica S. Ortiz[3], Fabricio Pérez[1], David Rivas[1], Jorge S. Sánchez[1],
and Paúl Canseco[2]

[1] Universidad de las Fuerzas Armadas ESPE, Sangolquí, Ecuador
victorhandaluz@uta.edu.ec, {vhandaluz1,hcteran,
obarteaga,mfperez3,drrivas,jssanchez}@espe.edu.ec
[2] Universidad Técnica de Ambato, Ambato, Ecuador
jazjose@hotmail.es
[3] Escuela Superior Politécnica de Chimborazo, Riobamba, Ecuador
andaluzgaby_4@yahoo.com, jessortizm@outlook.com

Abstract. This paper presents the development of algorithm nonlinear predictive controller based on the model, NMPC, trajectory tracking applied to a wheelchair. Also it illustrated the kinematic and dynamic modeling of human-wheelchair system is considered in which the deviation of the center of gravity of the system due to postural problems of man, limb amputations, or obesity user. In NMPC they considered restrictions on control measures, as well as in the states of the system. The results show that the movements of the wheelchair converge to the desired trajectory in accordance with the provisions of the theoretical design; with analysis times NMPC calculation algorithm concludes that the controller can be implemented in real time.

Keywords: Dynamic modeling · Control algorithms · Model predictive control · Wheelchair · Trajectory tracking · Helpful robots

1 Introduction

In this last times, research in advancing robotic is focusing on developing functional robots [1–4], these robots are used in industry in environments structured or partially structured for mass production, applications in welding, cut, in mechanization among others; whereas in environments unstructured, we have space applications with robots sent to explore Mars and space in general [5]; military applications in land, air and water mobile robots for mine detection, exploration of hostile environments, lookouts robots and rescue robots used for military personnel [6]; agricultural applications oriented precision farming robots used to: optimize crops, pest control and spraying, soil preparation [7] among others; medical applications consigned to better quality of life through tele-oriented robots robotics, rehabilitation [8], robots storage and distribution of drugs; among other applications.

© Springer International Publishing Switzerland 2016
N. Kubota et al. (Eds.): ICIRA 2016, Part II, LNAI 9835, pp. 179–190, 2016.
DOI: 10.1007/978-3-319-43518-3_18

Robotic oriented medicine can be classified according to their application (i) surgeon robot is intended for classroom and remote surgeries, exams through cameras, among these robots can highlight: Robotic Catheter Sensei X, NeuroArm Surgical System Da Vinci, among others (ii) robots prosthesis intended to replace or restore the function of a limb of people who have disabilities, among these are: prosthetic hand as the hand of Canterbury, artificial legs and arms [9]; therapeutic robots are aimed at facilitating the physical, motor, verbal and psychological rehabilitation of patients, they help in the movement of a person so that they can develop their daily activities [10]; care robots are aimed at improving the quality of life for people with special needs, senior citizens and people who have suffered an accident that limits their mobility whether partial or total, among which may be mentioned Manus' 80s, the Movaid, Helpmate, including that over time have improved their human service-oriented applications. In this context, in recent years there is great interest from the scientific community in the development of control algorithms to provide a stable and robust in each of robotic applications in medicine, so handling these autonomous robots become tools able to fulfill the tasks in a reliable manner. A wheelchair is designed to allow the movement of people with mobility problems or reduced mobility due to a physical injury or psychological illness (paraplegia, quadriplegia, etc.). There are two types of chairs manual and electric wheelchairs, the manual wheelchair depend directly the men for their movement, and depending on the disability a person needs for handling, while the electric chairs movement is performed by electric motors and handling is generally via a joystick.

In this context what is sought is to develop robotic chairs, so that their movement and management is an autonomous way, for this you have different types of control that can be applied in these chairs, within the algorithms used are: controllers based on fuzzy logic [11], the fuzzy logic is based on the decision from a vague, ambiguous, inaccurate or incomplete information; this methodology has the disadvantage of requiring a high computational effort [12], algorithms based on a method of multiple hypotheses to predict the driver's intentions [13], so that the chair control safer; PID controllers this driver is most commonly used in industrial robots, their algorithm is based on modeling in order to compensate the dynamic terms present in performance; other controller is based predictive algorithms, this method involves predicting future response and calculated by an optimization procedure, the actions with the process to a desired target [14].

The commercial electric wheelchairs are not designed exclusively for each disability that can occur in person, in its most general form all have a common operation, and this is to facilitate mobility. In this respect the scientific community seeks to expand the autonomy of wheelchairs so that the chair is the most functional for the patient independently regardless of the type of disability that is present.

It is important to note that the center of gravity of the human-wheelchair system changes due to postural problems of person, limb amputations, or obesity. In this context, this paper presents the autonomous control of human-wheelchair system based on dynamic model of the system. The dynamic model of human-wheelchair system is developed considering the lateral deviation of the center of mass caused by the movement of human.

This paper is organized into five sections, including the Introduction. In Sect. 2 the kinematic and dynamic modeling of human-wheelchair system is presented. The conceptualization of predictive control model is presented in Sect. 3; while in Sect. 4, the results of the implementation of the algorithm applied to NMPC wheelchair shown. Finally, conclusions are presented in Sect. 5.

2 Kinematic and Dynamic Modeling

This work is based on a wheelchair similar to a unicycle type mobile robot. A unicycle type mobile robot can rotate about its axis. It is assumed that the human-wheelchair system moves in on a flat horizontal surface. It is considered as a reference frame fixed vertically R(X,Y,Z). Traditionally the control of a robotic wheelchair is considered a point located in the middle of the virtual axis connecting the two main wheels. However, this work is considered to point to follow a predetermined path is located in front of the virtual axis (the point $h(x, y)$ of Fig. 1). Figure 1 illustrates the configuration of the wheelchair considered in this work. Where G represents the center of mass; B is the central axis of two wheels; is the point that is required to follow a trajectory in R; $h(x, y)$ the positions of the axes of the wheelchair; ψ it is the orientation of the wheelchair; ω is the angular velocity; u', $\bar{u}$ are the longitudinal and lateral velocities of the center of mass; $F_{rrx'}$, $F_{rly'}$ are the longitudinal and lateral forces of the left wheel; $F_{rlx'}$, $F_{rly'}$ are the longitudinal and lateral forces of the right wheel; $F_{hx'}$, $F_{hy'}$ are the longitudinal and lateral external forces produced by the human in H; τ_h It is the momentum exerted by the human; $F_{cx'}$, $F_{cy'}$, $F_{dx'}$, $F_{dy'}$ are the longitudinal and lateral forces exerted on C and D by subsequent castor wheels; $F_{ex'}$, $F_{ey'}$, $F_{fx'}$, $F_{fy'}$ are the longitudinal and lateral forces exerted on E and F by the front castor wheels; d, b_1, b_2, a, c they are distances.

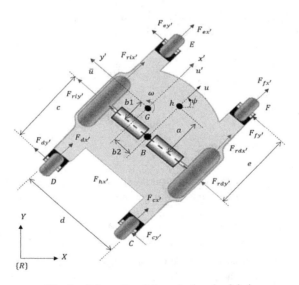

Fig. 1. Schematic of the robotic wheelchair

The strength and momentum equations for the mobile robot are:

$$\sum F_{x'} = m(\dot{u}' - \bar{u}\omega) = F_{rlx'} + F_{rrx'} + F_{cx'} + F_{dx'} + F_{ex'} + F_{fx'} \tag{1}$$

$$\sum F_{y'} = m(\dot{\bar{u}} - u'\omega) = F_{rly'} + F_{rry'} + F_{cy'} + F_{dy'} + F_{ey'} + F_{fy'} \tag{2}$$

$$\sum M_z = I_z\dot{\omega} = \frac{d}{2}(F_{rrx'} - F_{rlx'}) + b_2(F_{rrx'} + F_{rlx'}) - b_1(F_{rly'} + F_{rry'}) - (c - b_1)(F_{dy'} + F_{cy'})$$
$$+ (e - b_1)(F_{fy'} + F_{ey'}) + \left(\frac{d}{2} + b_2\right)(F_{cx'} + F_{fx'}) - \left(\frac{d}{2} - b_2\right)(F_{dx'} + F_{ex'}) \tag{3}$$

where $m = m_h + m_w$ that is of human-wheelchair system masses which m_h is the human mass and m_w the mass of the wheelchair; and I_z it is the moment of inertia about the vertical axis located in G the human-wheelchair system.

In general, most robots available on the market have a low level of PID controllers reference speed for monitoring input speeds and do not allow the motor voltage is proportional directly. Therefore, it is useful to express the model of the robotic wheelchair in a manner suitable considering the linear and angular velocity as input signals. So the model of the wheelchair can be expressed as,

$$\begin{bmatrix} \dot{x} \\ \dot{y} \\ \dot{\psi} \\ \dot{u} \\ \dot{\omega} \end{bmatrix} = \begin{bmatrix} u\cos\psi - a\omega\sin\psi \\ u\sin\psi + a\omega\cos\psi \\ \omega \\ \frac{\varsigma_3}{\varsigma_1}\omega^2 - \frac{\varsigma_4}{\varsigma_1}u \\ -\frac{\varsigma_5}{\varsigma_2}u\omega - \frac{\varsigma_1}{\varsigma_2}\omega \end{bmatrix} + \begin{bmatrix} 0 & 0 \\ 0 & 0 \\ 0 & 0 \\ \frac{1}{\varsigma_1} & 0 \\ 0 & \frac{1}{\varsigma_2} \end{bmatrix} \begin{bmatrix} u_{ref} \\ \omega_{ref} \end{bmatrix} \tag{4}$$

Where $\varsigma \in \Re^l$ con $l = 8$ y $\varsigma = [\varsigma_1 \quad \varsigma_2 \quad \cdots \quad \varsigma_l]^T$ is the vector of the dynamic parameters of human-wheelchair system. The dynamic parameters of the model of human-chair wheel system are defined as [3, 4, 16]:

$$\varsigma_1 = \frac{\frac{2I_e}{r} + \frac{2k_a}{R_a}k_{DT} + rm}{2\frac{k_a}{R_a}k_{PT}}; \qquad \varsigma_2 = \frac{rmb_2}{2\frac{k_a}{R_a}k_{PT}};$$

$$\varsigma_3 = \frac{\frac{2B_e}{r} + \frac{k_a}{rR_a}k_b + 2\frac{k_a}{R_a}k_{PT}}{2\frac{k_a}{R_a}k_{PT}}; \qquad \varsigma_4 = \frac{rmb_1}{2\frac{k_a}{R_a}k_{PT}};$$

$$\varsigma_5 = \frac{rmb_2}{\frac{k_a}{R_a}k_{PR}}; \qquad \varsigma_6 = \frac{\frac{dI_e}{r} + 2\frac{k_ak_{DR}}{R_a} + 2r\frac{m\left(\frac{b^2}{2} + b_1^2\right) + I_z}{d}}{2\frac{k_a}{R_a}k_{PR}};$$

$$\varsigma_7 = \frac{\frac{dB_e}{r} + \frac{dk_ak_b}{2rR_a} + \frac{2k_ak_{PR}}{R_a}}{2\frac{k_a}{R_a}k_{PR}}; \qquad \varsigma_8 = \frac{rmb_1}{\frac{k_a}{R_a}k_{PR}};$$

The identification of the dynamic parameters of human-wheelchair system is obtained from through the least squares estimation and applying a filter to the regression model [15, 16]. The identification of the dynamic parameters of human-wheelchair system with a human of 76 kg are: $\varsigma_1 = 0.4903$, $\varsigma_2 = 0.2112$, $\varsigma_3 = -0.0011$, $\varsigma_4 = 1.0203$, $\varsigma_5 = 0.0346$, $\varsigma_6 = 0.9766$, $\varsigma_7 = -0.0016$ y $\varsigma_8 = -0.1001$. Figure 2 shows the proposed dynamic model validation, where you can observe the good performance of the mathematical model obtained.

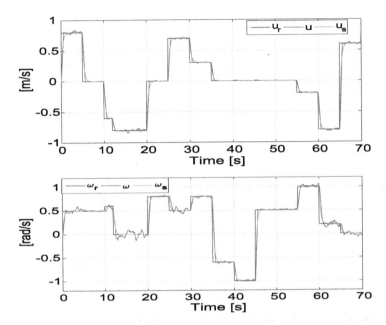

Fig. 2. Validation of the proposed dynamic model. The blue line are the velocities reference; while the blue line represents the dynamic model proposed and the red line are the signs of the experimental velocities human-wheelchair system. (Color figure online)

Therefore the (4) can be written compactly as,

$$\dot{x}(t) = f(x(t)) + g(v(t)) \tag{5}$$

where, $x(t) = [x \ \ y \ \ \psi \ \ u \ \ \omega]^T$ is the state vector of the system, $v(t) = [u_{ref} \ \ \omega_{ref}]^T$ is de control vector.

3   Model Predictive Control of Human-Wheelchair System

A general nonlinear deterministic model in discrete time of a system can be expressed as,

$$x(k+1) = f(x(k), v(k)) \tag{6}$$

$$y(k) = h(x(k)) \tag{7}$$

where, $x(k)$, $v(k)$ and $y(k)$ are the vectors of state, control input and output respectively.

Most MPC methods are based on a common schema [17]. It defines function J, which is often a quadratic function of the sum of the standard for future tracking errors of the trajectory.

$$e(k+i|k) = y(k+i|k) - y_d(k+i) \tag{8}$$

Preceded on a prediction horizon N plus the sum of the standard of predicted increases in the control action on a control horizon N_u,

$$J = \sum_{i=1}^{N} \delta_i \|e(k+i|k)\|_Q^2 + \sum_{i=1}^{N_u} \lambda_i \|\Delta v(k+i-1|k)\|_R^2 \tag{9}$$

$$\Delta v(k+i|k) = v(k+i|k) - v(k+i-1|k)$$

where, δ_i and λ_i are sequences that are usually chosen penalty constant, $yd(k+i)$ is the desired output and notation $y(k+i|k)$ indicates that $y(k+i)$ is calculated with the information known at the time k. The future system outputs $y(k+i|k)$ for $i = 1, \ldots, N$, are predicted by a model of the process from previous entries and exits instant k, and from the predicted future control actions, $v(k+i|k)$ for $i = 0, \ldots, Nu - 1$, which are those that are to be calculated. Additionally it has to $\|x\|_Q^2 = x^T Qx$ y $Q > 0$.

This form J can be expressed as a function depending only on the actions of future control. The objective of predictive control is to obtain a sequence of future control actions $[v(k), v(k+1|k), \ldots, v(k+Nu-1|k)]$, so that the outputs predicted using the model system $y(k+i|k)$, as close as possible to the reference $yd(k+i|k)$. This is achieved by minimizing J respect to control variables. After this sequence is obtained, receding horizon strategy, which consists of only applying the first control action $v(k)$ calculated is used. This process is repeated at each sampling time.

When nonlinear models are used, the MPC depends on finding a solution to a problem of nonlinear programming in each sampling step. To solve this, it is necessary to solve the optimization and system model. These two problems can be implemented in two ways: sequentially or simultaneously.

3.1   Sequential Optimization Algorithms

For sequential implementation, a solution is in each iteration of the optimization routine. The controls are the decision variables entering the algorithm, which computes the solution to the model. Then, this solution to evaluate objective and the calculated value is delivered to program optimization function is used. Then, the optimization variable takes the form,

$$z = [v(k) \quad v(k+1|k) \quad \cdots \quad v(k+N_u-1|k)]^T \tag{10}$$

The function that calculates the functional, you must first solve the system model with z values with the current state vector $x(k)$ applying N times (6). This vector sequence is obtained $[x(k+1|k) \quad x(k+2|k) \quad \cdots \quad x(k+N|k)]$. With (7) the values of the sequence set $[y(k+1|k) \quad y(k+2|k) \quad \cdots \quad y(k+N|k)]$ output of these values and the cost functional given by (9) is evaluated.

The cost functional J depends on the predicted outputs, which in turn are a function of the state vector and it shares control (optimization variables). Therefore, it follows that to obtain the gradient of the functional outputs should arise regarding control actions from k to $(k + Nu - 1)$. This is complicated and not always has a solution. Therefore, in the sequential solution there is no gradient and the information must be obtained by numerical differentiation, which is computationally negative, because a higher cost calculation and convergence problems is generated.

3.2   Sequential Optimization Algorithms

Unlike the sequential solution, the solution and simultaneous optimization include states and controls model as decision variables. The model equations are added to the optimization problem as a constraint equations. So is the optimization variable,

$$z = [x(k+1) \quad x(k+1|k) \quad \cdots \quad x(k+N|k) \\ v(k) \quad v(k+1|k) \quad \cdots \quad v(k+N_u-1|k)]^T \tag{11}$$

that is, considering the states and control actions as optimization variables. The dimension of this vector is $(eN + pNu)$, which it is greater than that resulting in the sequential proposal (pNu), where, e and p are the sizes of the state vectors and control input, respectively. This entails a considerable increase in the size of the optimization variable in relation to the sequential proposal. The model equations as equality constraints appear as shown in (12).

$$R = \begin{cases} x(k+1|k) = f(x(k), v(k)) \\ x(k+2|k) = f(x(k+1), v(k+1)) \\ \quad \vdots \\ x(k+N|k) = f(x(k+N-1), v(k+N_u)) \end{cases} \tag{12}$$

In this proposal, obtaining analytically gradient is simpler, so it can be incorporated explicitly the optimization algorithm. For the case of functional (9) is,

$$\nabla J = \begin{bmatrix} 2\delta_1 G^T_{x_{k+1}} Q[h(x(k+1|k)) - y_d(k+1)] \\ 2\delta_2 G^T_{x_{k+2}} Q[h(x(k+2|k)) - y_d(k+2)] \\ \vdots \\ 2\delta_N G^T_{x_{k+N}} Q[h(x(k+N|k)) - y_d(k+N)] \\ 2\lambda_1 R\Delta v(k) - 2\lambda_2 R\Delta v(k+1) \\ 2\lambda_2 R\Delta v(k+1) - 2\lambda_3 R\Delta v(k+2) \\ \vdots \\ 2\lambda_{N_u} R\Delta v(k+N_u-1) \end{bmatrix} \tag{13}$$

$$G^T_{x_i} = \left.\frac{\partial h(x(k), v(k))}{\partial x(k)}\right|_{k=i}$$

The gradient of equality constraints is a sparse matrix and takes the form of,

$$\nabla R = \begin{bmatrix} I_{e\times e} & -Fx(k+1) & \cdots & 0_{e\times e} & 0_{e\times e} \\ 0_{e\times e} & I_{e\times e} & \cdots & 0_{e\times e} & 0_{e\times e} \\ \vdots & \vdots & \ddots & \vdots & \vdots \\ 0_{e\times e} & 0_{e\times e} & \cdots & I_{e\times e} & -Fx(k+N-1) \\ 0_{e\times e} & 0_{e\times e} & \cdots & 0_{e\times e} & I_{e\times e} \\ -Fv(k) & 0_{e\times e} & \cdots & 0_{e\times e} & 0_{e\times e} \\ 0_{e\times e} & -Fv(k+1) & \cdots & 0_{e\times e} & 0_{e\times e} \\ \vdots & \vdots & \ddots & \vdots & \vdots \\ 0_{e\times e} & 0_{e\times e} & \cdots & 0_{e\times e} & -Fv(k+N_u-1) \end{bmatrix} \tag{14}$$

Where, $F_{x_{k+i}} = \left.\frac{\partial f(x(k),v(k))}{\partial x(k)}\right|_{k=i}$, $F_{v_{k+i}} = \left.\frac{\partial f(x(k),v(k))}{\partial v(k)}\right|_{k=i}$, I_{exe} It is the identity matrix of dimension e and 0_{pxe} It is a zero matrix of dimension pxe.

For small problems with a few states and a small prediction horizon, the sequential method is probably more effective [18]. For large problems, generally the simultaneous proposal is more robust, this is less likely to fail. In the sequential proposal, the addition of restrictions on states or outputs it is more complicated. In addition to restrictions derived from the model, you can add restrictions on the control action, at steady state, etc.

4 Experimental Results

In this work presents a wheelchair robot which was developed at the Universidad Técnica de Ambato (see Fig. 3). The wheelchair has two independently driven wheels by two direct current motors (in the center part), and four caster wheel around the central axis conferring greater stability to the human-wheelchair system (two in the rear part and two in the front part). Encoders installed on each one of the motor shafts allow knowing the relative position and orientation of the wheelchair.

In this section several experiments were executed for show the performance of the proposed controller and dynamic modelling of the human-machine system. Some of the results experiments using the wheelchair presented on Fig. 1 are reported in this section.

The maximum absolute values of the linear and angular velocities of the wheelchair for the test were 0.8 [m/s] and 0.5 [rad/s], respectively. It used a sampling time for testing of $To = 0.1$[s] and he worked with a prediction horizon $N = N_u = 7$. Is additionally selected the arrays $Q = diag[1, 1, 0.0000001]$ and $R = I$, and parameters $\delta = 92$ y $\lambda = 0.8$. The linear reference speed 0.25 [m/s] was used in the reference trajectory. To check the performance control the wheelchair, starts at position:

Fig. 3. Autonomous wheelchair robotic

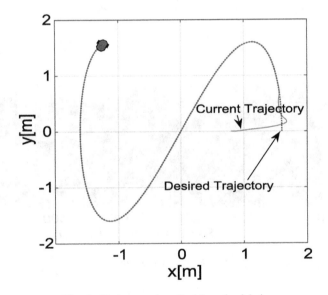

Fig. 4. Trajectory described by wheelchair.

$x = 0$ [m], $y = 0$ [m], $\psi = 0$ [rad] and must follow a circular path of 1 [m] radio, as it is shown in Fig. 4.

The trajectories in the x and y axis of the desired position and the wheelchair described are shown in Fig. 5, where you can see that the square error of control is limited in ± 0.1 [m].

Finally, the calculation time of the algorithm are smaller than the sampling period of the wheelchair, see Fig. 6.

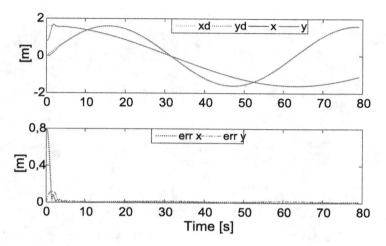

Fig. 5. Desired and actual positions of the wheelchair

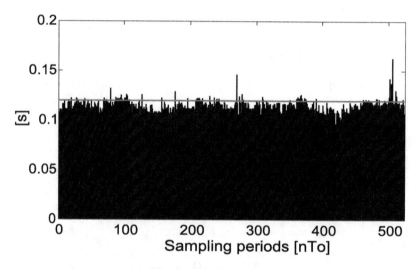

Fig. 6. Calculation time

5   Conclusions

Has solved the problem of driving a wheelchair through paths previously calculated by NMPC strategy as robot navigation algorithm, working with the dynamic model with deviation from the center of gravity of the human-wheelchair system. The results presented were obtained by sequential NMPC proposal, the times calculated with this algorithm were better than the times with simultaneously NMPC algorithm, as shown in the tests.

Acknowledgment. The authors would like to thanks to Universidad Técnica de Ambato for financing the project *Robotic Assistance for Persons with Disabilities* (Resolution: 1151-CU-P-2012). Also to the Universidad de las Fuerzas Armadas ESPE and to the Escuela Superior Politécnica de Chimborazo for the support to develop of the Master's Thesis *Modelación y Control Predictivo de un Robot Móvil con Centro de Masa Desplazado.*

References

1. Victores, J.G., Morante, S., Jardón, A., Balaguer, C.: Creating robotic assistance force by multimodal interaction. In: Iberoamerican Congress on Technology Support Disabilities, pp. 21–24 (2013)
2. Bastos-Filho, T., et al.: Towards a new modality-independent interface for a robotic wheelchair. IEEE Trans. Neural Syst. Rehabil. Eng. **22**(3), 567–584 (2014)
3. Andaluz, V.H., Canseco, P., Varela, J., Ortiz, J.S., Pérez, M.G., Morales, V., Robertí, F., Carelli, R.: Modeling and control of a wheelchair considering center of mass lateral displacements. In: Liu, H., Kubota, N., Zhu, X., Dillmann, R. (eds.) ICIRA 2015. LNCS, vol. 9246, pp. 254–270. Springer, Heidelberg (2015)
4. Andaluz, V.H., Canseco, P., Varela, J., Ortiz, J.S., Pérez, M.G., Roberti, F., Carelli, R.: Robust control with dynamic compensation for human-wheelchair system. In: Zhang, X., Liu, H., Chen, Z., Wang, N. (eds.) ICIRA 2014, Part I. LNCS, vol. 8917, pp. 376–389. Springer, Heidelberg (2014)
5. Ortiz, J., Morales, J., Peréz, M., Andaluz, V.: Bilateral teleoperation handlers phones. Polytech. Mag. **34**(2), 50–57 (2015)
6. Voth, D.: A new generation of military robots intelligent systems. IEEE, pp. 2–3 (2005)
7. Tu, C., Wyk, B.J., Djouani, K., Hamam, Y.: An efficient crop row detection method for agriculture robots. In: IEEE-CISP Image and Signal Processing, pp. 655–659 (2014)
8. Lo, A.C., Guarino, P.D.: Robot-assisted therapy for long-term upper-limb impairment after stroke. N. Engl. J. Med. **362**(19), 1772–1783 (2010)
9. Krusienski, D.J., Shit, J.J., Wolpaw, J.R.: Brain-computer interfaces in medicine. In: Mayo Clinic Proceedings, vol. 87, no. 3, pp. 268–279. IEEE Computer Society (2013)
10. Carrera, J.L., Guzmán, C.H., Oliver, M.A, Ortega, A.B.: Kinematic analysis of a therapeutic robot for rehabilitation of lower limbs. J. Ind. Eng., pp. 21–30 (2013)
11. Wu, B.F., Jen, C.L., Tsou, T.Y., Li, W.F., Tseng, P.Y.: Accompanist detection and following for wheelchair robots with fuzzy controller. In: Advanced Mechatronic Systems (ICAMechS), pp. 18–21 (2012)
12. Jamali, P., Tabatabaei, S.M., Sohrabi, O., Seifipour, N.: Software based modeling, simulation and fuzzy control of a Mecanum wheeled mobile robot. In: First RSI/ISM International Conference on Robotics and Mechatronics (ICRoM), pp. 200–204 (2013)
13. Carlson, T., Demiris, Y.: Collaborative control for a robotic wheelchair: evaluation of performance, attention, and workload. IEEE Trans. Syst. Man Cybern. Part B Cybern. **42**(3), 876–888 (2012)
14. Widyotriatmo, A., Rauzanfiqr, S.K., Suprijanto, S.: A modified PID algorithm for dynamic control of an automatic wheelchair. In: Control, Systems and Industrial Informatics (ICCSII), 23–26 September 2012, pp. 64–68 (2012)
15. Aström, K.J., Wittenmark, B.: Adaptive Control. Addison-Wesley, Reading (1995)
16. Reyes, F., Kelly, R.: On parameter identification of robot manipulator. In: IEEE International Conference on Robotics and Automation, pp. 1910–1915 (1997)

17. Ramírez, D., Limón-Marruedo, D., Gómez-Ortega, J., Camacho, E.: Application of Predictive Control for Nonlinear Model of a Mobile Robot Navigation using Genetic Algorithms, Numerical Methods in Engineering (1999)
18. Peña, M.: Model-based control blurs. Doctoral Thesis – INAUT – UNSJ (2002)

FESleeve: A Functional Electrical Stimulation System with Multi-electrode Array for Finger Motion Control

Tianqu Shao[1], Xiang Li[1], Hiroshi Yokoi[2], and Dingguo Zhang[1(✉)]

[1] School of Mechanical Engineering, Shanghai Jiaotong University,
Dongchuan Road 800, Shanghai, 200240, China
dgzhang@sjtu.edu.cn
[2] Faculty of Informatics and Engineering, University of Electro-Communications,
Tokyo, Japan

Abstract. Functional electrical stimulation (FES) is frequently used for recovering upper limb functions in patients with central nervous system lesions. However, some typical problems exist such as comfort, selectivity, fatigue and convenience when applying this technology. Targeting these problems, we developed a functional electrical stimulation system with multi-electrode array, called FESleeve, for finger motion control. With smaller size in electrode design, the system was capable of targeting and stimulating muscles related with specific finger movements more precisely, meanwhile improving selectivity. For easy use, the overall system was designed to be sleeve-shape and can be easily fixed on the forearm. Special stimulation configuration was designed to alleviate sting feelings and muscle fatigue. Experiments were conducted on 4 healthy subjects and 3 stroke patients with upper limb deficits. Results show that the system was able to induce desirable finger movements, such as single finger flexing and finger bending. No negative effects of the system were found in the experiments.

Keywords: Functional electrical stimulation · Multi-electrode array · Hand motion · Finger muscle · Stimulation selectivity

1 Introduction

Functional electrical stimulation (FES) is a common technique in the recovery of upper limb function. It can help the patients increase muscle strength in voluntary movement [1]. By inducing continuous electrical pulse trains, the technique generates muscle contraction when needed. To achieve a continuous contraction, at least 20 action potentials would be required. Otherwise, only twitches would occur [2].

Some widely known neuroprosthesis devices such as Handmaster [3] and Bionic Glove [4] were designed to restore or improve grasping function with surface electrodes. To realize more intricate finger motions, FES electrodes have to be attached to specific muscles and more electrodes would be needed. A better solution is to use multi-electrode array, which have the advanced features of dynamic changes of electrode position [5]. By altering only a subset of electrodes on the array, the stimulation can be made highly selective and avoid undesired muscle activation [6].

© Springer International Publishing Switzerland 2016
N. Kubota et al. (Eds.): ICIRA 2016, Part II, LNAI 9835, pp. 191–199, 2016.
DOI: 10.1007/978-3-319-43518-3_19

Some related work with multi-electrode array has been published. Professor Emi Tamaki and his team designed a Possessedhand [7], which stimulated muscles through 28 electrode pads (array of 2*14). The size of the electrodes is 10*30 mm. By varying the stimulation level, muscles at different depths in the forearm can be stimulated. It can control the proximal interphalangeal joints of index finger, middle finger, ring finger and middle finger independently. But the probability of little finger movement is low. They have also successfully applied the device in assisting playing musical instrument. Malešević and his team brought about an FES system [8] with a 4 × 4-electrode array each electrode of which can be activated separately. It allowed the targeting of motor neurons that activate synergistic muscles and produce a functional movement, such as grasping. Results of tests on stroke patients showed that the multi-pad electrodes provided the desired level of selectivity and could be used for generating a functional grasp.

In this paper, we present herein our design—FESleeve, which adopts smaller electrodes and smaller gaps between each electrode than that of previous works. By increasing the number of electrodes and decreasing the size of electrodes, selectivity for single finger movement is increased, so the FES system can control more muscles. Also, it can control the extension and flexion of the fingers since the FESleeve can cover both sides of forearm.

2 Methodology

2.1 System Structure

Figure 1 shows the structure of the FES system, which included a PC terminal, a FES device, a switching board and the FESleeve. A FES device was used to generate low-frequency pulses. The switching board has 48 channels, and each channel connects to an electrode of the FESleeve. The switch of channels takes orders from the PC terminal. The figure of the overall FESleeve system is shown in Fig. 2.

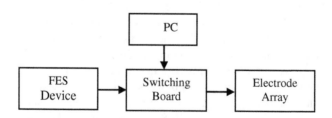

Fig. 1. System structure of the proposed FES system

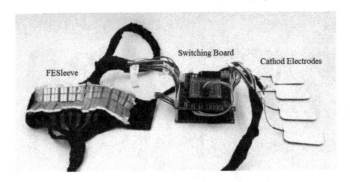

Fig. 2. FESleeve system

2.2   FESleeve Design

The material of electrode should provide the preferred performance with the least pain, and at the meantime not cause permanent skin damage or irritation [5]. Typically, there are three types of material: hydrogel, metal and fiber. In the design of FESleeve, we used the combination of sheet copper and hydrogel. The main reason was that it can be easily fixed on the sleeve and be connected to the switching board. Two electrodes were placed on one PCB board. In order to avoid sting during stimulation, a layer of sticky gel on the metal electrodes was glued so as to reduce uneven attachment on skin. In later experiments, for the consideration of convenience, we put a large piece of gel on the electrodes, and it did not affect much.

Electrode size is an important aspect in determining the stimulation effect. The larger the muscle, the larger size is required. As for the flexion or extension of fingers, relatively small electrode size is required. The application of small electrodes will result in a localized electrical field which decreases spreading of currents to motor neurons of adjacent muscles. In order to increase the selectivity, smaller size is preferred. In the experiments, we applied 30*30 mm, 25*25 mm, 20*20 mm, 15*15 mm, 10*10 mm, 8*8 mm. When the size is smaller than 10*10 mm, single finger movements without coupling can be achieved. But as the size goes smaller than 8*8 mm, subjects start to feel sting. Besides, when the size is too small, the current may spread within the fat layer and prevents the current to reach the motor nerves.

The FESleeve consisted of an array of 12 × 4 electrodes, with the gap of 1 mm between each one. That is to say, 24 PCB boards were fixed on the sleeve (Fig. 3). When wore on the forearm, such array can cover flexor digitorum superficialis muscle, flexor digitorum profundus muscle, extensor digitorum superficialis muscle and extensor digitorum profundus muscle. These muscles can be stimulated in experiment. This design can bring great convenience to subjects. All they should do is to place the FESleeve on the approximate position of forearm and fix it firmly with magic stripes. The calibration process can automatically select the preferred electrode for stimulation with high selectivity.

Fig. 3. Multi-electrode array of FESleeve with 48 channels

2.3 Electrical Parameters

A commercial FES device called Compex Motion II (Compex, Switzerland) was used. The stimulation frequency was 25 Hz in order to avoid the appearance of fatigue and the pulse width was set to 300 us, which was preferred by most subjects [9]. Since different people require different level of stimulation to realize certain functional movement, pulse amplitude should be set at each experiment according to the situation of different subjects. Normally, the pulse amplitude would be 7–15 mA for healthy people and higher for patients. When the electrodes get smaller, it requires smaller amplitude to achieve the finger movements.

2.4 Calibration

Multi-pad is an approach to dynamically change the size and position of the active region. Calibration is a method for determining optimal electrode positions and size, which shortens the time of conducting this procedure.

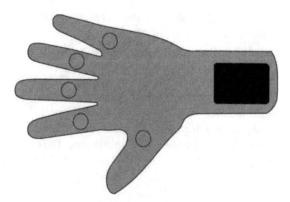

Fig. 4. The five measured joint angles (marked in blue circles) (Color figure online)

5DT Data Glove 14 Ultra (5DT, USA) was used in the process of calibration to acquire finger joint angles. Five channels of the data glove were used to differentiate functional movement patterns. As shown in Fig. 4, the five joints we measured were the metacarpophalangeal joint (MCP) of thumb and little finger, and proximal interphalangeal (PIP) joint of index finger, middle finger and ring finger. The data glove should be calibrated at the first place. For the measurement of flexion, the subjects should first clench the fist and the value of flexion would be set to "1", which denotes the maximum degree, then the subjects should open the hand and the value of flexion would be set to "0", which denotes the minimum degree.

During the calibration procedure, each electrode or a series of electrodes was activated in chronological order, and the degree of flexion or extension was recorded at the meantime. After the stimulation, an addressing algorithm ran automatically to find the optimal electrode positions for certain movement pattern. When the value of degree was less than 0.80, the algorithm set it to 0; otherwise the algorithm set it to 1. The values were stored in an n*5 matrix (n is the number of times of stimulation). For example, to identify the preferred electrode position for bending the ring finger, the algorithm searched for the row: [0 0 0 0 1].

2.5 Software

The calibration algorithm and the stimulation program were implemented in C# environment. The software provided two modes for stimulation. One was to manually choose the activated electrode position, and the other was to automatically activate each electrode in chronological order. The stimulation time for each electrode was set to 3 s, and the interval time was set to 1 s. The total time for one automatic calibration procedure was approximately three minutes.

3 Experiment

3.1 Subjects

Four able-bodied subjects (four males; aged 21–24) and three stroke patients (two males and one female; aged 45–65) participated in the experiment. All of the stroke patients

Fig. 5. A subject wearing the data glove and FESleeve

cannot move their fingers voluntarily, but their motor nerve is still intact. One thing to notice is that some patients' hands always gripped into fists because of muscle atrophy or spasticity.

3.2 Experimental Paradigm

Subjects were instructed to be seated in a chair with his/her arm on the desk. First, the FESleeve was wore on the lower part (at a distance of 1/3 the full length of forearm to the wrist) of the forearm. After that, the data glove was wore on the same hand (Fig. 5). Subjects were requested to keep the position of the arm at their best. The electrode array was positioned based on a previous report found in [10]. The electrode array should be placed on the lower side of the forearm, measured from the condyles of the elbow. One or more large cathode electrodes should be placed near the wrist. The pulse amplitude was adjusted according to the feeling and stimulation effect. Then the calibration procedure was started. For subjects of larger size, the FESleeve may not cover enough area. The slight change of electrode position may result in different stimulation effect. Normally, several times of calibration would be needed to find the optimal electrode positions. If calibration cannot find several functional movement patterns at one time,

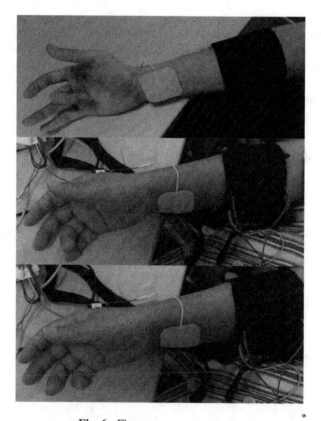

Fig. 6. Finger movement patterns

then the position of the FESleeve should be adjusted to fit the muscle shape better. Figure 6 shows some finger movement patterns that appeared in experiment.

3.3   Results

In order to examine the accuracy and stability of FESleeve, we recorded the patterns of functional finger movements that may appear in the calibration process and the possibility of appearance in one calibration process. Results of healthy subjects and stroke patients are presented in Tables 1 and 2, separately. The desired finger movement patterns are listed in the first column of each table.

Table 1. Possibility of each movement pattern in one calibration process (Healthy subjects)

Movement pattern	Healthy subjects			
	Subject A	Subject B	Subject C	Subject D
Index	75 %	25 %	25 %	25 %
Middle	100 %	75 %	75 %	75 %
Ring	100 %	100 %	100 %	100 %
Little	100 %	100 %	75 %	50 %
Ring&Little	100 %	100 %	100 %	75 %

Table 2. Possibility of each movement pattern in one calibration process (Patients)

Movement pattern	Patients		
	Subject E	Subject F	Subject G
Index	75 %	50 %	0
Middle	0	0	0
Ring	75 %	75 %	100 %
Little	50 %	50 %	100 %
Ring&Little	25 %	0	75 %

Experiencing up to four times of calibration, all healthy subjects could perform the desired finger motions. For stroke patients, only finger movement patterns that relate with the index finger, the ring finger and the little finger could be realized. The possibility of middle finger movement and ring finger movement was relatively higher in healthy people experiment. However, no stroke patients could move their middle fingers. After stroke, the impairment in hand motor control affected the movement of all the fingers to some extent, and the middle finger was worst affected. For all subjects, actuation of little finger and ring finger were found with high accuracy, while some coupling was present.

Experiments have also shown that even slight change of the electrode position could alter the stimulation effect, which proved the significance of adopting smaller electrodes and gaps.

4   Conclusion

In this paper, we developed an FES system, called FESleeve, with multi-electrode array for more precise finger motion control. Multiple electrodes with smaller size and gap can stimulate different muscles and improve the selectivity. With the adoption of material which combines hydrogel and sheet copper, FESleeve can provide both convenience and comfort for subjects. While tested on healthy subjects, the system works well and can easily realize separated motion of each finger. However, the overall success rate on stroke patients is lower than that on healthy subjects. They also performed less functional movement patterns. The reason is that they were not capable of exercising the muscles, and therefore muscles and tendons had atrophied. For all subjects, movement patterns, such as bending of the index finger is difficult to achieve, which stemmed from the regular configuration of electrode array. To improve the selectivity, we have developed another type of configuration (Fig. 7) and its effect will be tested in future work.

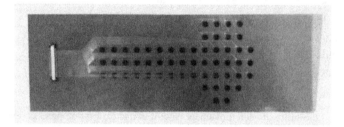

Fig. 7. The T-shape configuration of electrode array

Results of the experiment show that this new FES rehabilitation system can dynamically change the optimal electrode position for performing specific finger movements. The FESleeve system can be further improved by allowing the change of pulse amplitude during the calibration procedure so that muscles at different depth will be easier to activate. However, the change of current amplitude will result in much longer calibration time and may induce muscle fatigue. In the future, we will try to realize the change of current amplitude by optimizing the calibration algorithm.

Acknowledgement.   The authors would like to thank all subjects for participating experiments in this work. This research is supported by the National Natural Science Foundation of China (No. 51475292) and the National High Technology Research and Development Program (863 Program) of China (No. 2015AA020501).

References

1. Bajd, T., Kralj, A., Turk, R., et al.: Use of functional electrical stimulation in the rehabilitation of patients with incomplete spinal cord injuries. Int. J. Biomed. Eng. **11**(2), 96–102 (1989)
2. Popovic, M.R., Keller, T., Papas, I.P.I., Dietz, V., Morari, M.: Surface-stimulation technology for grasping and walking neuroprostheses. IEEE Eng. Med. Biol. Mag. **20**(1), 82–93 (2001)

3. Ijzerman, M., Stoffers, T., Groen, F., Klatte, M., Snoek, G., Vorsteveld, J., Nathan, J., Hermens, H.: The NESS handmaster orthosis: restoration of hand function in C5 and stroke patients by means of electrical stimulation. J. Rehab. Sci. **9**(3), 86–89 (1996)
4. Prochazka, A., Gauthier, M., Wieler, M., Kanwell, Z.: The Bionic Glove: an electrical stimulator garment that provides controlled grasp and hand opening in quadriplegia. Arch. Phys. Med. Rehab. **78**, 1–7 (1997)
5. Kapkin, O., Satar, B., Yetiser, S.: Electrodes for transcutaneous (surface) electrical stimulation. J. Autom. Control **70**(2), 35–45 (2008)
6. Hoffmann, U., Deinhofer, M., Keller, T.: Automatic determination of parameters for multipad functional electrical stimulation: application to hand opening and closing. In: 2012 Annual International Conference of the IEEE on Engineering in Medicine and Biology Society (EMBC), San Diego, CA, pp. 1859–1863 (2012)
7. Tamaki, E., Miyaki, T., Rekimoto, J.: PossessedHand: techniques for controlling human hands using electrical muscles stimuli. In: Computer Human Interaction, pp. 543–552 (2011)
8. Stavric, V.A., Mcnair, P.J.: Optimizing muscle power after stroke: a cross-sectional study. J. Neuroeng. Rehabil. **9**(1), 2633–2637 (2012)
9. Bowman, B.R., Baker, L.L.: Effects of waveform parameters on comfort during transcutaneous neuromuscular electrical stimulation. Ann. Biomed. Eng. **13**(1), 59–74 (1985)
10. Electrode Placement and Functional Movement, PALS Clinical Support. http://www.palsclinicalsupport.com/

A Calibration Method for Interbody Distance in Lumbosacral Alignment Estimation

Yoshio Tsuchiya[1]([✉]), Takashi Kusaka[1], Takayuki Tanaka[1],
Yoshikazu Matsuo[2], Makoto Oda[3], Tsukasa Sasaki[3],
Tamotsu Kamishima[4], and Masanori Yamanaka[4]

[1] Information Science and Technology, Hokkaido University, Sapporo, Japan
{tsuchiya,kusaka,tanaka}@hce.ist.hokudai.ac.jp
[2] National Institute of Technology, Hakodate College, Hakodate, Japan
matsuo@hakodate-ct.ac.jp
[3] Hokkaido University Hospital, Sapporo, Japan
{makoto,tsukasa}@huhp.hokudai.ac.jp
[4] Faculty of Health Sciences, Hokkaido University, Sapporo, Japan
ktamotamo2@yahoo.co.jp, yamanaka@hs.hokudai.ac.jp

Abstract. Anteflexion of the spine is essential for many physical activities of daily living. However, this motion places the lumbar discs under heavy loading because of changes in the shape of the lumbar spine, possibly leading to low back pain. With the aim of reducing low back pain, here we developed a wearable sensor system that can estimate lumbosacral alignment and lumbar load by measuring the shape of the lumbar skin as the lumbosacral alignment changes. The shape of the lumbar skin and posture angle are measured by using curvature sensors and accelerometers. In addition, the wearers physique must be considered for the system to be usable by a variety of people. We developed this system by measuring the body parameters associated with anteflexion and studied the changes in the dimensions of the lumbar spine with changes in posture. By measuring the dimensions of the lumbosacral spine on X-ray images, the posture angle, body surface area, and the dimensions of the lumbosacral spine have relevance. A calibration method for lumbosacral dimensions was developed using this relation. However, the estimation method for lumbosacral alignment could not maintain good calibration of interbody distance. Therefore, we further developed the estimation to include a calibration method for interbody distance, thereby improving the estimation accuracy.

Keywords: Wearable sensor · Lumbar spine · Lumbar load constitutional difference · Skin deformation · Individual difference correction

1 Introduction

Anteflexion of the spine is a crucial motion necessary to perform many tasks in work and daily living. It is particularly important in tasks such as caregiving and

N. Kubota et al. (Eds.): ICIRA 2016, Part II, LNAI 9835, pp. 200–210, 2016.
DOI: 10.1007/978-3-319-43518-3_20

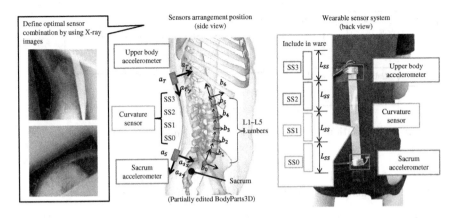

Fig. 1. Wearable sensor system developed by using X-ray image

carrying objects. However, anteflexion causes considerable loading of the lumbar spine [1] and deforms the lumbar discs [2].

European evidence-based guidelines for the prevention of low back pain were published in *Low back pain: Guideline for its management* [3], which is intended for prevention low back pain. Nachemson studied lumbar loading in relation to posture [1] and found that lumbar loading increases with the degree of anteflexion. In addition, National Institute of Occupational Safety and Health (NIOSH) guidelines specify an maximum allowable compressive force of 3400 N on lumbar discs. In Elfeituri et al. study, the lumbar load was average 3685 N at the time of lifting a weight of 23 kg [4]. To devise measures for preventing low back pain, it is necessary to identify high-risk postures. Shibata et al. [5] estimated lumbosacral alignment by using three-dimensional measurements. The lumbar load was then calculated from the estimation results. However, the methods of Shibata et al. and Nachemson are impractical for routine measurement of lumbar load. To reduce the risk of low back pain, it is necessary to identify postures that cause increased lumbar load. Thus, a system is needed that can measure lumbar load easily and routinely. Developing such a system was the aim of this study. Toward this end, a wearable sensor system was constructed to enable measurement of lumbar load on a daily basis. The ultimate goal of this research is to develop a system for identifying postures associated with a high risk of low back pain. For such a system, it is necessary to calculate lumbar loading from movement and posture parameters. In previous work toward this goal, the relation between posture and intervertebral loading was clarified, and both the center of gravity in the upper-body and the waist shape were estimated [6]. An estimation method for lumbosacral alignment has been developed to consider individual differences and thereby to improve the accuracy of the estimation [7]. In particular, the lumbosacral dimensions were calibrated by using posture angle and body surface area [8]. However, the estimation method could not maintain good calibration of interbody distance. In this study, we extend the estimation method to maintain

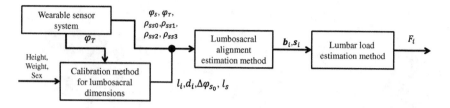

Fig. 2. Estimation algorithm for the posture and shape of lumbar spine

calibration of interbody distance in order to improve the estimation accuracy. A coordinate transformation matrix is used for this purpose.

2   Wearable Sensor System

Figure 1 shows the wearable sensor system for estimating lumbosacral alignment and lumbar load. The system, composed of three sets of flex sensors and two accelerometers, was developed based on X-ray image analysis. The flex sensors are embedded into compression sportswear. By using a pair of flex sensors, the shape of the lumbar skin is measured during retroflexion and anteflexion. The flex sensor obtains the skin curvature. The accelerometers are located at the top and bottom of the system and are referred to as the upper-body accelerometer and sacrum accelerometer, respectively. The upper-body is regarded as a rigid body for measuring its posture. The upper-body accelerometer is used to estimate the posture of the upper-body, and the sacrum accelerometer to estimate the posture of the pelvis. The measurement system is designed such that the sensor positions do not change during anteflexion. The device is affixed to the thighs with rubber bands underneath the clothes. In addition, since the flex sensors may come into contact with the waist during anteflexion, their tightness is reduced, and a belt-type wearable jig is used to prevent the sensors from moving upward. The sacrum accelerometer is placed on the sacrum below the flex sensors, and the upper-body accelerometer is placed above them. In this paper, the upper-body indicates the region at the thoracic vertebrae and above.

3   Estimation of Lumbosacral Alignment

Figure 2 shows estimation flowchart of lumbosacral alignment(b_i, s_i), and lumbar loadF_i. A calibration method for lumbosacral dimensions [7] calibrate 4 parameters, that are Interbody distance l_I, distance d_i between skin and vertebra, correction angle $\Delta\varphi_{s_0}$ of sacral, and distance l_s correction of coordinates. This estimation method of lumbosacral alignment uses the lumbar skin curvature values $\rho_{SS_j}(j = 0\text{--}3)$, which are obtained by the flex sensors, together with the components of gravitational acceleration a_{S_x} and a_{S_y}, as shown in Fig. 1. Using data from these sensors, we represent the shape of the skin as the "skin line" and the shape of the lumbar spine as the "spine line". Figure 3 shows estimation

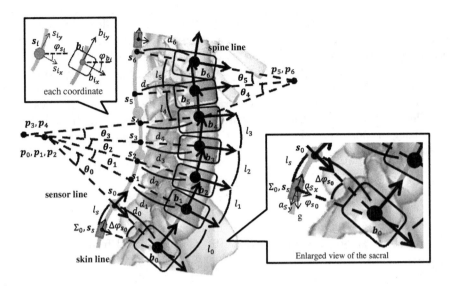

Fig. 3. Lumbosacral coordinates. This model is based on the lumbosacral alignment estimation. The method for calculating the posture angle from the accelerometer data is also shown. Partially edited BodyParts3D [9]

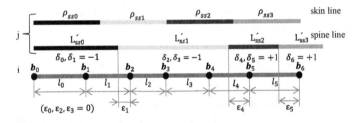

Fig. 4. Measurement area of curvature sensors in the partition method for lumbar curvature

example of lumbosacral alignment. The skin line is a curve connecting point s_0 to point s_6, and the spine line is a curve connecting point b_0 to point b_6. The curves from c_{s_0} to c_{s_6} and c_{b_0} to c_{b_6} contain inflection points in the lumbar spine shape model and the posture estimation model, respectively. The slope of the tangent to each curve is assumed to change continuously.

The pelvic tilt angle φ_{s_0} is calculated from the following equation by using sacrum accelerometer value.

$$\varphi_{s_0} = \tan^{-1}\left(\frac{a_{Sx}}{a_{Sy}}\right) \tag{1}$$

Parameters for distance between the skin and vertebrae d_0 are calculated from Eq. (2).

$$b_0 = \begin{pmatrix} d_0 \cos(\varphi_{s_0} + \Delta\varphi_{s_0}) \\ d_0 \sin(\varphi_{s_0} + \Delta\varphi_{s_0}) \end{pmatrix} \tag{2}$$

Matsumura reported that the pelvic tilt differs between male and female [10]. In addition, the actual pelvic tilt angle and the pelvic tilt measured from the sacrum acceleration sensor are different. Therefore, pelvic tilt angle is taken as $\Delta\varphi_{s_0}$. Figure 4 shows how the lumber curvature value is calculated by using the curvature sensor value.

The curvature sensor length is constant ($L_{ss} = 50\,\text{mm}$), but the measurement length of one curvature sensor is changed by changes in the spine line with changing posture. Spine line length L_{SS_j} is calculated from the following equation.

$$L'_{ss_j} = \frac{\rho_{SS_j} L_{SS}}{1/\rho_{SS_j} + \acute{d}} \tag{3}$$

Here, $\acute{d}$ is the average of d_i in the measurement area. The difference between the sum of the sensor measurement lengths ε_i is calculated from the following equation ($\varepsilon_i > 0$).

$$\varepsilon_i = \sum_{k=0}^{i} l_k - \sum_{k=0}^{j} L'_{SS_k} \tag{4}$$

Here, j is the curvature sensor number and j is selected by minimizing ε_i. The following formula is used for conversion from the curvature sensor value to the curvature for each vertebral body.

$$\rho_{L_i} = \delta_i \frac{\left|\rho_{SS_j}\right| \left|\rho_{SS_{j+1}}\right|}{(1 - \alpha_{l_e}) \left|\rho_{SS_{j+1}}\right| + \alpha_{l_{e_i}} \left|\rho_{SS_j}\right|} \tag{5}$$

α_{l_e} is the ratio of the relative range of interbody distance, the ratio is calculated by measurement values between two curvature sensors. δ_i is calculated from the following equation.

$$\delta_i = \begin{cases} \text{sign}(\rho_{SS_j}) & (\varepsilon_i < \frac{1}{2}) \\ \text{sign}(\rho_{SS_j+1}) & (\varepsilon_i > \frac{1}{2}) \end{cases} \tag{6}$$

Based on these results, $\alpha_{l_{e_i}}$ is calculated.

$$\alpha_{l_{e_i}} = \begin{cases} \frac{l_i - \varepsilon_i}{l_i} & (\varepsilon_i > 0) \\ 1 & (\text{otherwise}) \end{cases} \tag{7}$$

The value of ρ_i is negative for retroflexion and positive for anteflexion. Each angle θ_i in the figure is calculated by using the following formula.

$$\theta_i = \frac{l_i}{1/\rho_{L_i} + d_i} \tag{8}$$

For counterclockwise rotation, the angle is taken to be positive. The origin of the global coordinate system is Σ_0, as shown in Fig. 3. The origin is a point on the skin corresponding to s_0 on the sacrum. The arrows at points s_0 to s_6 and b_0 to b_6 in the figure are the normal and tangential lines. The normal is defined to be along the x-axis, and the tangent is defined to be along the y-axis. The center of curvature p_i is calculated using the following formula.

$$p_i = \frac{1}{\rho_{L_i}} \begin{pmatrix} s_{ix} + \cos\varphi_{b_i} \\ s_{iy} + \sin\varphi_{b_i} \end{pmatrix} \tag{9}$$

Equation (10) shows the transformation matrix used for s_i and b_i, which are calculated using Eqs. (11) and (12).

$$^iT_{i+1} = \begin{pmatrix} \cos\theta_i & -\sin\theta_i & p_{i_x} - p_{i_x}\cos\theta_i + p_{i_y}\sin\theta_i \\ \sin\theta_i & \cos\theta_i & p_{i_y} - p_{i_x}\cos\theta_i - p_{i_y}\cos\theta_i \\ 0 & 0 & 1 \end{pmatrix} \tag{10}$$

$$\begin{pmatrix} s_{i+1} \\ 1 \end{pmatrix} = {}^iT_{i+1} \begin{pmatrix} s_i \\ 1 \end{pmatrix} \tag{11}$$

$$\begin{pmatrix} b_{i+1} \\ 1 \end{pmatrix} = {}^iT_{i+1} \begin{pmatrix} b_i \\ 1 \end{pmatrix} \tag{12}$$

In the above, $\rho_{SS_j}(j = 0-3)$, a_{Sx}, and a_{Sy} are updated to reflect changes in posture. Then, a new lumbosacral alignment is estimated by this method.

4   Calibration Method for Interbody Distance in Lumbosacral Alignment Estimation

The estimation method described above did not consider individual differences in lumbosacral dimensions. To address this, we proposed a calibration method for lumbosacral dimensions, and could be estimating lumbosacral alignment with higher accuracy [7]. In the lumbosacral alignment estimation, we consider individual differences in three parameters: distance between vertebrae (interbody distance l_i), distance between the skin and vertebrae, and corrected pelvic tilt angle. The correction method for each parameter uses the upper-body posture angle and body surface area. This study calculated the body surface area by following the method of Kurazumi [8]. Our previous estimation method has the problem that the interbody distance does not remain calibrated. We therefore consider a lumbosacral alignment estimation method to maintain good calibration of interbody distance.

Figure 5 shows the calibration method for interbody distance in lumbosacral alignment estimation. P is always the origin in coordinate transformations. The coordinates of Q are calculated by the following equation.

$$Q = \begin{pmatrix} x_Q \\ y_Q \end{pmatrix} = \begin{pmatrix} l\cos(\pi - \psi - \Delta\psi) \\ l\sin(\pi - \psi - \Delta\psi) \end{pmatrix} \tag{13}$$

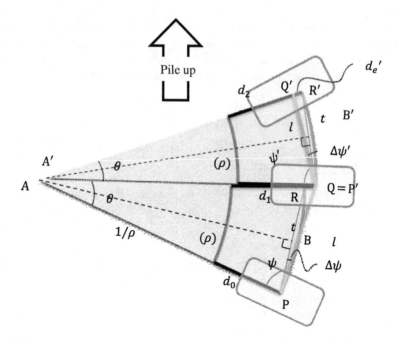

Fig. 5. Calibration method for interbody distance in lumbosacral alignment estimation. P is always the origin in coordinate transformations. Q is the next lumbar point in the calibration method.

Here, ψ and $\Delta\psi$ are needed in order to solve the equation. As shown in the figure, $\triangle ABP$ gives the following relation with the help of the additional line PR.

$$t = 2r \sin \frac{\theta}{2}$$
$$= r\theta \tag{14}$$

Next, θ is calculated by using the small-angle approximation of $\sin \theta$. Therefore, we can derive θ by applying the law of cosines to $\triangle AQP$ as follows.

$$\theta = \cos^{-1} \left(\frac{r^2 + (r + d_e)^2 - l^2}{2r(r + d_e)} \right). \tag{15}$$

By using θ, ψ is calculated as follows.

$$\psi = \frac{\pi}{2} - \frac{\theta}{2} \tag{16}$$

Finally, $\Delta\psi$ is calculated by applying the law of cosines to $\triangle PRQ$.

$$\Delta\psi = \cos^{-1} \left(\frac{t^2 + l^2 - d_e{}^2}{2tl} \right) \tag{17}$$

Q is transformed to the global coordinate system by using the following transformation matrix.

$$
{}^0Q = \begin{pmatrix} {}^0x_Q \\ {}^0y_Q \\ 1 \end{pmatrix} = \begin{pmatrix} \cos\theta_p & -\sin\theta_p & {}^0x_p \\ \sin\theta_p & \cos\theta_p & {}^0y_p \\ 0 & 0 & 1 \end{pmatrix} \begin{pmatrix} x_Q \\ y_Q \\ 1 \end{pmatrix} \tag{18}
$$

Here, θ_p are the coordinate angles from the lumbosacral alignment estimation, and it is same as φ_{b_i} in Eq. (18). Figure 5 shows the results of the calibration method for interbody distance of the lumbosacral alignment estimation.

5   Experiment on Lumbosacral Alignment Estimation

The objective of the experiment was to evaluate the accuracy of the corrected estimation of lumbosacral alignment. We estimated the lumbosacral alignment in various postures by using X-ray images of the sagittal plane. Posture 1 is a standing posture. Posture 2 is halfway to a stooped posture. Posture 3 is a stooped posture. Participants in the experiment were 4 men and 4 women. Table 1 shows each participant's weight, height, and body surface area. The lumbar skin shape was extracted from X-ray images to obtain the skin lumbar curvatures 1, 2, and 3. In the X-ray imaging experiment, a motion capture system was used to measure body posture. Then, the pelvic tilt angle and the upper-body posture angle were calculated from X-ray images and motion capture data. The obtained values were used in the method for estimating lumbosacral alignment with the correction of lumbosacral dimensions [7] and the calibration method for interbody distance as introduced in this paper. We then calibrated the method. As a target for comparison, a previous lumbosacral alignment estimation method [6] was given each average parameter value. The coordinates of lumbar vertebrae were estimated using the proposed method and conventional method The cumulative error (e [mm]) of all the vertebral body coordinates was determined by comparing the measured values (R_i) on X-ray images with the estimated vertebral body coordinates (s_i).

$$
e = \Sigma\sqrt{(R_{i_x} - s_{i_x})^2 + (R_{i_y} - s_{i_y})^2} \tag{19}
$$

The experimental results are shown in Fig. 6. This results shows the average error of all subjects. For all three postures, the error was smaller in the proposed method that in the conventional method As shown in Fig. 6, the error was reduced by up to 39.9 %. Figure 7 shows the experimental results for a participant (male 1) on an X-ray image. Figure 7 shows that estimation error is increased in top lumbar, because the origin of lumbosacral alignment estimation is the sacral coordinate s_0.

is increased in top lumbar. The error was smallest for male 1, for whom the average estimation error of a vertebral body was 9.9 mm. Estimation seems better with the proposed method than with the uncorrected method because the results are more accurate.

Table 1. Participant body data for lumbosacral alignment estimation experiment

	Height [cm]	Weight [kg]	Body surface area [m²]
Male 1	177.4	75.7	1.903
Male 2	168.3	66.6	1.739
Male 3	171.0	72.4	1.816
Male 4	179.1	67.0	1.835
Female 1	148.5	43.4	1.361
Female 2	154.8	50.0	1.488
Female 3	151.8	40.7	1.341
Female 4	160.4	51.5	1.541

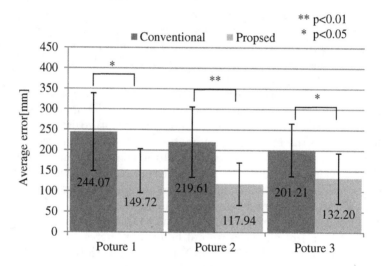

Fig. 6. Estimation results of the lumbosacral alignment, by using X-ray image

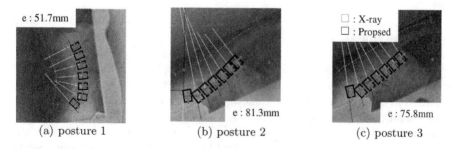

(a) posture 1   (b) posture 2   (c) posture 3

Fig. 7. Estimation of the lumbosacral alignment in each posture (male 1)

In general, lumbar load is estimated by using the position of center of gravity in the upper-body and weight. In this study, we defined that the error of the position of the center of gravity is increased by an increase in the estimated error of the lumbosacral vertebral alignment. Estimation alignment error of 100 mm corresponds to an error of 100 N in lumbar load [11]. The average error of lumbosacral alignment estimation is 133.29 mm when using proposed method, which would give an error of 130 N in estimated lumbar load. This error is satisfactorily small relative to the maximum permissible level of lumbar load (3400 N) set by NIOSH.

6   Conclusion and Future Works

We developed a wearable sensor system for estimation of lumbosacral alignment. Furthermore, we proposed a method for estimating lumbosacral alignment and posture from the pelvic tilt angle and lumbar skin curvature. We also included a calibration method for interbody distance. The error in lumbosacral alignment from the proposed method was lower than that from an uncorrected estimation method. It is necessary to further improve the estimation accuracy of lumbosacral alignment and to compare the lumbar loading estimated using the proposed method with that estimated by conventional methods.

References

1. Nachemson, A.L.: The lumbar spine: an orthopaedic challenge. Spine **1**(1), 59–71 (1976)
2. Tadano, S., Ishikawa, H., Itoh, M., Kaneda, K.: Strain distribution on the sagittal plane of an intervertebral disc. Trans. Jpn. Soc. Mech. Eng. A **57**(537), 1202–1207 (1991). (In Japanese)
3. Low Back Pain: Guidelines for Its Management. European Commission Research Directorate General (2004)
4. Elfeituri, F.E., Taboun, S.M.: An evaluation of the NIOSH lifting equation: a psychophysical and biomechanical investigation. Int. J. Occup. Saf. Ergon. **8**(2), 243–258 (2002)
5. Shibata, K., Inoue, Y., Iwata, Y., Katagawa, J., Fijii, R.: Study on noninvasive estimate method for intervertebral disk load at lumbar vertebrae. Trans. Jpn. Soc. Mech. Eng. A **78**(791), 2483–2495 (2012). (In Japanese)
6. Tsuchiya, Y., Yoshikazu, M., Takayuki, T.: Estimation of lumbar load by 2D reconstruction of spine line using wearable sensor system. In: 2014 IEEE International Conference on Systems Man and Cybernetics (SMC), pp. 3669–3674. IEEE (2014)
7. Tsuchiya, Y., Kusaka, T., Tanaka, T., Matsuo, Y., Oda, M., Sasaki, T., Kamishima, T., Yamanaka, M.: Calibration method for lumbosacral dimensions in wearable sensor system of lumbar alignment. In: 2015 37th Annual International Conference of the IEEE Engineering in Medicine and Biology Society (EMBC), pp. 3909–3912. IEEE (2015)
8. Kurazumi, Y., Horikoshi, T., Tsuchikawa, T., Matsubara, N.: The Body Surface Area

9. BodyParts3D, ©2008 Creative Commons Attribution- ShareAlike 2.1 Japan License

10. Matumura, S., Usa, H., Ogawa, D., Ichikawa, K., Hata, M., Mitomo, S., Takei, H.: Kenjoseijin niokeru kotsuban to kashi araimento no nendai hikaku to seisa no bunseki. In: Congress of the Japanese Physical Therapy Association 2012, pp. 48100760–48100760 (2013). (In Japanese). Japanese Society of Biometeorology **31**(1), 5–29 (1994). (In Japanese)

11. Tsuchiya, Y., Kusaka, T., Yoshikazu, M., Takayuki, T.: Wearable lumbar load estimation system for work load evaluation. J. Ergon. Occup. Saf. Health **16**, 68–71 (2014). (In Japanese)

Human-Wheelchair System Controlled by Through Brain Signals

Jessica S. Ortiz[3], Víctor H. Andaluz[1,2(✉)], David Rivas[1],
Jorge S. Sánchez[1], and Edison G. Espinosa[1]

[1] Universidad de las Fuerzas Armadas ESPE, Sangolquí, Ecuador
{vhandaluz1,drrivas,jssanchez,
egespinosal}@espe.edu.ec
[2] Universidad Técnica de Ambato, Ambato, Ecuador
victorhandaluz@uta.edu.ec
[3] Escuele Superior Politécnica de Chimborazo, Riobamba, Ecuador
jessortizm@outlook.com

Abstract. This work presents a dynamic controller for a robotic wheelchair, that allows people with lower and upper extremity impairments to move through of brain signals. The person receives visual feedback of the movement of the robot and it sends desired position-velocity commands through of the Emotiv EPOC device. The desired velocity of the wheelchair is considered as a function of the disregard of the person to move the robotic wheelchair. Additionally, the kinematic and dynamic modeling of a human-wheelchair system where it is considered that its mass center is not located at the wheels' axis center of the wheelchair. Finally, the results are reported to verify the performance of the proposed system.

Keywords: Wheelchair · Dynamic modeling · Cascade control · Lyapunov's method

1 Introduction

The integration of robotic issues into the medical field has become of great interest in recent years. Service, assistance, rehabilitation and surgery are the more benefited human health-care areas by the recent advances in robotics. Specifically, autonomous and safe navigation of wheelchairs inside known and unknown environments is one of the important goals in assistance robotics [1–7].

A robotic wheelchair can be used to allow people with lower and upper extremity impairments or severe motor dysfunctions overcome the difficulties in driving a wheelchair. The robotic wheelchair system integrates a sensory subsystem, a navigation and control module and a user-machine interface to guide the wheelchair in autonomous or semi-autonomous mode [4–6]. In autonomous mode, the robotic wheelchair goes to the chosen destination without any participation of the user in the control. This mode is intended for people who have great difficulties to guide the wheelchair. In the semi-autonomous mode the user shares the control with the robotic wheelchair. In this case only some motor skills are needed from the user.

N. Kubota et al. (Eds.): ICIRA 2016, Part II, LNAI 9835, pp. 211–222, 2016.
DOI: 10.1007/978-3-319-43518-3_21

On the other hand, the different architectures of control already proposed in the literature there is described the teleoperation which allows to govern the robot (slave) by means of the algorithms sent by the operator the same ones that will have to interact with the environment [8]. The disabled people have difficulty in moving his body freely, but his brain there issue signs electroencephalography -EEG-, the same ones that can be expressed so freely as they want with the suitable equipment (Emotiv), the major possible degree can obtain of telepresence, that is to say, that allows to the operator to realize tasks with so many skill as if it was manipulating directly the environment [9, 10]. The operator issues signs measured as mental commands, facial expressions or brain performance metrics known as EEG, is a non-invasive method to record electrical activity of the brain along the scalp. EEG measures voltage fluctuations resulting from ionic current flows within the neurons of the brain, this information combined with a good feedback of efforts allows him to realize his task of a more skillful way [11, 12].

In such context, this work proposes a bilateral teleoperation system in order to allow people with lower and upper extremity impairments or severe motor dysfunctions to overcome the difficulties in locomotion. It comprises a robotic wheelchair (slave) so that it can move on unstructured environments. The human operator receives visual signal and sends velocity and position commands generated by electromyogram signals through Emotiv EPOC haptic device (master) to the remote site. The desired velocity of the wheelchair is considered as a function of the disregard of the person to move the robotic wheelchair. On the other hand, is important to indicate that the wheelchair's center of gravity changes due to postural issues, limb amputations, or obesity [13]. Therefore, in the present work a dynamic model of the human-wheelchair system is developed considering lateral deviations of the center of mass originated in user's movement, limb amputations, or obesity. Furthermore, in this work it is proposed a method to solve the path following problem and positioning for a wheelchair robot to assist persons with severe motor diseases. The proposed control scheme is divided into two subsystems, each one being a controller itself: (i) the first on is a kinematic controller with saturation of velocity commands, which is based on the wheelchair robot's kinematic. The path following problem is addressed in this subsystem. It is worth noting that the proposed controller does not consider $s(t)$ as an additional control input as it is frequent in literature; and (ii) an dynamic compensation controller that considered the human-wheelchair system dynamic model, which are directly related to physical parameters of the system. In addition, both stability and robustness properties to parametric uncertainties in the dynamic model are proven through Lyapunov's method. To validate the proposed control algorithm, experimental results are included and discussed.

2   Human-Wheelchair System

The robotic wheelchair used in this work presents similar characteristics to that of a unicycle-like mobile robot, because it has two driven wheels which are controlled independently by two direct current motors and four caster wheel to maintain balance, while the unicycle-type mobile robots have a caster wheel to maintain stability. The kinematic modeling of the human-wheelchair system is developed in the next sub-section, considering a horizontal work plane where the wheelchair moves. The wheelchair type unicycle-like mobile robot presents the advantages of high mobility, high traction with pneumatic tires, and a simple wheel configuration.

2.1 Kinematic Modeling

Based on what was written in previous paragraphs, this work is based on unicycle-like wheelchair. A unicycle wheelchair is a driving robot that can rotate freely around its axis. The term unicycle is often used in robotics to mean a generalized cart or car moving in a two-dimensional world; these are also often called unicycle-like or unicycle-type vehicles.

It is assumed that the human-wheelchair system moves on a planar horizontal surface. Let $\mathcal{R}(\mathcal{X}, \mathcal{Y}, \mathcal{Z})$ be any fixed frame with $\mathcal{Z}$ vertical. Traditionally, in the motion control of wheelchair robots, the wheelchair is considered as a point located at the middle of the virtual axle. However, in this work, the point that should follow a predetermined trajectory is located in front of the virtual axle (point $h(x, y)$ of Fig. 1). Such point is herein after named as the point of interest. G is of the center of mass of the wheelchair. Figure 1 illustrates the wheelchair considering in this work.

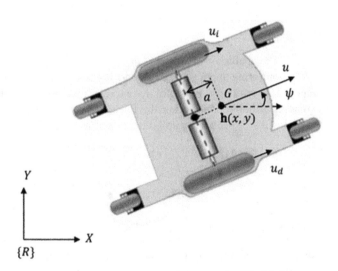

Fig. 1. Schematic of the autonomous wheelchair robotic

The configuration instantaneous kinematic model of the holonomic wheelchair is defined as,

$$\begin{cases} \dot{x} = u\cos\psi - a\omega\sin\psi \\ \dot{y} = u\sin\psi + a\omega\cos\psi \\ \dot{\psi} = \omega \end{cases} \tag{1}$$

also the equation system (1) can be written in compact form as

$$\dot{\mathbf{h}} = J(\psi)\mathbf{v}$$
$$\dot{\psi} = \omega \tag{2}$$

where $\dot{\mathbf{h}} = [\dot{x} \quad \dot{y}]^T \in \Re^2$ represents the vector of axis velocity of the $\mathcal{R}(\mathcal{X}, \mathcal{Y}, \mathcal{Z})$ system; $J(\psi) = \begin{bmatrix} \cos\psi & -a\sin\psi \\ \sin\psi & a\cos\psi \end{bmatrix} \in \Re^{2\times 2}$ is a singular matrix; and the control of maneuverability of the wheelchair is defined $\mathbf{v} \in \Re^n$ and $\mathbf{v} = [u \quad \omega]^T \in \Re^2$ in which u and ω represent the linear and angular velocities of the wheelchair, respectively.

On the other side, of (1) is determined the non-holonomic velocity constraint of the wheelchair robotic which determines that it can only move perpendicular to the wheels axis,

$$\dot{x}\sin\psi - \dot{y}\cos\psi + a\omega = 0 \tag{3}$$

2.2   Dynamic Model

In general, most robots available on the market have a low level of PID control-lers reference speed for monitoring input speeds and do not allow the motor voltage is proportional directly. Therefore, it is useful to express the model of the robotic wheelchair in a manner suitable considering the linear and angular velocity as input signals. So the model of the wheelchair can be expressed as [14, 15].

$$\mathbf{M}(\varsigma)\dot{\mathbf{v}} + \mathbf{C}(\varsigma, \mathbf{v})\mathbf{v} = \mathbf{v}_{ref} \tag{4}$$

where, $\mathbf{M}(\varsigma) \in \Re^{n\times n}$ with $n = 2$ and $\mathbf{M}(\varsigma) = \begin{bmatrix} \varsigma_1 & -\varsigma_7 \\ -\varsigma_8 & \varsigma_2 \end{bmatrix}$ represents the human-wheelchair system's inertia; $\mathbf{C}(\varsigma, \mathbf{v}) \in \Re^{n\times n}$ and $\mathbf{C}(\varsigma, \mathbf{v}) = \begin{bmatrix} \varsigma_4 & -\varsigma_3\omega \\ \varsigma_5\omega & \varsigma_6 \end{bmatrix}$ represents the components of the centripetal forces; $\mathbf{v} \in \Re^n$ and $\mathbf{v} = [u \quad \omega]^T$ is the vector of system's velocity; $\mathbf{v}_{ref} \in \Re^n$ and $\mathbf{v}_{ref} = [u_{ref} \quad \omega_{ref}]^T$ is the vector of velocity control signals for the wheelchair; and $\varsigma \in \Re^l$ with $l = 8$ and $\varsigma = [\varsigma_1 \quad \varsigma_2 \quad \cdots \quad \varsigma_l]^T$ is the vector of system's dynamic parameters.

3 Problem Formulation

The human operator controls the wheelchair by sending position commands to the system: x_d, and y_d, one for each axis in respect to the inertial frame $\mathcal{R}(\mathcal{X}, \mathcal{Y}, \mathcal{Z})$, using a haptic device.

$$P_{\mathbf{d}} = [x_d \ \ y_d]^T$$

The human operator commands are generated with the use of the Emotiv EPOC haptic device from Emotiv Systems Electronics Company [9] as indicated in Fig. 2. Its positions P_x, and P_y are translated into position commands x_d, and y_d for the locomotion of the wheelchair, through the following rotation matrix,

$$\begin{bmatrix} x_d \\ y_d \end{bmatrix} = \begin{bmatrix} \cos(\psi) & -\sin(\psi) \\ \sin(\psi) & \cos(\psi) \end{bmatrix} \begin{bmatrix} P_x \\ P_y \end{bmatrix}$$

where ψ represents the orientation of the wheelchair that rotates about the axis $\mathcal{Z}$.

Fig. 2. Emotiv EPOC haptic device

The other hand, the desired velocity of the wheelchair will depend on the task, the control error, the angular velocity, etc. In this case, it is considered that the reference velocity depends on the control errors and the angular velocity. It is defined as,

$$|\boldsymbol{v}_P(s_D, h)| = v_{P\max} \frac{1}{1 + k_i i_{\mathcal{P}} + k_\rho \|\rho\|} \tag{5}$$

where, $\upsilon_{P\max}$ represents the desired maximum velocity on the path $\mathcal{P}(\mathcal{S})$; k_i and k_ρ are positive constants that weigh of inattention level on path and control error, respectively; i_P is the inattention of moving of the wheelchair, and is defined as:

$$i_P(t) = 1 - \frac{U(t)}{U_{\max}}$$

where $U_{\max}$ is the maximum power of concentration of the human operator.

3.1   Problem of Motion Control of the Wheelchair

As represented in Fig. 3, the path to be followed is generated by human operator through the velocity and position commands generated by electromyogram signals through Emotiv EPOC haptic device. In this context the desire path is denoted as $\mathcal{P}(\mathcal{S})$, where $\mathcal{P}(\mathcal{S}) = (x_P(s), y_P(s))$; the actual desired location $P_d = (x_P(s_D), y_P(s_D))$ is defined as the closest point on $\mathcal{P}(\mathcal{S})$ to the human-wheelchair system, with s_D being the curvilinear abscissa defining the point P_d; the unit vector tangent to the path in the point P_d is denoted by $\mathbf{T}$; θ_T is the orientation of $\mathbf{T}$ with respect to the inertial frame $\mathcal{R}(\mathcal{X}, \mathcal{Y}, \mathcal{Z})$; $\tilde{x} = x_P(s_D) - x$ is the position error in the $\mathcal{X}$ direction; $\tilde{y} = y_P(s_D) - y$ is the position error in the $\mathcal{Y}$ direction; ρ represents the distance between the wheelchair position $h(x, y)$ and the desired point P_d, where the position error in the ρ direction is $\tilde{\rho} = 0 - \rho = -\rho$, i.e., the desired distance between the wheelchair position $h(x, y)$ and the desired point P_d must be zero; and θ_ρ is the orientation of the error $\tilde{\rho}$ with respect to the inertial frame $\mathcal{R}(\mathcal{X}, \mathcal{Y}, \mathcal{Z})$.

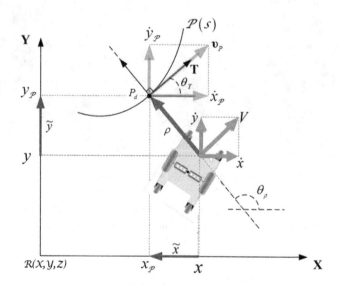

Fig. 3. The orthogonal projection of the point of interest over the path.

The path-following problem is solved by a control law capable of making the point of interest to assume a desired velocity equal to

$$V = \mathbf{v}_P(s_D, h) = |\mathbf{v}_P(s_D, h)| \angle \theta_T \tag{6}$$

besides making the robot to stay on the path, that is, $\tilde{x} = 0$ and $\tilde{y} = 0$. Therefore, if $\lim_{t \to \infty} \tilde{x}(t) = \mathbf{0}$ and $\lim_{t \to \infty} \tilde{y}(t) = \mathbf{0}$ then $\lim_{t \to \infty} \rho(t) = 0$ and $\lim_{t \to \infty} \tilde{\psi}(t) = 0$, being $\tilde{\psi}$ the orientation error of the wheelchair, defined as $\tilde{\psi} = \theta_T - \psi$.

Worth noting that the reference desired velocity $\mathbf{v}_P(s_D, h)$ of the wheelchair during the tracking path need not be constant, with is common in the literature [1, 3, 13–15],

$$\mathbf{v}_P(s_D, h) = f(k, s_D, \rho(t), \omega(t), \dots) \tag{7}$$

the wheelchair's desired velocity can be expressed as: constant function, curvilinear abscissa function of the path, position error function, angular velocities function of the wheelchair; and the others consideration.

4   Controllers Design

The design of the controller is based mainly on two cascaded subsystems: (1) *Kinematic controller* where the control errors $\rho(t)$ and $\tilde{\psi}(t)$ may be calculated at every measurement time and used to drive the mobile robot in a direction which decreases the errors; and (2) *Dynamic compensation controller*, which main objective is to compensate the dynamics of the human-wheelchair system, thus reducing the velocity tracking error.

4.1   Kinematic Controller

The proposed kinematic controller is based on the kinematic model of the wheelchair (2), *i.e.*, $\dot{\mathbf{h}} = f(\psi)\mathbf{v}$. Hence following control law is proposed,

$$\begin{bmatrix} u_c \\ \omega_c \end{bmatrix} = \mathbf{J}^{-1} \left(\begin{bmatrix} \dot{x}_P \\ \dot{y}_P \end{bmatrix} + \begin{bmatrix} \rho_x \\ \rho_y \end{bmatrix} \right) \tag{8}$$

with

$$\dot{x}_P = |\mathbf{v}_P| \cos(\theta_T) \quad \text{and} \quad \dot{y}_P = |\mathbf{v}_P| \sin(\theta_T) \tag{9}$$

where u_c and ω_c are the velocities outputs of the kinematic controller, $\mathbf{v}_P$ is the reference velocity input of the wheelchair for the controller, $\dot{x}_P$ is the projection of $\mathbf{v}_P$ in the $\mathcal{X}$ direction, $\dot{y}_P$ is the projection of $\mathbf{v}_P$ in the $\mathcal{Y}$ direction, $\mathbf{J}^{-1}$ is the matrix of inverse kinematics for the wheelchair, and ρ_x and ρ_y are the position error in the $\mathcal{X}$ and $\mathcal{Y}$ direction, respectively, respect to the inertial frame $\mathcal{R}(\mathcal{X}, \mathcal{Y}, \mathcal{Z})$, In order to include an analytical saturation of velocities in the wheelchair, the **tanh(.)** function, which limits the errors ρ_x and ρ_y, is proposed. Hence it is defined as,

$$\rho_x = l_x \tanh\left(\frac{k_x}{l_x}\tilde{x}\right) \quad \text{and} \quad \rho_y = l_y \tanh\left(\frac{k_y}{l_y}\tilde{y}\right). \tag{10}$$

Now, the behaviour of the control position error of the wheelchair is now analysed assuming -by now- perfect velocity tracking *i.e.*, $u(t) \equiv u_c(t)$ and $\omega(t) \equiv \omega_c(t)$. Hence manipulating (2) and (8), is can be written the behavior of the velocity of the point of interest of the wheelchair for the closed-loop system, that is given by

$$\begin{bmatrix} \dot{x} \\ \dot{y} \end{bmatrix} = \begin{bmatrix} \dot{x}_P \\ \dot{y}_P \end{bmatrix} + \begin{bmatrix} l_x \tanh\left(\frac{k_x}{l_x}\tilde{x}\right) \\ l_y \tanh\left(\frac{k_y}{l_y}\tilde{y}\right) \end{bmatrix}. \tag{11}$$

The analysis of the stability of the closed-loop system is represented in [15]; hence, it can now be concluded that $\lim_{t\to\infty} \tilde{\rho}(t) \to 0$, *i.e.*, $\tilde{x}(t) \to 0$ and $\tilde{y}(t) \to 0$ with $t \to \infty$ asymptotically. Therefore, it can be concluded that the final velocity of the point of interest will be $V = |v_P(s_D, h)| \angle \theta_T$ hence $\tilde{\psi}(t) \to 0$ for $t \to \infty$ asymptotically.

4.2 Dynamic Compensation Controller

If not considered the perfect velocity tracking in kinematic controller design, *i.e.*, $u(t) \neq u_c(t)$ and $\omega(t) \neq \omega_c(t)$. This velocity error motivates to design of an dynamic compensation controller; the objective of this controller is to compensate the dynamic of the human and of the wheelchair, thus reducing the velocity tracking error, hence the following control law dynamic model based (4) is proposed,

$$\begin{bmatrix} u_{ref} \\ \omega_{ref} \end{bmatrix} = \mathbf{M}\left(\begin{bmatrix} \dot{u}_c \\ \dot{\omega}_c \end{bmatrix} + \begin{bmatrix} \sigma_u \\ \sigma_\omega \end{bmatrix} \right) + \mathbf{C}\begin{bmatrix} u \\ \omega \end{bmatrix} \tag{12}$$

with

$$\sigma_u = l_u \tanh\left(\frac{k_u}{l_u}\tilde{u}\right) \quad \text{and} \quad \sigma_\omega = l_\omega \tanh\left(\frac{k_\omega}{l_\omega}\tilde{\omega}\right)$$

where $\tilde{u}(t) = u_c(t) - u(t)$ *and* $\tilde{\omega}(t) = \omega_c(t) - \omega(t)$ are the linear and angular velocity errors, respectively; $l_u > 0$, $k_u > 0$, $l_\omega > 0$ and $k_\omega > 0$ are positive gain constants that weigh the control error.

Now manipulating (4) and (12), have the behavior of the velocity errors of the human-wheelchair for the closed-loop system,

$$\begin{bmatrix} \dot{u} \\ \dot{\omega} \end{bmatrix} = \begin{bmatrix} \dot{u}_c \\ \dot{\omega}_c \end{bmatrix} + \begin{bmatrix} l_u \tanh\left(\frac{k_u}{l_u}\tilde{u}\right) \\ l_\omega \tanh\left(\frac{k_\omega}{l_\omega}\tilde{\omega}\right) \end{bmatrix}. \tag{13}$$

Next, a Lyapunov candidate function and its time derivative on the system trajectories are introduced in order to consider the corresponding stability analysis $V(\tilde{u}, \tilde{\omega}) = \frac{1}{2}(\tilde{u}^2 + \tilde{\omega}^2) > 0$; the time derivative of the Lyapunov candidate function is,

$$\dot{V}(\tilde{u}, \tilde{\omega}) = \tilde{u}\dot{\tilde{u}} + \tilde{\omega}\dot{\tilde{\omega}} \tag{14}$$

After introducing the derivate of (13) in (14), the time derivative $\dot{V}(\tilde{u}, \tilde{\omega})$ is now

$$\dot{V}(\tilde{u}, \tilde{\omega}) = -\tilde{u}l_u \tanh\left(\frac{k_u}{l_u}\tilde{u}\right) - \tilde{\omega}l_\omega \tanh\left(\frac{k_\omega}{l_\omega}\tilde{\omega}\right) < 0 \tag{15}$$

Hence, from (15) it can now be concluded that $\tilde{u}(t) \to 0$ and $\tilde{\omega}(t) \to 0$ with $t \to \infty$ asymptotically.

5 Experimental Results

In this section the performance of the proposed controllers and dynamic modeling of human wheel-chair system is demonstrated. Some of the results of the wheelchair shown in Fig. 4, in order to evaluate the performance of the proposed controller.

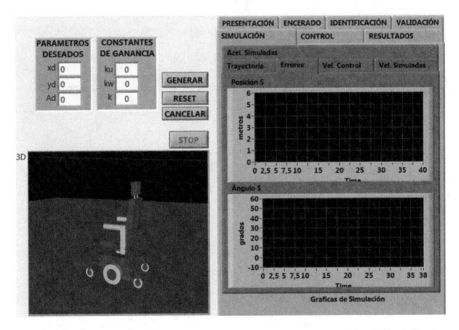

Fig. 4. Human Machine Interface developed for the analysis of the model and the performance of the controller proposed

The experiment corresponds to the performance of the proposed controller for path following problem. Note that for the path following problem the desired velocity of the wheelchair will depend on the task, the control error, the angular velocity, etc. For this case, it is consider that the reference velocity depends on the control errors, the angular velocity and inattention of moving of the wheelchair. Figures 5, 6 and 7 show the results of the experiment. Figure 5 shows the movement of the wheelchair on the X-Y space of the path following problem experiment, and finally the Figs. 6 and 7 present the linear and angular velocities of the wheelchair. It can be seen that the proposed controller works correctly.

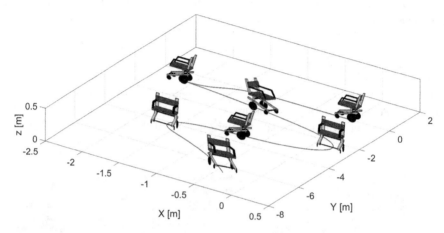

Fig. 5. Stroboscopic movement of the wheelchair in the path following experiment.

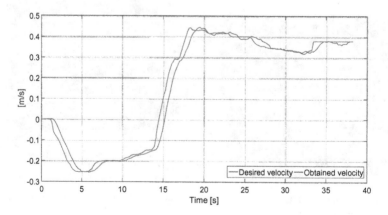

Fig. 6. Linear velocity of the Human-Wheelchair System

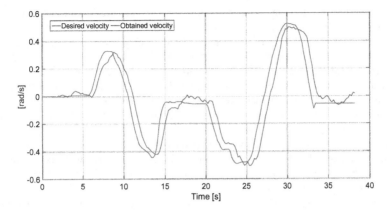

Fig. 7. Velocity and angular of the Human-Wheelchair System

6   Conclusions

In this paper the control for a human-wheelchair system, considering the brain signals sent commands remote site, position and speed of movement is taken for wheelchair. The controller proposed resolved the path following problem for wheelchair robot, which is also capable of positioning the robot. Human sends velocity and position commands generated by electromyogram signals through Emotiv EPOC haptic device (master) to the remote site (wheelchair). The desired velocity of the wheelchair is considered as a function of the disregard of the person to move the robotic wheelchair. Finally, the stability and robustness are proved by considering the Lyapunov's method, and the performance of the proposed controller is shown through real experiments.

Acknowledgment. The authors would like to thanks to the Universidad Técnica de Ambato for financing the project *Robotic Assistance for Persons with Disabilities* (Resolution: 1151-CU-P-2012). Also to the Universidad de las Fuerzas Armadas ESPE and to the Escuela Superior Politécnica de Chimborazo for the support to develop of the Master's Thesis *Control de una silla de ruedas a través de señales cerebrales*.

References

1. Bastos-Filho, T.F.: Towards a new modality independent interface for a robotic wheelchair. IEEE Trans. Neural Syst. Rehabil. Eng. **22**(3), 567–584 (2014)
2. Wang, Y., Chen, W.: Hybrid map-based navigation for intelligent wheelchair. In: IEEE International Conference on Robotics and Automation, China, pp. 637–642 (2011)
3. Andaluz, V.H., Ortiz, J.S., Sanchéz, J.S.: Bilateral control of a robotic arm through brain signals. In: De Paolis, L.T., Mongelli, A. (eds.) AVR 2015. LNCS, vol. 9254, pp. 355–368. Springer, Heidelberg (2015)
4. Mazo, M.: An integral system for assisted mobility. IEEE Robot. Autom. Mag. **8**(1), 46–56 (2001)

5. Zeng, Q., Teo, C., Rebsamen, B., Burdet, E.: A collaborative wheelchair system. IEEE Trans. Neural Syst. Rehabil. Eng. **16**(2), 161–170 (2008)

6. Parikh, S.P., Grassi, V., Kumar, V., Okamoto, J.: Integrating human inputs with autonomous behaviors on an intelligent wheelchair platform. IEEE Intell. Syst. **22**(2), 33–41 (2007)

7. Biswas, K., Mazumder, O., Kundu, A.S.: Multichannel fused EMG based biofeedback system with virtual reality for gait rehabilitation. In: IEEE Proceedings in International Conference on Intelligent Human Computer Interaction, India (2012)

8. Hirche, S., Buss, M.: Human-oriented control for haptic teleoperation. Proc. IEEE **100**(3), 623–647 (2012)

9. Emotiv Systems Electronics Company. http://emotiv.com/

10. Jang, W.A., Lee, S.M., Lee, D.H.: Development BCI for individuals with severely disability using EMOTIV EEG headset and robot. In: 2014 International Winter Workshop on Brain-Computer Interface (BCI) (2014)

11. Enache, A., Cepisca, C., Paraschiv, M., Banica, C.: Virtual instrument for electroencephalography data acquisition. In: 7th International Symposium on Advanced Topics in Electrical Engineering (ATEE), pp. 1–4 (2011)

12. Huang, D., Qian, K., Fei, D.-Y., Jia, W.: Electroencephalography (EEG)-based Brain-Computer Interface (BCI): a 2-D virtual wheelchair control based on event-related desynchronization/synchronization and state control. IEEE Trans. Neural Syst. Rehabil. Eng. **20**, 379–388 (2011)

13. Sapey, B., Stewart, J., Donaldson, G.: Increases in wheelchair use and perceptions of disablement. Disabil. Soc. **20**(5), 489–505 (2005)

14. Andaluz, V.H., Canseco, P., Varela, J., Ortiz, J.S., Pérez, M.G., Roberti, F., Carelli, R.: Robust control with dynamic compensation for human-wheelchair system. In: Zhang, X., Liu, H., Chen, Z., Wang, N. (eds.) ICIRA 2014, Part I. LNCS, vol. 8917, pp. 376–389. Springer, Heidelberg (2014)

15. Andaluz, V.H., Canseco, P., Varela, J., Ortiz, J.S., Pérez, M.G., Morales, V., Robertí, F., Carelli, R.: Modeling and control of a wheelchair considering center of mass lateral displacements. In: Liu, H., Kubota, N., Zhu, X., Dillmann, R. (eds.) ICIRA 2015. LNCS, vol. 9246, pp. 254–270. Springer, Heidelberg (2015)

Adaptive Control of the Human-Wheelchair System Through Brain Signals

Víctor H. Andaluz[1,2(✉)], Jessica S. Ortiz[3], Fernando A. Chicaiza[1], José Varela[1], Edison G. Espinosa[1], and Paúl Canseco[2]

[1] Universidad de las Fuerzas Armadas ESPE, Sangolquí, Ecuador
{vhandaluz1,fachicaiza,egespinosa1}@espe.edu.ec
[2] Universidad Técnica de Ambato, Ambato, Ecuador
victorhandaluz@uta.edu.ec
[3] Escuela Superior Politécnica de Chimborazo, Riobamba, Ecuador
jessortizm@outlook.com

Abstract. This work presents an adaptive dynamic control to solve the path following problem for the human-wheelchair system, allowing people with lower and upper extremity impairments to move a wheelchair through brain signals. The desired velocity of the wheelchair is considered as a function of the disregard of the person to move the robotic wheelchair. Additionally, the kinematic and dynamic modeling of a human-wheelchair system where it is considered that its mass center is not located at the wheels' axis center of the wheelchair. This controller design is based on two cascaded subsystems: a kinematic controller with command saturation, and an adaptive dynamic controller that compensates the dynamics of the human-wheelchair system. Stability and robustness are proved by using Lyapunov's method. Experimental results show a good performance of the proposed controller as proved by the theoretical design.

Keywords: Wheelchair · Dynamic modeling · Cascade control · Lyapunov's method

1 Introduction

A robotic wheelchair can be used to allow people with lower and upper extremity impairments or severe motor dysfunctions overcome the difficulties in driving a wheelchair. The robotic wheelchair system integrates a sensory subsystem, a navigation and control module and a user-machine interface to guide the wheelchair in autonomous or semi-autonomous mode [1–6]. In autonomous mode, the robotic wheelchair goes to the chosen destination without any participation of the user in the control. This mode is intended for people who have great difficulties to guide the wheelchair. In the semi-autonomous mode the user shares the control with the robotic wheelchair. In this case only some motor skills are needed from the user.

In order to reduce performance degradation, on-line parameter adaptation is relevant in applications where the mobile robot dynamic parameters may vary, such as load transportation. It is also useful when the knowledge of the dynamic parameters is limited. As

© Springer International Publishing Switzerland 2016
N. Kubota et al. (Eds.): ICIRA 2016, Part II, LNAI 9835, pp. 223–234, 2016.
DOI: 10.1007/978-3-319-43518-3_22

an example, the trajectory tracking task can be severely affected by the change imposed to the robot dynamics. Hence, some path following control architectures already proposed in the literature have considered the dynamics of the mobile robots [7, 8].

In such context, this work proposes a bilateral teleoperation system in order to allow people with lower and upper extremity impairments or severe motor dysfunctions to overcome the difficulties in locomotion. It comprises a robotic wheelchair so that it can move on unstructured environments. The human operator receives visual signal and sends velocity and position commands generated by electromyogram signals through Emotiv EPOC device to the wheelchair. The desired velocity of the wheelchair is considered as a function of the disregard of the person to move the robotic wheelchair. On the other hand, is important to indicate that the wheelchair's center of gravity changes due to postural issues, limb amputations, or obesity [9]. Therefore, in the present work a dynamic model of the human-wheelchair system is developed considering lateral deviations of the center of mass originated in user's movement, limb amputations, or obesity. Furthermore, in this work it is proposed a method to solve the path following problem and positioning for a wheelchair robot to assist persons with severe motor diseases. The proposed control scheme is divided into two subsystems, each one being a controller itself: (i) the first on is a kinematic controller with saturation of velocity commands, which is based on the wheelchair robot's kinematic. The path following problem is addressed in this subsystem. It is worth noting that the proposed controller does not consider $s(t)$ as an additional control input as it is frequent in literature; and (ii) an dynamic compensation controller that considered the human-wheelchair system dynamic model, which are directly related to physical parameters of the system. In addition, both stability and robustness properties to parametric uncertainties in the dynamic model are proven through Lyapunov's method. To validate the proposed control algorithm, experimental results are included and discussed.

2   Problem Formulation

The human operator controls the movement the wheelchair by sending position commands to the robot: x_d, and y_d, one for each axis in respect to the inertial frame $\mathcal{R}(\mathcal{X}, \mathcal{Y}, \mathcal{Z})$, using a EEG sensor.

$$P_{\mathbf{d}} = \begin{bmatrix} x_d \, y_d \end{bmatrix}^T$$

The human operator commands are generated through of the Emotiv EPOC device from Emotiv Systems Electronics Company [10] as indicated in Fig. 1. Its positions P_x, and P_y are translated into position commands x_d, and y_d for the locomotion of the wheelchair, through the following rotation matrix,

$$\begin{bmatrix} x_d \\ y_d \end{bmatrix} = \begin{bmatrix} \cos \psi & -\sin \psi \\ \sin \psi & \cos \psi \end{bmatrix} \begin{bmatrix} P_x \\ P_y \end{bmatrix}$$

where ψ represents the orientation of the wheelchair that rotates about the axis $\mathcal{Z}$.

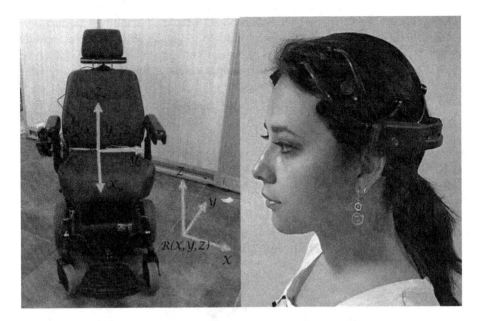

Fig. 1. Emotiv EPOC haptic device

The other hand, the desired velocity of the robotic wheelchair will depend on the task, path desired, control error, angular velocity, among others. In this work, it is considered that the reference velocity depends on the control errors and the angular velocity. It is defined as,

$$\left|\mathbf{v}_P(s_D, h)\right| = v_{P\max} \frac{1}{1 + k_i i_p + k_\rho \|\rho\|}$$

where, $v_{P\max}$ represents the desired maximum velocity on the path $\mathcal{P}(s)$; k_i and k_ρ are positive constants that weigh of inattention level on path and control error, respectively; i_p is the inattention of moving of the wheelchair, and is defined as:

$$i_p(t) = 1 - \frac{U(t)}{U_{\max}}$$

where $U_{\max}$ is the maximum power of concentration of the human operator.

Remark 1. In the literature it is common to find that considers ways to track the speed is constant, which restricts movement to perform a task.

3   Human – Wheelchair System

The robotic wheelchair used in this work presents similar characteristics to that of a unicycle-like mobile robot, because it has two driven wheels which are controlled independently by two direct current motors and four caster wheel to maintain balance, while the unicycle-type mobile robots have a caster wheel to maintain stability. The kinematic modeling of the human-wheelchair system is developed in the next subsection, considering a horizontal work plane where the wheelchair moves. The wheelchair type unicycle-like mobile robot presents the advantages of high mobility, high traction with pneumatic tires, and a simple wheel configuration.

3.1 Kinematic Modeling

It is assumed that the human-wheelchair system moves on a planar horizontal surface. Let $\mathcal{R}(\mathcal{X}, \mathcal{Y}, \mathcal{Z})$ be any fixed frame with $\mathcal{Z}$ vertical. Traditionally, in the motion control of wheelchair robots, the wheelchair is considered as a point located at the middle of the virtual axle. However, in this work, the point that should follow a predetermined trajectory is located in front of the virtual axle (point $h(x, y)$ of Fig. 2). Such point is herein after named as the point of interest. G is of the center of mass of the wheelchair. Figure 2 illustrates the wheelchair considering in this work [11].

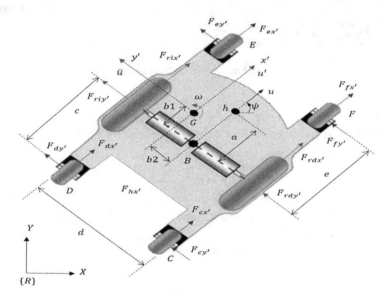

Fig. 2. Schematic of the autonomous wheelchair robotic

The configuration instantaneous kinematic model of the holonomic wheelchair is defined as,

$$\begin{cases} \dot{x} = u \cos \psi - a\omega \sin \psi \\ \dot{y} = u \sin \psi + a\omega \cos \psi \\ \dot{\psi} = \omega \end{cases} \tag{1}$$

also the equation system (1) can be written in compact form as

$$\dot{\mathbf{h}} = \mathbf{J}(\psi)\mathbf{v}$$
$$\dot{\psi} = \omega \tag{2}$$

where $\dot{\mathbf{h}} = [\dot{x}\ \dot{y}]^T \in \mathfrak{R}^2$ represents the vector of axis velocity of the $\mathcal{R}(\mathcal{X}, \mathcal{Y}, \mathcal{Z})$ system; $\mathbf{J}(\psi) = \begin{bmatrix} \cos\psi & -a\sin\psi \\ \sin\psi & a\cos\psi \end{bmatrix} \in \mathfrak{R}^{2\times2}$ is a singular matrix; and the control of maneuverability of the wheelchair is defined $\mathbf{v} \in \mathfrak{R}^n$ and $\mathbf{v} = [u\ \omega]^T \in \mathfrak{R}^2$ in which u and ω represent the linear and angular velocities of the wheelchair, respectively.

On the other side, of (1) is determined the non-holonomic velocity constraint of the wheelchair robotic which determines that it can only move perpendicular to the wheels axis,

$$\dot{x}\sin\psi - \dot{y}\cos\psi + a\omega = 0 \tag{3}$$

3.2 Dynamic Model

In general, most robots available on the market have a low level of PID controllers reference speed for monitoring input speeds and do not allow the motor voltage is proportional directly. Therefore, it is useful to express the model of the robotic wheelchair in a manner suitable considering the linear and angular velocity as input signals. So the model of the wheelchair can be expressed as [11, 12].

$$\mathbf{M}(\varsigma)\dot{\mathbf{v}} + \mathbf{C}(\varsigma, \mathbf{v})\mathbf{v} = \mathbf{v}_{\mathbf{ref}} \tag{4}$$

where, $\mathbf{M}(\varsigma) \in \mathfrak{R}^{n\times n}$ with $n = 2$ and $\mathbf{M}(\varsigma) = \begin{bmatrix} \varsigma_1 & -\varsigma_7 \\ -\varsigma_8 & \varsigma_2 \end{bmatrix}$ represents the human-wheelchair system's inertia; $\mathbf{C}(\varsigma, \mathbf{v}) \in \mathfrak{R}^{n\times n}$ and $\mathbf{C}(\varsigma, \mathbf{v}) = \begin{bmatrix} \varsigma_4 & -\varsigma_3\omega \\ \varsigma_5\omega & \varsigma_6 \end{bmatrix}$ represents the components of the centripetal forces; $\mathbf{v} \in \mathfrak{R}^n$ and $\mathbf{v} = [u\ \omega]^T$ is the vector of system's velocity; $\mathbf{v}_{ref} \in \mathfrak{R}^n$ and $\mathbf{v}_{\mathbf{ref}} = [u_{ref}\ \omega_{ref}]^T$ is the vector of velocity control signals for the wheelchair; and $\varsigma \in \mathfrak{R}^l$ with $l = 8$ and $\varsigma = [\varsigma_1\ \varsigma_2\ \dots\ \varsigma_l]^T$ is the vector of system's dynamic parameters.

4 Controllers Design

The proposed control scheme to solve the path following problem is shown in Fig. 3, the design of the controller is based mainly on two cascaded subsystems.

(1) *Kinematic Controller* with saturation of velocity commands, where the control errors $\rho(t)$ and $\tilde{\psi}(t)$ may be calculated at every measurement time and used to drive the mobile robot in a direction which decreases the errors. Therefore, the control aim is to ensure that $\lim_{t\to\infty}\rho(t) = 0$ and $\lim_{t\to\infty}\tilde{\psi}(t) = 0$.

(2) *Adaptive Dynamic Compensation Controller*, which main objective is to compensate the dynamics of the mobile robot, thus reducing the velocity tracking error. This controller receives as inputs the desired velocities $\mathbf{v_c} = [\, u_c \;\; \omega_c \,]^T$ calculated by the kinematic controller, and generates velocity references $\mathbf{v_{ref}}(t)$ for the mobile robot. The velocity control error is defined as $\tilde{\mathbf{v}} = \mathbf{v_c} - \mathbf{v}$. Hence, the control aim is to ensure that $\lim_{t \to \infty} \tilde{\mathbf{v}}(t) = \mathbf{0}$.

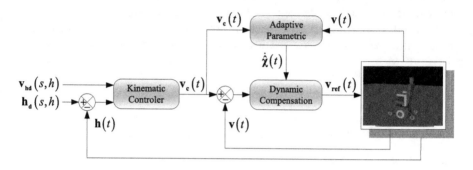

Fig. 3. Adaptive dynamic controller: block diagram.

4.1 Kinematic Controller

The desired operational motion of the wheelchair is an application $P_d = \big(x_P(s_D), \, y_P(s_D) \big)$ is the actual desired location generated by human through of the Emotiv EPOC device. The problem of control is to find the control vector of maneuverability $\big(\mathbf{v_c}(t) | t \in [t_0, t_f] \big)$ to achieve the desired operational motion. Thus, the proposed kinematic controller is based on the kinematic model of the wheelchair (1), *i.e.*, $\dot{\mathbf{h}} = f(\mathbf{h})\mathbf{v}$. Hence following control law is proposed,

$$\begin{bmatrix} u_c \\ \omega_c \end{bmatrix} = \mathbf{J}^{-1} \left(\begin{bmatrix} \dot{x}_P \\ \dot{y}_P \end{bmatrix} + \begin{bmatrix} \rho_x \\ \rho_y \end{bmatrix} \right) \tag{5}$$

with

$$\dot{x}_P = |\boldsymbol{v}_P| \cos \left(\theta_T \right) \text{ and } \dot{y}_P = |\boldsymbol{v}_P| \sin \left(\theta_T \right)$$

where u_c and ω_c are the velocities outputs of the kinematic controller, $\boldsymbol{v}_P$ is the reference velocity input of the wheelchair for the controller, $\dot{x}_P$ is the projection of $\boldsymbol{v}_P$ in the $\mathcal{X}$ direction, $\dot{y}_P$ is the projection of $\boldsymbol{v}_P$ in the $\mathcal{Y}$ direction, $\mathbf{J}^{-1}$ is the matrix of inverse kinematics for the wheelchair, and ρ_x and ρ_y are the position error in the $\mathcal{X}$ and $\mathcal{Y}$ direction, respectively, respect to the inertial frame $\mathcal{R}(\mathcal{X}, \mathcal{Y}, \mathcal{Z})$. In order to include an analytical saturation of velocities in the wheelchair, the **tanh**(.) function, which limits the errors ρ_x and ρ_y, is proposed. Hence it is defined as,

$$\rho_x = l_x \tanh\left(\frac{k_x}{l_x}\tilde{x}\right) \text{ and } \rho_y = l_y \tanh\left(\frac{k_y}{l_y}\tilde{y}\right). \qquad (6)$$

Now, the behaviour of the control position error of the wheelchair is now analysed assuming - by now - perfect velocity tracking *i.e.*, $u(t) \equiv u_c(t)$ and $\omega(t) \equiv \omega_c(t)$. Hence manipulating (1) and (5), is can be written the behavior of the velocity of the point of interest of the wheelchair for the closed-loop system, that is given by

$$\begin{bmatrix} \dot{x} \\ \dot{y} \end{bmatrix} = \begin{bmatrix} \dot{x}_P \\ \dot{y}_P \end{bmatrix} + \begin{bmatrix} l_x \tanh\left(\frac{k_x}{l_x}\tilde{x}\right) \\ l_y \tanh\left(\frac{k_y}{l_y}\tilde{y}\right) \end{bmatrix}. \qquad (7)$$

The analysis of the stability of the closed-loop system is represented in [12]; hence, it can now be concluded that $\lim_{t\to\infty} \tilde{\rho}(t) \to 0$, *i.e.*, $\tilde{x}(t) \to 0$ and $\tilde{y}(t) \to 0$ with $t \to \infty$ asymptotically. Therefore, it can be concluded that the final velocity of the point of interest will be $V = |\mathbf{v}_P(s_D, h)| \angle \theta_T$ hence $\tilde{\psi}(t) \to 0$ for $t \to \infty$ asymptotically.

4.2   Adaptive Dynamic Compensation Controller

The objective of the adaptive dynamic compensation controller is to compensate the dynamic of the mobile robot, thus reducing the velocity tracking error. This subsystem receives the desired velocities $\mathbf{v_c} = [\, u_c \ \ \omega_c \,]^T$ and generates velocity references $\mathbf{v}_{ref} = [\, u_{ref} \ \ \omega_{ref} \,]^T$ to be sent to the mobile robot.

If there is not considering perfect velocity tracking in Subsect. 4.1, *i.e.*, $\mathbf{v} \neq \mathbf{v_c}$, the velocity error is defined as, $\tilde{\mathbf{v}} = \mathbf{v_c} - \mathbf{v}$. This velocity error motivates to design of an adaptive dynamic compensation controller with a robust parameter updating law. The dynamic compensation control law for the mobile robot is,

$$\mathbf{v_{ref}} = \boldsymbol{\eta}\hat{\boldsymbol{\chi}} = \boldsymbol{\eta}\boldsymbol{\chi} + \boldsymbol{\eta}\tilde{\boldsymbol{\chi}} = \mathbf{M}\boldsymbol{\sigma} + \mathbf{C}\mathbf{v} + \boldsymbol{\eta}\tilde{\boldsymbol{\chi}} \qquad (8)$$

where $\boldsymbol{\eta}(\mathbf{q}, \mathbf{v}, \boldsymbol{\sigma}) \in \mathfrak{R}^{2\times8}$, $\boldsymbol{\chi} = [\, \chi_1 \ \chi_2 \ \cdots \ \chi_8 \,]^T$ and $\hat{\boldsymbol{\chi}} = [\, \hat{\chi}_1 \ \hat{\chi}_2 \ \cdots \ \hat{\chi}_8 \,]^T$ are respectively the unknown vector, real parameters vector and estimated parameters vector of the robot, whereas $\tilde{\boldsymbol{\chi}} = \hat{\boldsymbol{\chi}} - \boldsymbol{\chi}$ is the vector of parameter errors and $\boldsymbol{\sigma} = \dot{\mathbf{v}}_c + \mathbf{L_v} \tanh\left(\mathbf{L_v^{-1}} \mathbf{K_v}\tilde{\mathbf{v}}\right)$. A Lyapunov candidate function is proposed as

$$V(\tilde{\mathbf{v}}, \tilde{\boldsymbol{\chi}}) = \tfrac{1}{2}\tilde{\mathbf{v}}^T \mathbf{M}\tilde{\mathbf{v}} + \tfrac{1}{2}\tilde{\boldsymbol{\chi}}^T \boldsymbol{\gamma}\tilde{\boldsymbol{\chi}}$$

where $\boldsymbol{\gamma} \in \mathfrak{R}^{8\times8}$ is a positive definite diagonal matrix and $\mathbf{M} \in \mathfrak{R}^{2\times2}$ is a positive definite matrix defined in (4). The time derivative of the Lyapunov candidate function is,

$$\dot{V}(\tilde{\mathbf{v}}, \tilde{\boldsymbol{\chi}}) = -\tilde{\mathbf{v}}^T \mathbf{M}\mathbf{L_v} \tanh\left(\mathbf{L_v^{-1}}\mathbf{K_v}\tilde{\mathbf{v}}\right) - \tilde{\mathbf{v}}^T \boldsymbol{\eta}\tilde{\boldsymbol{\chi}} + \tilde{\boldsymbol{\chi}}^T \boldsymbol{\gamma}\dot{\tilde{\boldsymbol{\chi}}} + \tfrac{1}{2}\tilde{\mathbf{v}}^T \dot{\mathbf{M}}\tilde{\mathbf{v}}. \qquad (9)$$

The robust updating law

$$\dot{\hat{\boldsymbol{\chi}}} = \boldsymbol{\gamma}^{-1}\mathbf{L}^T\tilde{\mathbf{v}} - \boldsymbol{\gamma}^{-1}\boldsymbol{\Gamma}\hat{\boldsymbol{\chi}} \qquad (10)$$

is adopted to update the parameter estimated, where $\Gamma \in \Re^{8\times8}$ is a diagonal positive gain matrix. Let us consider that the dynamic parameters can vary, *i.e.*, $\chi = \chi(t)$ and $\dot{\tilde{\chi}} = \dot{\hat{\chi}} - \dot{\chi}$. Now, substituting (10) in (9), the following expression it is obtained,

$$V(\tilde{\mathbf{v}}, \tilde{\chi}) = -\tilde{\mathbf{v}}^{\mathbf{T}}\mathbf{M}\mathbf{L}_{\mathbf{v}}\tanh\left(\mathbf{L}_{\mathbf{v}}^{-1}\mathbf{K}_{\mathbf{v}}\tilde{\mathbf{v}}\right) - \tilde{\chi}^{\mathbf{T}}\Gamma\tilde{\chi} - \tilde{\chi}^{\mathbf{T}}\Gamma\chi$$
$$- \tilde{\chi}^{\mathbf{T}}\gamma\dot{\chi} + \tfrac{1}{2}\tilde{\mathbf{v}}^{\mathbf{T}}\dot{\mathbf{M}}\tilde{\mathbf{v}} \tag{11}$$

In [13], it has been shown the stability of the adaptive dynamic compensation controller for a mobile manipulator, where it was proved that velocity error and parameter errors are ultimately bounded. Hence, we can conclude that $\tilde{\mathbf{v}}(t)$ and $\tilde{\chi}(t)$ are ultimately bounded.

4.3   Adaptive Dynamic Compensation Controller

The behaviour of the tracking error of the mobile robot is now analyzed relaxing the assumption of perfect velocity tracking.

$$\dot{\tilde{\mathbf{h}}} + \mathbf{L}\tanh\left(\mathbf{L}^{-1}\mathbf{K}\,\tilde{\mathbf{h}}\right) = \mathbf{J}\tilde{\mathbf{v}} + \Upsilon. \tag{12}$$

The Lyapunov candidate $V(\tilde{\mathbf{h}})$ is proposed, and its time derivative is $\dot{V}(\tilde{\mathbf{h}}) = \tilde{\mathbf{h}}^{\mathbf{T}}(\mathbf{J}\tilde{\mathbf{v}} + \Upsilon) - \tilde{\mathbf{h}}^{\mathbf{T}}\mathbf{L}\tanh(\mathbf{L}^{-1}\mathbf{K}\,\tilde{\mathbf{h}})$. A sufficient condition for $\dot{V}(\tilde{\mathbf{h}})$ to be negative definite is

$$\left|\tilde{\mathbf{h}}^{\mathbf{T}}\mathbf{L}\tanh\left(\mathbf{L}^{-1}\mathbf{K}\,\tilde{\mathbf{h}}\right)\right| > \left|\tilde{\mathbf{h}}^{\mathbf{T}}(\mathbf{J}\tilde{\mathbf{v}} + \Upsilon)\right|. \tag{13}$$

For large values of $\tilde{\mathbf{h}}$, it can be considered that: $\mathbf{L}\tanh\left(\mathbf{L}^{-1}\mathbf{K}\,\tilde{\mathbf{h}}\right) \approx \mathbf{L}$. Therefore, $\dot{V}(\tilde{\mathbf{h}})$ will be negative definite only if $\|\mathbf{L}\| > \|\mathbf{J}\tilde{\mathbf{v}} + \Upsilon\|$, thus making the velocity errors $\tilde{\mathbf{h}}$ to decrease.

Now, for small values of $\tilde{\mathbf{h}}$, it can be expressed that: $\mathbf{L}\tanh\left(\mathbf{L}^{-1}\mathbf{K}\,\tilde{\mathbf{h}}\right) \approx \mathbf{K}\tilde{\mathbf{h}}$, and (13) can be written as, $\|\tilde{\mathbf{h}}\| > \|\mathbf{J}\tilde{\mathbf{v}} + \Upsilon\|/\lambda_{\min}(\mathbf{K})$, thus implying that the error $\tilde{\mathbf{h}}$ is ultimately bounded, with bound,

$$\|\tilde{\mathbf{h}}\| \leq \frac{\|\mathbf{J}\tilde{\mathbf{v}} + \Upsilon\|}{\lambda_{\min}(\mathbf{K})} \tag{14}$$

5   Experimental Results

In this section the performance of the proposed controllers and dynamic modeling of human wheel-chair system is demonstrated. Some of the results of the wheelchair shown in Fig. 4, in order to evaluate the performance of the proposed controller.

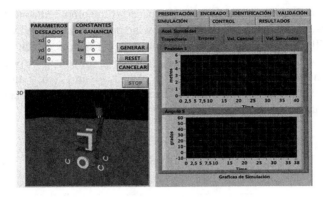

Fig. 4. Human Machine Interface developed for the analysis of the model and the performance of the controller proposed

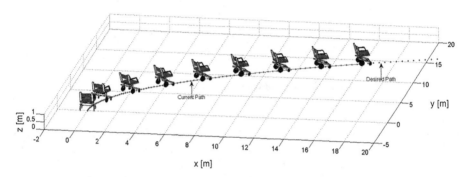

Fig. 5. Stroboscopic movement of the wheelchair in the path following experiment.

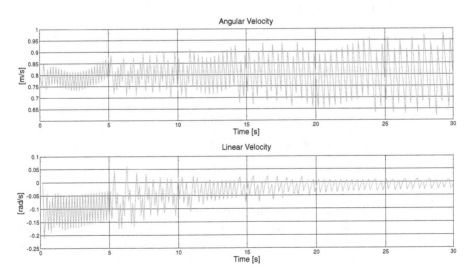

Fig. 6. Linear and angular velocity of the wheelchair

The experiment corresponds to the performance of the proposed controller for path following problem. Note that for the path following problem the desired velocity of the wheelchair will depend on the task, the control error, the angular velocity, etc. For this case, it is consider that the reference velocity depends on the control errors, the angular velocity and inattention of moving of the wheelchair. Figures 5, 6, 7 and 8 show the results of the experiment. Figure 5 shows the movement of the wheelchair on the *X-Y* space of the path following problem experiment.

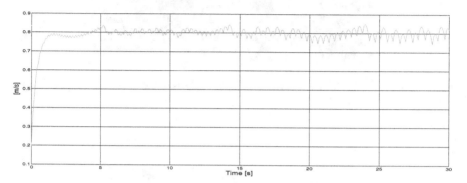

Fig. 7. Velocity of the human-wheelchair system

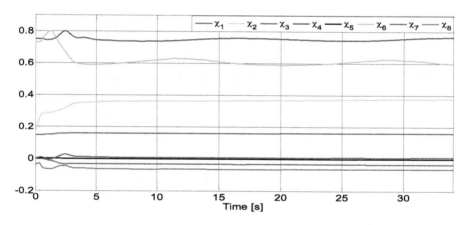

Fig. 8. Adaptive parameters evolution

While the Figs. 6 and 7 present the linear and angular velocities of the wheelchair. It can be seen that the proposed controller works correctly; and finally the Fig. 8 shows the evolution of the adaptive parameters, where it can be seen that all the parameters converge to fixed values.

6   Conclusions

In this paper the control for a human-wheelchair system, considering the brain signals sent commands remote site, position and speed of movement is taken for wheelchair. The controller proposed resolved the path following problem for wheelchair robot, which is also capable of positioning the robot. Human sends velocity and position commands generated by electromyogram signals through Emotiv EPOC haptic device (master) to the remote site (wheelchair). The desired velocity of the wheelchair is considered as a function of the disregard of the person to move the robotic wheelchair. Finally, the stability and robustness are proved by considering the Lyapunov's method, and the performance of the proposed controller is shown through real experiments.

Acknowledgment. The authors would like to thanks to the Universidad Técnica de Ambato for financing the project *Robotic Assistance for Persons with Disabilities* (Resolution: 1151-CU-P-2012). Also to the Universidad de las Fuerzas Armadas ESPE and to the Escuela Superior Politécnica de Chimborazo for the support to develop of the Master's Thesis *Control de una silla de ruedas a través de señales cerebrales*.

References

1. Bastos-Filho, T., et al.: Towards a new modality-independent interface for a robotic wheelchair. IEEE Trans. Neural Syst. Rehabil. Eng. **22**(3), 567–584 (2014)
2. Wang, Y., Chen, W.: Hybrid map-based navigation for intelligent wheelchair. In: IEEE International Conference on Robotics and Automation, China, pp. 637–642 (2011)
3. Cheng., W.-C., Chiang, C.-C.: The development of the automatic lane following navigation system for the intelligent robotic wheelchair. In: IEEE International Conference on Fuzzy Systems, Taiwan, pp. 1946–1952 (2011)
4. Mazo, M.: An integral system for assisted mobility. IEEE Robot. Autom. Mag. **8**(1), 46–56 (2001)
5. Zeng, Q., Teo, C., Rebsamen, B., Burdet, E.: A collaborative wheelchair system. IEEE Trans. Neural Syst. Rehabil. Eng. **16**(2), 161–170 (2008)
6. Parikh, S.P., Grassi, V., Kumar, V., Okamoto, J.: Integrating human inputs with autonomous behaviors on an intelligent wheelchair platform. IEEE Intell. Syst. **22**(2), 33–41 (2007)
7. Soeanto, D., Lapierre, L., Pascoal, A.: Adaptive non-singular path-following, control of dynamic wheeled robots. In: Proceedings of 42nd IEEE/CDC, Hawaii, USA, 9–12 December, pp. 1765–1770 (2003)
8. Xu, Y., Zhang, C., Bao, W., Tong, L.: Dynamic sliding mode controller based on particle swarm optimization for mobile robot's path following. In: International Forum on Information Technology and Applications, pp. 257–260 (2009)
9. Sapey, B., Stewart, J., Donaldson, G.: Increases in wheelchair use and perceptions of disablement. Disabil. Soc. **20**(5), 489–505 (2005)
10. Emotiv Systems Electronics Company. http://emotiv.com/
11. Andaluz, V.H., Canseco, P., Varela, J., Ortiz, J.S., Pérez, M.G., Roberti, F., Carelli, R.: Robust control with dynamic compensation for human-wheelchair system. In: Zhang, X., Liu, H., Chen, Z., Wang, N. (eds.) ICIRA 2014, Part I. LNCS, vol. 8917, pp. 376–389. Springer, Heidelberg (2014)

12. Andaluz, V.H., Canseco, P., Varela, J., Ortiz, J.S., Pérez, M.G., Morales, V., Robertí, F., Carelli, R.: Modeling and control of a wheelchair considering center of mass lateral displacements. In: Liu, H., Kubota, N., Zhu, X., Dillmann, R. (eds.) ICIRA 2015. LNCS, vol. 9246, pp. 254–270. Springer, Heidelberg (2015)
13. Andaluz, V., Roberti, F.: Robust control with redundancy resolution and dynamic compensation for mobile manipulators. In: IEEE-ICIT International Conference on Industrial Technology, pp. 1449–1454 (2010)

Intelligent Space

Acquiring Personal Attributes Using Communication Robots for Recommendation System

Aoi Suzuki[✉], Eri Sato-Shimokawara, Toru Yamaguchi, and Kentaro Ikehata

Faculty of System Design, Tokyo Metropolitan University, Tokyo, Japan
suzuki-aoi@ed.tmu.ac.jp, {eri,yamachan}@tmu.ac.jp,
ikehata@rcsc.co.jp

Abstract. It looks easy to continue conversations and make them fun, but it is difficult because of various factors. It is difficult to continue conversations with the common topics. Conversations aren't continued without pleasant topics. If pleasant topics are obtained from personal attributes, conversations become lively and he/she wants to talk about himself/herself. It is difficult to obtain pleasant topics from personal attributes. If robots research and analyze about them daily, personal attributes are obtained effectively. This paper describes dialog system to obtain personal attributes and recommend information based on personal attributes.

Keywords: Dialog system · Personal information · Recommendation

1 Introduction

It looks easy to continue conversations and make them fun, but it is difficult because of various factors. For example, gender, age, background and so on. If he/she is on good terms with a helper with a common hobby, his/her communicates with the helper easily. Conversations with robots are similar to conversations with human. If robots understand factors to make conversation fun, users talk with them positively. In addition, if users talk with them continuously, robot understands them more and more and conversations with robots will become lively.

Generally, person begins to talk from common topics such as weather and news for the first time. These topics are known by every people and people easy to talk them at any time. However, conversations about these topics tend not to continue long, because these topics aren't pleasant. It is difficult to continue conversations with the common topics. Conversations aren't continued without pleasant topics. Moreover pleasant topics depend on personal attributes. If pleasant topics are obtained from personal attributes, conversations become lively and he/she wants to talk about himself/herself.

It is difficult to obtain pleasant topics from personal attributes. Especially, when he/she meets him/her for the first time, there is no information about personal attributes. Personal attributes are able to be obtained from actions and conversations in daily life. However, it is difficult to research and analyze about them daily. If robots research and analyze about them daily, personal attributes are obtained effectively. That leads to

© Springer International Publishing Switzerland 2016
N. Kubota et al. (Eds.): ICIRA 2016, Part II, LNAI 9835, pp. 237–246, 2016.
DOI: 10.1007/978-3-319-43518-3_23

obtaining pleasant topics. Moreover if robots collect them and upload to servers, various robots are able to use them everywhere and recommend good information for users.

The rest of the paper is organized as follows. The motivation for this research is described in Sect. 2. Our dialog system is described in Sect. 3. The experiment results are shown in Sect. 4.

2 Motivation

The concrete example of the services of dialog system generally known at present is Siri [1] which was developed by Apple and shabette_concier [2] which was developed by NTT DOCOMO. Siri stores spoken words by user and operates device by the use of analysis result of spoken words. Accordingly, convenience of device is improved. However, Siri only works with utterances of user basically. Siri doesn't recommend information by use of user's information positively and purvey topics that user may like. In addition shabette_concier is much the same, but virtual robot is displayed in device. Therefore it easy to be intimate with robot compared with Siri.

The dialog systems generally known at present didn't have elements that let users to want to talk. The previous research shows the matching mechanism including a dialog system [3]. However, the dialog system wasn't able to obtain information well.

Moreover the way of generating dialogs is important. Inaba et al. [4] construct dialog agent using statistical response method and gamification. Tsukahara and Uchiumi [5] generate dialog from the label propagation over a correlation network of words and utterance patterns. Research of analysis of dialog for obtaining personal information is structure analysis. Higashinaka et al. [6] analyze predicate argument structure to obtain personal information. Kaneko et al. [7] recognize actions by ZigBee acceleration sensor, virtual robot on smartphone asked questions of them. The way not only to recognize actions but In addition to obtain it from dialog with a robot which is in the field of activities of daily life is thought to obtain that in daily life. Shinoda et al. [8] I research to obtain attributes of person from dialog with the robot by speech recognition. Matching at the place where people gather, for example community center by information of people about hobby and ability in daily life may make regional community active.

The example of recommend systems generally known at present is Amazon's system [9]. Amazon's system recommends information of items. The information is made from items that are similar to items user bought and search results of user. However, it isn't able to obtain personal information daily and promote users utilization of the system. If system promotes that, it is supported to obtain information easily and manage change of user.

Accordingly, we researched fitting conversations for users. If robots change topics, perhaps users want to continue to talk to them. Moreover if robots hear the topic users want to talk attentively, users will want to talk it positively. We did an experiment about them. As a result, to ask users about the topic in detail by robots has an effect on augmenting the number of times of users' utterance [10].

However, robots collect conversations with users, it is difficult to analyze them and obtain personal attributes. Moreover it is difficult to assess personal attributes based on only users' utterances. Accordingly, the dialog system with robot which is described in this paper obtains personal attributes from not only conversations but also using various sensors. Figure 1 shows Pepper [11] which was developed by Aldebaran. The previous research uses android smartphones. They easy to obtain conversations. However, it is difficult for them to obtain other information about personal attributes. For example, gender, age. Pepper has various sensors and obtains users' gender and age. Moreover pepper obtain users' expression. Using users' expression becomes easy to assess users' feelings about conversations. If users' feelings about conversations are obtained, robots are able to talk about the topic users become good feeling and recommend information that users want. Pepper also has arms like a human. Pepper is able to use body language and express feelings using arms. Therefore pepper is the robot has effects on bringing out positiveness of conversations in users. The dialog system with robot which is described in this paper uses pepper.

Fig. 1. Pepper which was developed by Aldebaran. Pepper has various sensors and obtain users' gender, age and expression. Pepper also has arms like human. Pepper is able to use body language and express feelings using arms.

The dialog system with robot which is described in this paper has 2 parts. They are called "Collect part" and "Recommend part" in this paper. Figure 2 shows image of "Collect part" in the dialog system with robot which is described in this paper. Devices obtain information from dialog with a user and upload them to the server. In previous research, Shimokawara et al. [12] asked users daily questions by smartphone and obtain information about daily life. In addition, Pepper obtains users' gender and age and upload them to the server. Moreover the system obtains users' expression finally. It is difficult to analyze conversations and obtain personal attributes from only text data. Accordingly the system analyses conversations and obtains personal attributes to use users' expressions.

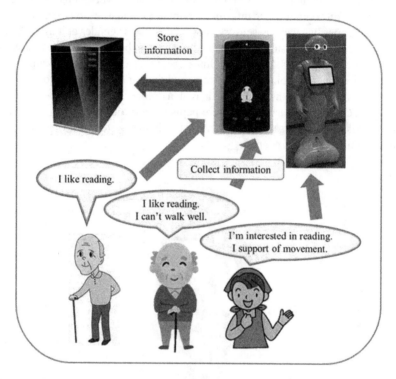

Fig. 2. Image of "Collect part" in the dialog system with robot which is described in this paper. Devices obtain information from dialog with a user and upload them to the server. In this paper, pepper obtain users' gender and age and upload them to the server. Moreover the system obtain users' expression finally.

Figure 3 shows image of "Recommend part" in the dialog system with robot which is described in this paper. Devices talk and recommend information based on information which is stored on the server with. The dialog system with robot which is described in this paper recommends information considering users' age. The system selects information to recommend using users' age. Moreover this part also obtains users' expression finally. Because we assess the information the system recommended.

Fig. 3. Image of "Recommend part" in the dialog system with robot which is described in this paper. Devices talk and recommend information based on information which are stored the server with. The dialog system with robot which is described in this paper recommend information considering users' age.

The dialog system with robot which is described in this paper aim to bring out the feeling that users want to talk. In order to do that, robots should understand personal attributes and have fit conversations for users. Therefore, users want to talk with robot positively and more personal attributes are obtained. It may also lead to generate user models. If user models are generated based on the information, the system recommends information to the user is similar to user models.

In this research, we construct the dialog system in order to obtain personal attributes and use them for generating fitting conversations with Pepper.

3   Dialog System

The dialog system is supposed to be able to respond to changes of users like talking with human. The dialog system aims to let users to talk. Therefore the dialog system needs to obtain users' information. In addition the dialog system needs to analyze them and be reflected in the conversation. Figure 4 shows the system was installed in pepper. Pepper is able to use a display and arms. Therefore pepper recommends information effectively using them.

Fig. 4. The system was installed in pepper. Pepper is able to use a display and arms. Therefore pepper recommend information effectively using them.

The dialog system with robot which is described in this paper recommends information of tourist spots. Pepper recommends information about tourist spots based on weather information from Web API [13] which was developed by RC Solution Co.

Figure 5 shows the flowchart of "Collect part". Firstly, Pepper obtains users' gender and age. Secondly, Pepper obtains weather information from Web API and searches recommended spots based on weather information. Thirdly, user chooses a recommended spot and pepper recommends information about the recommended spot. Finally, pepper obtains user's expression.

Figure 6 shows the flowchart of "Recommend part". This flowchart is similar to the flowchart of "Collect part". However, recommended spots are chosen not by users but by using users' age.

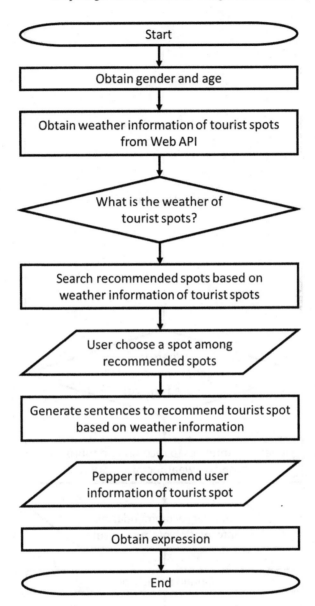

Fig. 5. The flowchart of "Collect part".

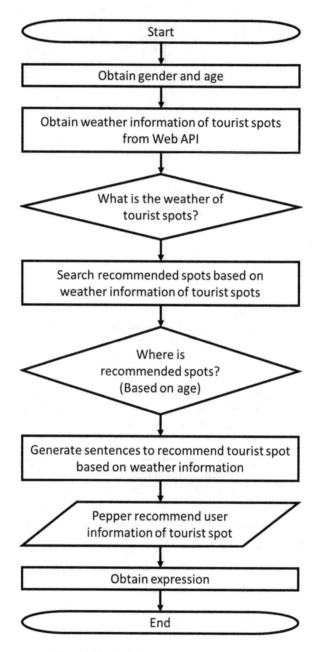

Fig. 6. The flowchart of "Recommend part".

"Collect part" is used to build a database from information about users. Accordingly, "Collect part" makes relationships between recommends information and information about users. "Recommend part" is used to recommend users to information using the database. Accordingly, "Recommend part" find fit information for users based on information about users.

4 Experiment and Result

Figure 7 shows the log of "Collect part". "Collect part" obtain users' gender and age. Moreover users' expressions are obtained after pepper recommend information. Therefore pepper is able to assess recommended information based on users' expressions.

```
[INFO ] behavior.box :onInput_message:27 _Behavior_lastUploadedChoregrapheBehaviorbehavior_1873921976:/Log_2: Message text: male
[INFO ] behavior.box :onInput_message:27 _Behavior_lastUploadedChoregrapheBehaviorbehavior_1873921976:/Log_2: Message text: 29
[INFO ] behavior.box :onInput_message:27 _Behavior_lastUploadedChoregrapheBehaviorbehavior_1873921976:/SayKanagawaSunny_51/Log_1:
Message text: 神奈川はいいお天気です。晴れた日は鎌倉散策はいかがですか？竹林が有名な報国じや鶴岡八幡宮がおすすめです。
[INFO ] behavior.box :onInput_message:27 _Behavior_lastUploadedChoregrapheBehaviorbehavior_1873921976:/Log_2: Message text: sad
```

Fig. 7. The log of "Collect part".

Figure 8 shows the log of "Recommend part". "Recommend part" recommend information based on users' age. Moreover users' expressions are obtained after pepper recommend information.

```
[INFO ] behavior.box :onInput_message:27 _Behavior_lastUploadedChoregrapheBehaviorbehavior_1845157336:/Log_2: Message text: male
[INFO ] behavior.box :onInput_message:27 _Behavior_lastUploadedChoregrapheBehaviorbehavior_1845157336:/Log_2: Message text: 41
[INFO ] behavior.box :onInput_message:27 _Behavior_lastUploadedChoregrapheBehaviorbehavior_1845157336:/SayTochigiSunny_59/Log_1:
Message text: お天気がいいので、日光東照宮に行ってみてはいかがですか？眠り猫や、鳴き竜が有名です。
[INFO ] behavior.box :onInput_message:27 _Behavior_lastUploadedChoregrapheBehaviorbehavior_1845157336:/Log_2: Message text: neutral
```

Fig. 8. The log of "Recommend part".

5 Conclusion

We develop a dialog system to obtain personal information and recommend information based on personal attributes. This paper shows the dialog system use pepper and obtain not only text data of conversations but also users' gender, age and expression. Moreover the dialog system chooses information that fits for users based on users' age. It is difficult to obtain personal attributes from only text data of conversations. However, using users' expression becomes easy to obtain information about personal attributes.

In this paper, "Collect part" collect personal information and "Recommend part" recommend information based on personal information. However, "Recommend part" don't use information that "Collect part" collect. Personal information is changed daily. Dialog systems should consider personal information. And then personal information should be reflected in generating conversations. Moreover the dialog system is run by pepper, we aim to run the dialog system by various devices. Personal information is upload to servers, various devices and robots should be able to use them. If various devices are able to use personal information in the server everywhere, users are supposed to want to use that.

References

1. Apple: Siri. https://www.apple.com/jp/ios/siri/
2. Uchida, W., Midori, O.: Shabette-Concier for Raku-Raku Smartphone: Improvements to Voice Agent Service for Senior Users (2013)
3. Suzuki, A., Gomi, R., Kaneko, T., Shimokawara, E., Yamaguchi, T.: Development of matching mechanism for co-occurrence and mutual assistance in community. In: System Integration of the 15th, 2G3-1, Japan, December 2014
4. Inaba, M., Iwata, N., Toriumi, F., Hirayama, T., Enokibori, Y., Takahashi, K., Mase, K.: Constructing a non-task-oriented dialogue agent using statistical response method and gamification. In: ICAART, vol. 1, pp. 14–21 (2014)
5. Tsukahara, H., Uchiumi, K.: Dialogue generation using the label propagation over a correlation network of words and utterance patterns. In: The 29th Annual Conference of the Japanese Society for Artificial Intelligence, 2L4-OS-07a-7, June 2015 (Japanese)
6. Higashinaka, R., Imamura, K., Meguro, T., Miyazaki, C., Kobayashi, N., Sugiyama, H., Hirano, T., Makino, T., Matsuo, Y.: Towards an open-domain conversational system fully based on natural language processing. In: Proceedings of the 25th International Conference on Computational Linguistics, pp. 928–939 (2014)
7. Tetsuya, K., Reona, G., Eri, S.-S., Toru, Y.: Acquisition of the human characteristics through dialogue with a robot at home. In: 25th 2014 International Symposium on Micro-NanoMechatronics and Human Science, Nagoya, Japan, TA2_2_3, 9–12 November 2014
8. Shinoda, Y., Nomura, S., Lee, H., Takatani, T., Wada, K., Shimokawara, E., Yamaguchi, T.: A dialogue analysis of elderly person with a chat robot. In: 2015 RISP International Workshop on Nonlinear Circuits, Communications and Signal Processing (NCSP 2015), 28PM1-2-2, Kuala Lumpur, Malaysia, February 27–March 2 2015
9. Amazon.co.jp. http://www.amazon.co.jp/
10. Suzuki, A., Gomi, R., Sato-Shimokawara, E., Yamaguchi, T.: Effectiveness of dialog contents for obtaining personal attribute. In: 2015 Conference on Technologies and Applications of Artificial Intelligence (TAAI 2015), Tayih Landis Hotel Tainan, Taiwan, 20–22 November 2015, pp. 200–205 (2015)
11. Softbank: Pepper. http://www.softbank.jp/robot/
12. Sato-Shimokawara, E., Suzuki, A., Gomi, R., Yamaguchi, T.: Obtaining user's preference and ability from human-robot conversation towards mutual assistance. In: The 41st Annual Conference of the IEEE Industrial Electronics Society (IECON), ss37 02 - Human Support and Monitoring Technology on Human Factors - Motion and Behavior II, Yokohama, Japan, 11 November 2015, pp. 3557–3560 (2015)
13. RC Solution Co. http://www.rcsc.co.jp/

Robot Control Interface System Using Glasses-Type Wearable Devices

Eichi Tamura$^{(\boxtimes)}$, Yoshihiro Yamashita, Yihsin Ho,
Eri Sato-Shimokawara, and Toru Yamaguchi

Faculty of System Design, Tokyo Metropolitan University,
Asahigaoka 6-6, Hino, Tokyo, Japan
{tamura-eichi,yamashita-yoshihiro}@ed.tmu.ac.jp,
{ho-yihsin,eri,yamachan}@tmu.ac.jp

Abstract. A hand gesture input interface system has been developed. Two kinds of hand gestures pointing and scrolling are able to be recognized from video images by glasses-type wearable device's camera. Using the system, it realized commanding to mobility robots by gestures in user's sights. Introducing pointing gesture, the current pointing rate increase from clicking gesture.

Keywords: Internet of Things · Human-computer interaction · Wearable devices · Hand gesture · Mobility robot

1 Introduction

Using wearable device is one of the hot topics in the Internet of Things (IoT). There are some types of wearable devices. Glasses-type wearable device is one of them. This paper focusses on the interface of glasses-type wearable devices. As it stands, general inputs of glasses-type wearable devices are controllers dedicated to the device, voice recognitions or smart phone applications. Using hand gestures is one of the attractive method in the human-computer interaction (HCI) field. If gesture recognition using the video form glasses camera is available, it is gesture from user's sight. Therefore, gesture inputs from glasses camera are suitable to direct the way. Using gesture in users' sight, direction becomes accurate.

Using depth sensors such as Microsoft Kinect is general way to recognize gestures. Kinect is able to recognize various gestures using depth sensors [1]. However, infrared needs some distance between sensor and gesture. In other words, it is unsuitable for wearable devices. Using video images is easy to adapt for wearable devices. To recognize hand gestures from video images, the system needs to extract users' hands from them. There are some methods to extract the objects from video frames. For example, optical flow [2], color band selection [3], or stereo-camera [4]. These methods need a lot of processing for wearable device use. Therefore, we introduced the Gaussian Mixture Model (GMM) foreground

© Springer International Publishing Switzerland 2016
N. Kubota et al. (Eds.): ICIRA 2016, Part II, LNAI 9835, pp. 247–256, 2016.
DOI: 10.1007/978-3-319-43518-3_24

segmentation method [5–8] for extracting hands and developed an interface system [9]. The developed system uses standard video images as input and run on real-time processing.

In the research areas of HCI and IoT, automated driving is also one of the hot topic. The National Highway Traffic Safety Administration (NHTSA) defined levels of vehicle automation [10]. These levels are 0 to 4. Level 0 is No-Automation and level 4 is Full Self-Driving Automation. Although this level 4 seems to perfect, we focus on this level 3 (Limited Self-Driving Automation). There are more interactions of human and robot in the level 3 than the level 4. Interacting with robots and taking responsibility of the driving, people are able to enjoy it. There are some approaches to interact with robots. One of popular methods is voice recognition. It is a close way of the human-human interaction. However, for example, when you direct robot that a way you want to go, using gesture is more attractive method than voice recognition.

The rest of this paper is organized as follows. The developed system is described in Sect. 2. The experiments and the result are described in Sect. 3. Section 4 is conclusion.

2   Robot Control Interface System

Figure 1 shows overview of the system. The inputs of the system is video images from the camera mounted on glasses-type wearable devices. The output is commands for the mobility robot. There are three steps to decide the command; hand extraction, fingertip detection, and motion analysis.

2.1   Hand Extraction

The hand extraction is realized by the GMM foreground segmentation. The details of the developed foreground segmentation are described in [6,7]. Using RGB value of input images, system extracts foreground area from the images. These foreground area are users' hands in input images from the camera on glasses. Figure 2 shows a result of the GMM foreground segmentation system. In Fig. 2(b), the black areas are defined as background areas and the others are defined as foreground areas by the system. Mostly hand areas are defined as foreground. It means the GMM foreground segmentation realizes hand extraction.

The GMM foreground segmentation is able to work at indoors or outdoors environments. The weak point is moving the camera. For glasses-type wearable devices, it needs correcting camera shakes.

2.2   Fingertip Detection and Motion Analysis

The fingertip detection is based on the system described in [9]. The algorithm is simple; basically rule is "the far-end point of gravity point of foreground is

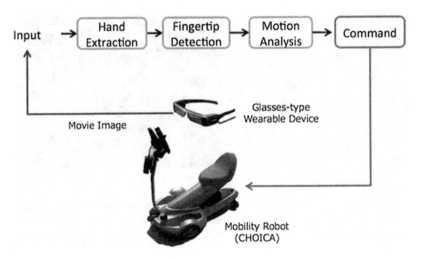

Fig. 1. Overview and system flow. Hand gestures are captured by the camera of glasses-type wearable device. The system receives the images as inputs. The first step of the system is hand extraction using GMM foreground segmentation. The second step is fingertip detection. Using result of second step, the system recognizes the gesture and sends the command to the mobility robot.

fingertip point." The system uses these positions of fingertip point and gravity point in motion analysis part.

In motion analysis part, the system recognizes two kinds of gestures. One is pointing, and the other is scrolling. Pointing gesture is putting the fingertip on place user wants to select. Scrolling gesture is moving the finger right side to left side (or left side to right side).

The algorithm of recognizing pointing gesture is organized as follows. The numbers are fixed for system running on 5 fps.

1. If there is a hand in the frame, go the next.
2. If the current fingertip position nearly the previous fingertip position, the previous position is marked.
3. If the current fingertip position nearly the marked position, pointing check stacks are added.
4. When the stacks become three, the current fingertip position is defined as pointing point.

To recognize where is pointed, the system introduces fuzzy inference method described in [9]. This method adjusts pointed area for users using initial setup data.

The algorithm of recognizing scrolling (right side to left side) gesture is organized as follows.

1. If fingertip and gravity point of foreground points do not move to rise or fall in last five frames, go the next.

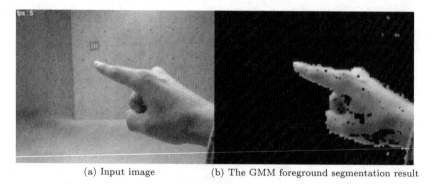

(a) Input image   (b) The GMM foreground segmentation result

Fig. 2. An example of the GMM foreground segmentation result from one frame of video images. (a) is the input image (original), (b) is the result of the GMM foreground segmentation. The foreground area means hand area in the images form the camera on glasses.

2. When all of fingertip position in five frames move to left side, the system defined that as scrolling

2.3 Command Decision

Basic commands for control mobility robots are "go," "turn," and "stop." The system uses pointing gesture for "go" command. The pointing areas are divided three (left, front, right). That means, pointing at the way want to go, the mobility robot goes there. Scrolling gestures are used for "turn" command. If you scroll right to left, mobility robot turns to left and vice versa. "Stop" command is the most important command. If "stop" command does not work, it causes accidents. Therefore, the gesture corresponds stop is defined as put finger on camera. When camera is covered, nothing is appeared in the input image. It is able to response quick and reliable.

3   Experiments

Experiments were carried out to test the developed gesture recognition system. To test blurring by head shake, two pattern experiments were prepared. Figure 3 shows them. One pattern is using a USB camera on a tripod shown in Fig. 3(a). In this pattern, the camera was set up in front of the participants. The other pattern is using a camera mounted on glasses shown in Fig. 3(b). In this pattern, participants wore the glasses mounting a camera and put their chin a tripod. The reason for putting the chin on the tripod is to simulate head shake correcting which is developing. The experiments flow are shown in follows.

1. Participants test the system. (Checking how to recognize gestures the system). All participants gestured with their right hand.

(a) Using fixed camera

(b) Using glasses camera

Fig. 3. Two patterns of experiments. One is using a USB camera fixed by a tripod due to avoid shaking the camera shown in (a). The other is using a camera mounted on glasses shown in (b). In (b) pattern, participants put their chin on a tripod due to avoid head shaking.

2. Participants gesture in front of a camera in one of the patterns shown in Fig. 3. Pointing gestures are five times each way (left, front, and right). Scrolling gesture is five times each way (left to right and right to left).
3. Participants gesture in front of a camera in the other pattern shown in Fig. 3. The number of gesturing times is same as the former.

Participants are five students (early twenties). For easy explanation, participants are named as participant-a to participant-e. Furthermore, to avoid ordering effect, the orders of pattern shown in Fig. 3 and the orders of pointing area are shuffled.

Figure 4 shows one example of recognizing pointing gesture. Orange circles are fingertip points and blue boxes are gravity points of foregrounds. The finger is not moving in these frames. At the frame 565 shown in Fig. 4(g), the system recognized the pointing gesture using the transition of the fingertip moves. Though foreground segmentation did not work correct (almost area was defined foreground wrongly) at frame 562 shown in Fig. 4(c), the system caught the fingertip position.

Figure 5 shows one example of recognizing left to right scrolling gesture. Orange circles are fingertip points and blue boxes are gravity points of foregrounds. At the frame 1522 shown in Fig. 5(i), scrolling gesture was recognized. Figure 5(i) shows the traces of fingertip as small orange circles.

Table 1 shows the rates of correct pointing in the pattern shown in Fig. 3(a). Three participants' rates reach 90 % and total is 84 %. In the previous research [9],

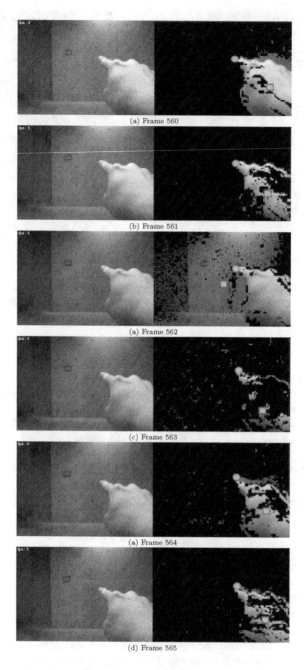

Fig. 4. One example of recognizing pointing gesture. Left side pictures are input and right side pictures are output which shows except the background area. Orange circles show fingertips, blue boxes show the center of gravity. The system recognized pointing gesture using fingertip positions of (a) to (g) at (g). (Color figure online)

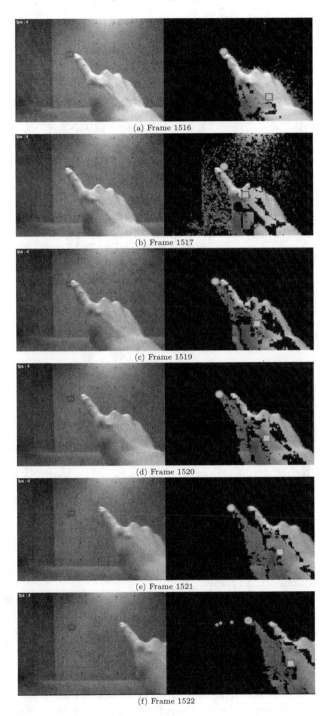

(a) Frame 1516

(b) Frame 1517

(c) Frame 1519

(d) Frame 1520

(e) Frame 1521

(f) Frame 1522

Fig. 5. One example of recognizing scrolling gesture. Left side pictures are input and right side pictures are output which shows except the background area. Orange circles show fingertip, blue boxes show the center of gravity. (Color figure online)

Table 1. Correct pointing rates using a fixed camera

Participants	Left	Front	Right	All
a	60 %	80 %	80 %	73.3 %
b	60 %	60 %	80 %	66.7 %
c	80 %	100 %	100 %	93.3 %
d	100 %	100 %	80 %	93.3 %
e	80 %	100 %	100 %	93.3 %
Total	76 %	88 %	88 %	84 %

Table 2. Correct pointing rates using glasses camera

Participants	Left	Front	Right	All
a	60 %	80 %	80 %	73.3 %
b	60 %	100 %	40 %	66.7 %
c	80 %	100 %	100 %	80.0 %
d	20 %	60 %	80 %	53.3 %
e	40 %	40 %	80 %	53.3 %
Total	56 %	68 %	72 %	65.3 %

Table 3. Correct scrolling rates using a fixed camera

Participants	Left to right	Right to left	All
a	80 %	100 %	90 %
b	100 %	100 %	100 %
c	40 %	100 %	70 %
d	100 %	60 %	80 %
e	40 %	20 %	30 %
Total	72 %	76 %	74 %

the recognition of clicking gesture was 68.3 % as total. In comparison with this, there are clear differences in these.

Table 2 shows the rates of correct pointing in the pattern shown in Fig. 3(b). The current rates decrease from using a fixed camera. In this case, the problem is noise generation. These noises be tendency seem to depend on participants. For example, more noises occurred when participant-d pointing in left side area.

Table 3 shows the rates of correct scrolling in the pattern shown in Fig. 3(a). The total rate is 74 %, however, participants can be divided in two types; good for scrolling gesture or not. Therefore, it expects practice effects.

Table 4 shows the rates of correct scrolling in the pattern shown in Fig. 3(b). This result is not good at using. From this result, it needs strong shake correcting function to use scrolling gesture.

Table 4. Correct scrolling rates using glasses camera

Participants	Left to right	Right to left	All
a	0 %	40 %	20 %
b	20 %	20 %	20 %
c	60 %	100 %	80 %
d	40 %	60 %	50 %
e	20 %	0 %	10 %
Total	28 %	44 %	36 %

4 Conclusion

The finger gesture input interface system for wearable device has been developed. The system controls a mobility robot using hand gesture. The two kinds of gestures (pointing and scrolling) are captured by the camera on glasses and recognized by the GMM foreground segmentation and simple algorithm of detecting fingertip. The current rate of recognizing pointing is 84 % and the current rate of recognize scrolling is 74 % when stable image processing works.

In this paper, the current rates of the gesture for the stop command are not measured. We have to carry out more experiments in the future.

References

1. Ren, Z., Yuan, J., Meng, J., Zhang, Z.: Robust part-based hand gesture recognition using kinect sensor. IEEE Trans. Multimedia **15**(5), 1110–1120 (2013)
2. Horn, B.K., Schunck, B.G.: Determining optical flow. In: 1981 Technical Symposium East. International Society for Optics and Photonics (1981)
3. Sato, E., Nakajima, A., Yamaguchi, T.: Nonverbal interface for user-friendly manipulation based on natural motion. In: IEEE International Symposium on Computational Intelligence in Robotics and Automation (CIRA 2005) We-C1: Intelligent Systems, Espoo, Finland (2005)
4. Fukusato, Y., Sakurai, S., Sato-Shimokawara, E., Yamaguchi, T.: Multi phase environment information interface by using "kukanchi: interactive human-space design and intelligence". In: Proceedings of the 17th IEEE International Symposium on Robot and Human Interactive Communication (ROMAN), Cape Town, pp. 526–531. Technische Universitat Munchen, Munich (2008)
5. Stauffer, C., Grimson, W.E.L.: Adaptive background mixture models for real-time tracking. In: IEEE Computer Society Conference on Computer Vision and Pattern Recognition, Fort Collins, CO, vol. 2, pp. 246–252. IEEE (1999)
6. Tezuka, H., Nishitani, T.: Multiresolutional Gaussian mixture model for precise and stable foreground segmentation in transform domain. IEICE Trans. Fundam. Electron. Commun. Comput. Sci. **92**(3), 772–778 (2009)
7. Yamashita, Y., Nishitani, T., Yamaguchi, T., Bunken, O.: Software implementation approach for fingertip detection based on Color Multi-Layer GMM, JeJu Island. In: The 18th IEEE International Symposium on Consumer Electronics (ISCE 2014). IEEE, JeJu Island (2014)

8. Matsui, S., Yamashita, Y., Yamacuchi, T., Nishitani, T.: Robust finger motion interface for IT terminals on GMM foreground segmentation. In: SICE System Integration (SI 2013), CD-ROM 1K4-3 Japan (2013). (In Japanese)
9. Tamura, E., Yamashita, Y., Ho, Y., Sato-Shimokawara, E., Nishitani, T., Yamaguchi, T.: Wearable finger motion input interface system with GMM foreground segmentation. In: 2015 Conference on Technologies and Applications of Artificial Intelligence (TAAI 2015), Taiwan, pp. 213–220 (2015)
10. National Highway Traffic Safety Administration: Preliminary Statement of Policy Concerning Automated Vehicles

Relationship Between Personal Space and the Big Five with Telepresence Robot

Yoshifumi Kokubo$^{(\boxtimes)}$, Eri Sato-Shimokawara, and Toru Yamaguchi

Faculty of System Design, Tokyo Metropolitan University, Tokyo, Japan
kokubo-yoshifumi@ed.tmu.ac.jp,
{eri,yamachan}@tmu.ac.jp

Abstract. The robot called "Telepresence Robot" is on business in recent years. However, the robot does not consider user's personal space when the robot approaches the user. Telepresence robot might give users fear if the robot unintentionally invades the user's personal space when the user have conversation with the robot at the first time. To better develop the system of the robot, here we study about how near the robot approaches a user in advance. We verified the relationship between personal space and the big five by using neural network. It is revealed that the most influential factor for the neural network is agreeableness. By confirming user's personal space in advance, the conversation between Telepresence Robot and human becomes natural and interactive.

Keywords: Telepresence robot · Personal space · The Big Five

1 Introduction

Nowadays, while industrial robot is developed, development of communication robot comes up. Even in the communication robots, the robot called *Telepresence Robot* is on business in recent years [1]. Telepresence robot is the robot used by a user for communication with another person who is in remote place. People using the robot are able to look at the user's face who is in a remote place and talk with the person without going to see him or her. Working people do not need to go to their office or make a business trip by using the robot. The robot has two main functions: remote mobile control and videophone such as Skype. Most of the robot is for office use [2]. However, general telepresence robots do not consider a user's personal space when the robot approaches a user.

Personal Space is the territory that person likes to keep a specific distance from other persons in order to keep him or her comfortable. Although it surrounds a person, it is not always the constant area. It depends on a lot of factors such as your mental condition, the relationship with the person who you talk with and etc. Between you and someone, for example, a stranger approaches you, you do not want him or her to approach you, so that the personal space becomes wide. However, your family or friend approaches you, you do not mind the person approaches you more than a stranger, so that the personal space becomes small.

N. Kubota et al. (Eds.): ICIRA 2016, Part II, LNAI 9835, pp. 257–265, 2016.
DOI: 10.1007/978-3-319-43518-3_25

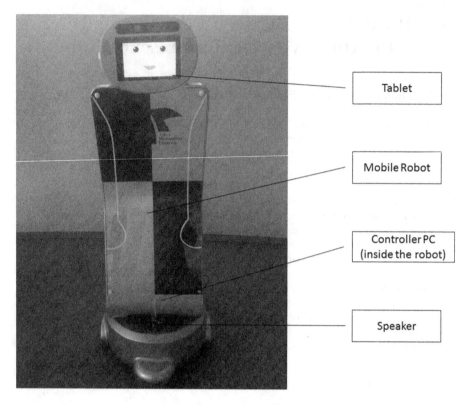

Fig. 1. The image of the telepresence robot. This robot developed by VECTOR Inc. This robot mounts the tablet, controller PC and the speaker.

Besides, personal space in human-human and human-robot is different. Personal space between robot and human is already researched in various ways [3–8]. User's personal space with a robot is measured by an experiment at the present time. Telepresence robot might give users fear if the robot unintentionally invades the user's personal space when the user has conversation with the robot at the first time. To accomplish fluent conversation, optimal distance has to be held between a robot and a user in advance. To make the optimal distance clear in advance, the present study investigated the relationship between personal space and an user's characteristic.

The rest of the paper is organized as follows. The detail is described in each experiment. The experiment about how the robot decelerates is described in Sect. 2. The relationship between personal space and the big five is explained in Sect. 3.

2   Experiment

2.1   Concept

To clarify the relationship between personal space and a user's characteristic, the experiment measuring personal space with a telepresence robot was carried out at the first. Figure 1 shows the image of the telepresence robot.

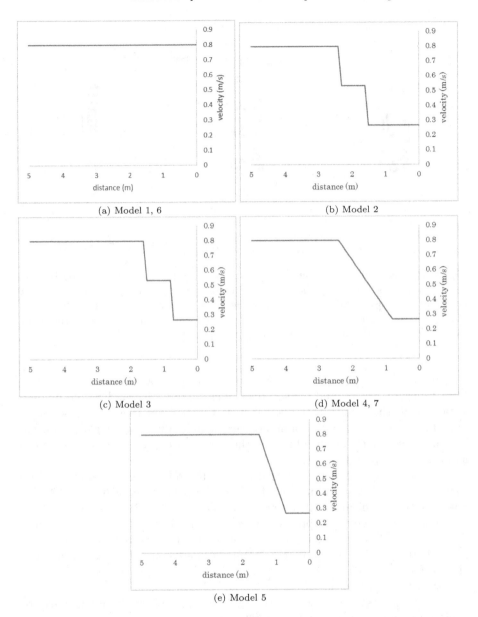

Fig. 2. The models robot approaches to person. Those models are designed in consideration of the two factors. The first factor is the position when the robot starts to decelerate. One position is set from 160 [cm] in front of the person as twice as the distance of the personal space [10]. The other is set to 240 [cm] as three times from the distance of the personal space. One pattern is that the speed changes two phased slowdown. The other is the speed rate declines linearly. The second factor is the method the robot decelerate. One pattern is that the speed changes two phased slowdown. The other is the speed rate declines linearly. The third factor is whether or not which have caution. In the model 6, the utterance *"I approach."* is notified. In the model 7, the utterance *"I decelerate and stop after."* was notified.

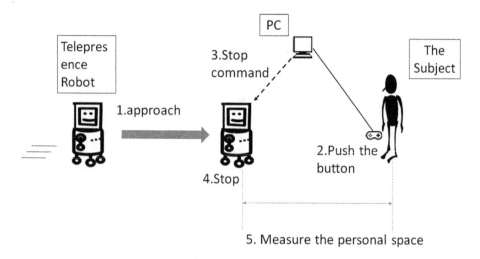

Fig. 3. The method the subject make the robot stop: (1) The robot approaches the subject. (2) the subject stop the robot at the position where he or she does not want the robot to approach him or her anymore by using the game pad. (3) The Master PC receive the command value from the game pad and send a stop command to the robot through the network. (5) Measure the personal space.

The robot starts to approach a subject from the position distant 5 [m] in front from the subject, and the subject facing the robot is standing and holding a game pad, the subject stop the robot at the position where he or she does not want the robot to approach him or her anymore by using the game pad. We measured the direct distance between the head of the robot and the toe of the subject in each pattern.

We prepared the 5 approaching models of the robot. The experiments were carried out a total of 7 times by changing the pattern in each time. The all patterns are shown in Fig. 2. At the first, the model 1 (Fig. 2(a)) which does not change the speed to approach a user was considered. Further three factors were considered to influence the scope of personal space as follows. The first factor is the position when the robot starts to decelerate. The second factor is the method the robot decelerates. The last factor is whether or not utterance as the pre-action exists or not.

About the first factor, we set the two positions where the robot starts to decelerate. One position is set from 160 [cm] in front of the person as twice as the distance of the personal space [10]. The other is set to 240 [cm] as three times from the distance of the personal space. About the second factor, we set the two methods the robot approaches a person with. One pattern is that the speed changes two phased slowdown. The other is the speed rate declines linearly. We combine these two factors, therefore, we made 4 models (Fig. 2(b) to (e)). Moreover about the third factor, the model 6 and 7 which have caution was prepared. In the model 6, the utterance *"I approach."* was notified in the

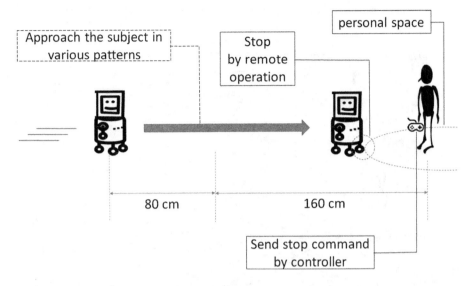

Fig. 4. The image of the experiment. The robot approaches the subject in various patterns. The model 2 and model 4 in Fig. 2 starts to decelerate distant 240 [cm] from the subject. The model 3 and model 5 in Fig. 2 starts to decelerate distant 160 [cm] from the subject.

position distant 320 [cm] from the subject. in the model 7, the utterance *"I decelerate and stop after."* was notified 1 s before starting deceleration, thus in the position distant 320 [cm] from the subject as well as model 6.

In this experiment, we set the maximum velocity at 0.8 [m/s], that is the limit speed of the robot. Minimum speed was set at 0.33 [m/s] which the subjects can immediately escape in front of the robot. In the model 2 and 3, the velocity one step decelerated is set at 0.67 [m/s] and two set decelerated is set at

Table 1. The experimental order in the group 1 and 2 for model 1 to 7. The group 1 is arranged in the order which is supposed to give a sense of fear the most to not. The group 2 is arranged by reverse of the first order.

Group 1	Group 2
Model 1	Model 4
Model 3	Model 5
Model 2	Model 2
Model 5	Model 3
Model 4	Model 1
Model 6	Model 7
Model 7	Model 6

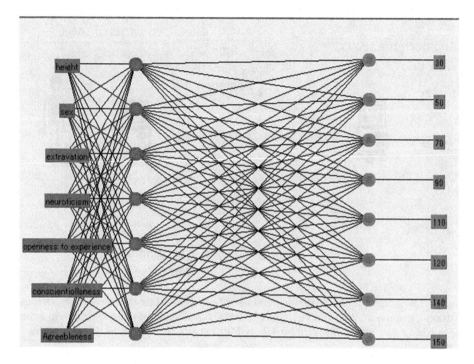

Fig. 5. Capture of execution of neural network on Weka. Input data is height, sex, extraversion, neuroticism, openness to experience concientiollsness and agreebleness. Models generated from Table 2 which values are rounded off to the nearest 10.

0.33 [m/s]. In model 4 and 5, the speed after deceleration was set at 0.33 [m/s]. The demonstration movie of the all patterns is uploaded on Youtube[1].

Before the experiment was carried out, the preliminary experiment was carried out. The subjects controlled the robot until the subjects judged that they are accustomed to the robot. The subjects let the robot to approach them and move front, back, left and right. Therefore, the experiment was carried out in the situation the subjects were accustomed to the robot.

To avoid order effect, the two order of the patterns were prepared (Table 1). One group is arranged in the order which is supposed to give a sense of fear the most to not. The other group is arranged by the reverse of the first order, that is the order starts from the model which supposed to the safest likely model. These order based on the preliminary experiment among us.

The experiment was carried out with the real environment using telepresence robot (Fig. 1). The subjects was males and females, a total of ten people, in their twenties.

---

[1] https://www.youtube.com/watch?v=3sQ388kR8FE\&feature=youtu.be.

Table 2. The data of 10 person's personal space associated with each model.

Subject	Model 1	Model 2	Model 3	Model 4	Model 5	Model 6	Model 7
Subject 1	117	89	127	81	76	105	89
Subject 2	150	158	153	140	166	86	125
Subject 3	106	119	148	126	124	84	88
Subject 4	31	57	51	46	42	59	45
Subject 5	73	53	59	45	42	76	51
Subject 6	111	83	75	81	86	64	74
Subject 7	138	99	107	132	150	118	113
Subject 8	92	117	116	117	137	153	95
Subject 9	53	60	58	80	53	61	53
Subject 10	121	120	103	141	158	110	87

2.2 Experimental Results

In this experiment, the data of 10 person's personal space associated with each model was obtained as shown in Table 2.

2.3 Big Five

Lewis Goldberg proposed that *the Big Five personality traits* that are used to describe human personality [11]. The traits of personality can be reduced to five independent dimensions, Extraversion, Neuroticism, Openness to Experience, Conscientiollsness and Agreeableness, that is called the Big Five. The each question of the test consisting of 60 questions was each personal traits terms for consisting the big five dimensions. The test was carried out for the same 10 subjects in the experiment, and the subjects evaluated each question on 7-point scale in Japanese.

Table 3. The score of each dimension in the personality test for each subject

	Subject									
	1	2	3	4	5	6	7	8	9	10
Extraversion	45	48	52	52	48	51	49	49	49	46
Neuroticism	55	38	36	69	48	54	46	61	37	68
Openness to experience	36	54	48	29	51	41	44	51	51	30
Conscientiollsness	48	53	55	49	43	48	51	57	55	56
Agreeableness	46	42	48	35	50	44	51	54	47	46

3   Relationship Between Personal Space and the Big Five

The neural network on Weka [12] was used for analysis of the relationship between personal space and the big five. As being supervised learning, height, sex, and each big-five data was input and each distance which was measured in the experiment was set as model. Model data from Table 2 is rounded off to the nearest 10. About a variable of the neural network, the unit of the hidden layer is set as 6, learning rate is 0.3, momentum is 0.2 and epoch is 500.

The each probability to be correctly classified model. In item All, height, sex, extraversion, neuroticism, openness to experience concientiollsness and agreeableness is used as input. In item extraversion, height, sex and extraversion is used as input. The same applies hereafter. It is to be noted that average of item All is almost 100 %. Therefore, the training with using all factors is presumed effectively. Comparing each factor of the big five in Table 3, it is presumed that agreeableness effects the most to select the correct model (Table 4).

Table 4. The accuracy rate about each patterns. In item all, height, sex, extraversion, neuroticism, openness to experience concientiollsness and agreeableness is used as input. In item extraversion, height, sex and extraversion is used as input. The same applies hereafter. And the average of each item is shown on right side.

	Model 1	Model 2	Model 3	Model 4	Model 5	Model 6	Model 7	Average
All	100	100	100	100	100	100	90	98.6
Extraversion	90	60	50	60	50	50	30	55.7
Neuroticism	60	50	80	60	50	50	50	57.1
Openness to experience	60	60	60	60	60	50	60	58.6
Conscientiollsness	60	90	60	50	50	40	40	55.7
Agreebleness	60	70	70	70	50	70	50	62.9

4   Conclusion

We verified the relationship between personal space and the big five. The training with using all factors is presumed effectively. Particularly, it is possible that agreeableness effects the most to select the correct model. Continuously, we considered analysis of the relationship in detail. We would confirm which model incorrect output choices. Even if the system provides incorrect answer, if its error would be small, it is possible for data to be made use of research users personal space in advance.

References

1. Nakanishi, H., Murakami, Y., Nogami, D., Ishiguro, H.: Minimum movement matters: impact of robot-mounted cameras on social telepresence. In: Proceedings of the 2008 ACM Conference on Computer Supported Cooperative Work. ACM (2008)

2. Kristoffersson, A., Coradeschi, S., Loutfi, A.: A review of mobile robotic telepresence. Adv. Hum. Comput. Interact. **2013**, 3 (2013)
3. Nakauchi, Y., Simmons, R.: A social robot that stands in line. Auton. Robots **12**(3), 313–324 (2002)
4. Bainbridge, W.A., Hart, J., Kim, E.S., Scassellati, B.: The effect of presence on human-robot interaction. In: The 17th IEEE International Symposium on Robot and Human Interactive Communication, RO-MAN 2008. IEEE (2008)
5. Walters, M.L., Dautenhahn, K., Te Boekhorst, R., Koay, K.L., Kaouri, C., Woods, S., Nehaniv, C., Lee, D., Werry, I.: The influence of subjects' personality traits on personal spatial zones in a human-robot interaction experiment. In: IEEE International Workshop on Robot and Human Interactive Communication: ROMAN 2005. IEEE (2005)
6. Mumm, J., Mutlu, B.: Human-robot proxemics: physical and psychological distancing in human-robot interaction. In: Proceedings of the 6th International Conference on Human-Robot Interaction. ACM (2011)
7. Yasumoto, M., Kamide, H., Mae, Y., Ohara, K., Takubo, T., Arai, T.: Personal space for the humanoid robot and a presentation method. In: The Robotics and Mechatronics Conference, pp. 2A2-D18(1)–2A2-D18(4) (2010)
8. Kokubo, Y., Yamaguchi, Y., Sato-Shimokawara, E., Yamaguchi, T.: Influence of approaching patterns of telepresence robot for personal space. In: 2015 Conference on Technologies and Applications of Artificial Intelligence (TAAI 2015), Tayih Landis Hotel Tainan, Taiwan, 20–22 November 2015, pp. 221–226 (2015)
9. Mitsumura, S., Fujimoto, Y., Yamaguchi, T.: Development of a walking rehabilitation robot based on technique of physical therapist. In: 2014 RISP International Workshop on Nonlinear Circuits, Communications and Signal Processing NCSP 2014, Honolulu, Hawaii, USA, 28 February–3 March, pp. 629–632 (2014)
10. Yamaguchi, Y., Mitsumura, S., Kokubo, Y., Shimokawara, E., Yamaguchi, T.: Proposal of home-use telepresence robot system considering user's personal space. In: 2015 RISP International Workshop on Nonlinear Circuits, Communications and Signal Processing (NCSP 2015), Kuala Lumpur, Malaysia, 28AM2-2-3, 27 February–2 March (2015)
11. Goldberg, L.R.: An alternative description of personality: the big-five factor structure. J. Pers. Soc. Psychol. **59**(6), 1216 (1990)
12. Hall, M., et al.: The WEKA data mining software: an update. ACM SIGKDD Explor. Newsl. **11**(1), 10–18 (2009)

Development of Behavior Observation Robot Ver.2

Motoyasu Tooyama[✉], Kazuyoshi Wada, and Mutsuki Yageta

Graduate School of System Design, Tokyo Metropolitan University,
6-6 Asahigaoka, Hino, Tokyo 191-0065, Japan
m.tooyama0104@gmail.com

Abstract. In field studies, the observation is often used. However, it requires huge burden to observers. In order to solve the problem, we have proposed "Behavior observation robot" which can be a substitute for human observer. The robot should avoid receiving much attention from the target subjects. However, previously developed robot could not follow the object when the robot lost the target object at once, and generated loud noise from the movement mechanism. In this study, we propose a new "Behavior observation robot" and describe its functions.

Keywords: Personal space · Field study · Robot therapy · Observation

1 Introduction

Observation is often used in many fields such as psychology, sociology, human engineering and so on. The field observation is one of the observation methods. Observers go to target subjects' working environment and observe target subjects in order to evaluate and analyze the natural state of human. There are two observation methods, direct and indirect observation method. Direct observation can record situations and target subjects' behavior and evaluate complex factors such as impressions of target subjects. However, evaluation criterion is unclear and observers bear huge burden. Moreover, there is another important problem. In observation field, observer's existence affects subjects, and then changes subject's behavior. This is called "observer effect". This effect causes declining to a reliance of observation. On the other hand, indirect observation (video recording etc.) can reduce observer effect. However, there are other problems such as restriction of target subjects' behavior and occurrence of occlusions. Multiple sensor system avoids these problem, however it has other problems. Installing sensor devices in facility is difficult to obtain permission.

To solve these problems, we have proposed "Behavior Observation Robot" which observes target subject as a substitute of human observer. In the prior study, the necessary functions of Behavior Observation Robot are; (1) minimize observer effects of the robot; (2) track the subject's posture and position changes; and (3) occlusion avoidance. Figure 1 shows the specification of the Prototype Behavior Observation Robot which is composed of Network camera, Laser Range Scanner, Small omnidirectional robot, LaptopPC. Then, we minimized observer effects of the robot in terms of the location of Behavior Observation Robot. However, the current system of Behavior

N. Kubota et al. (Eds.): ICIRA 2016, Part II, LNAI 9835, pp. 266–275, 2016.
DOI: 10.1007/978-3-319-43518-3_26

Network Camera SNC-RX550N	
Effective pixels	380,000pixels
Zoom ratio	Optics26times/Digital12times
Pan drive range	360℃ endless rotaing
Tilt drive range	−90～0℃
Laser Range Scanner UST-30LX	
Maximum detection distance	0.1～30m
Ranging accuracy	±50mm
Scanning angle	270℃
Small omnidirectional robot MDT-RO-02	
Height[mm]	860
Diameter[mm]	390
Weight[kg]	8
LaptopPC	
CPU	Core i7
memory	4.00GB

Fig. 1. Prototype behavior observation robot

Observation Robot had various problems; (1) when the robot lost the target object at once, it could not follow the object. (2) loud noise from the movement mechanism.

In this paper, we discuss necessary functional elements for new Behavior Observation Robot to solve the above problems, and then develop new robot.

Section 2 explains target environment of Behavior Observation Robot, Sect. 3.1 describes concept of new robot, Sect. 3.2 explains its specification and functions and Finally, Sect. 4 offers conclusions.

2 Target Environment

As a model case, Behavior Observation Robot targets the field study of robot therapy at a facility for the elderly [2]. Robot therapy is a new method of psychotherapy for the elderly and patients suffering from dementia. Psychophysiological and social effects are expected to emerge through interaction with animal type robots. The therapeutic seal robot PARO is designed for such therapy and is widely used around the world. Typically, PARO robot therapy is facilitated by a trained caregiver and conducted as a group activity, with elderly people sitting around a table interacting freely with PARO (Fig. 2).

The Behavior Observation Robot records a seated subject, particularly his/her facial expression, in the above mentioned therapy scenario. Within the environment, which is a relatively large space (e.g., dining hall) with a flat floor, a variety of either seated or walking people (caregivers, elderly people, etc.) are present.

Fig. 2. A scene of robot therapy (target working scene of behavior observation robot).

3 Behavior Observation Robot

3.1 Concept of Behavior Observation Robot

As described in the last chapter, when a place changes fluidly like facilities observation in an observation experiment, it is trouble very much for plural people to visit the field each time from a point of the time and labor. So, in this target environment the observation system which have three concept have been thought about to perform the substitute of conventional observer.

- Behavior Observation Robot records subject behaviors autonomously.
- The robot does not interrupt the natural behaviors of the subject as much as possible.
- Any other equipment (CCD sensor, laser sensor and etc.) is not installed within the working environment.

3.2 Required Specification for New Robot

Based on this three concept and the problem in the current Behavior Observation Robot, we considered the required function for the new robot as follows.

- The height of the robot is 1,000 mm.
  - The height of the robot match the eye level of the sitting target subject. And, in this target environment there are lots of chair and desk, so not to disturb observation the height of the robot needs to be 1,000 mm.
- The noise level of robot is under 40 dB.
  - The noise shows up the existence of the robot. The prototype Behavior Observation Robot generates the noise of 68 dB which is the same as noisy street, therefore its noise level should be under 40 dB which is the same as quiet library.

- The robot can observe target subject continuously over 30-minute.
  - The experiment of robot therapy using PARO is conducted generally for 30 min. So, the robot needs to observe during the period.
- The observable distance is more than 3.5 m.
  - All people have personal space where they feel uncomfortable when other people enter in. Considering personal space, the distance between robot and target subject should be more than 3.5 m in order to minimize observer effects.

3.3   Behavior Observation Robot Ver.2

We developed a new Behavior Observation Robot which realized required functions. Figure 3 shows its appearance and equipment.

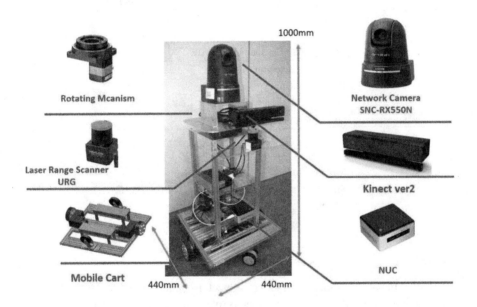

Fig. 3. Appearance and equipped device

Kinect v2, facial recognition device, acquires target's face coordinate, and transmits the coordinate to network camera. To the facial expression analysis and behavior analysis, the network camera acquires an image of the appropriate angle of view by using its pan, tilt and zooming function. Even though the network camera lose target subject in the angle of view, the camera can follow target subject by the target's face coordinate data from Kinect.

In order to keep appropriate view direction, Kinect v2 and the camera were placed on rotating table to cancel the rotation of robot.

The prototype robot used omnidirectional wheel, therefore it caused loud mechanical noise. New robot adopts simple two-wheel type mobile robot to reduce the noise (Table 1).

Table 1. Specification of new behavior observation robot

Network camera SNC-RX550 N	
Effective pixels	380,000 pixels
Zoom ratio	Optics 26×/Digital 12×
Pan drive range	360 °C endless rotating
Tilt drive range	—90–0 °C
Kinect ver2	
Human detection range	0.5–4.5 m
Color(resolution)	1920 × 1080
depth (re solution)	512 × 424
Human posture	Max 6 person
Laser Range Scanner UST-10LX	
Maximum detection distance	10 m
Ranging accuracy	±40 mm
Scanning angle	270 °C
NUC 5i7RYH	
CPU	Core i7-5557U
Memory	16 GB
Mobile cart	
Size [mm]	W440 × D440 × H130
Weight	12 [kg]
Equipped with sensors	Rotary encoder

3.4 Acquire Appropriate View Angle

At first, the facial coordinates of target subject is acquired by the face detection function of Kinect. Then, rotation mechanisms rotates the cameras (Kinect and network camera) to adjust its direction to the subject. The network camera performs tilt movement based on the subject's facial coordinate, and then, controls its angle of view based on the distance to the nose of target subject by zooming function.

Figure 4 shows the configuration of Kinect and the network camera. The center of each camera are aligned in vertical. Horizontal distance of the network camera and Kinect is 120 mm, and the vertical distance is 250 mm.

The tilt angle (SNC tilt) of network camera is calculated based on the target's coordinates obtained by Kinect (Fig. 5). The angle is obtained by the following equations.

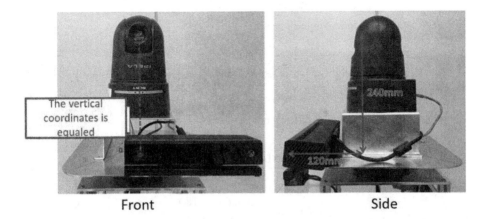

Fig. 4. Configuration of Kinect and network camera

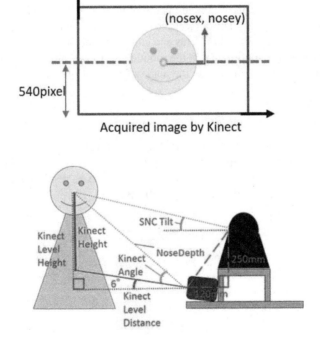

Fig. 5. Relationship among Kinect, network camera and a subject

$$\text{KinectHeight} = (540 - \text{nosy}) \times (1.0 \times \text{NoseDepth} \times 0.001) \tag{1}$$

$$\text{KinectAngle} = \arcsin\left(\frac{\text{KinectHeight}}{\text{NoseDepth}}\right) \times \frac{180}{\pi} \tag{2}$$

$$\text{KinectLevelHeight} = \text{NoseDepth} \times \sin\left(\frac{(\text{KinectAngle}) + 6 \times \pi}{180}\right) \tag{3}$$

$$\text{KinectLevelDistance} = \text{NoseDepth} \times \left|\cos\left(\frac{(\text{KinectAngle} + 6) \times \pi}{180}\right)\right| \tag{4}$$

$$\alpha = \frac{\text{KinectLevelHeight} - 250}{\text{KinectLevelDistance} + 120} \tag{5}$$

$$\text{SNC Tilt} = \frac{\arctan \alpha \times 180}{\pi} \tag{6}$$

To acquire appropriate angle of view, we obtained experimentally the relationship between a zoom level of the network camera and the NoseDepth.

$$\text{zoom level} = \frac{\text{NoseDepth [mm]} \times 0.001}{0.70} \tag{7}$$

Figure 6 shows the result of simple operation test. (i) Kinect and network camera captured a target subject. (ii) The subject moved to another place, the network camera lost him. (iii) The system rotated cameras and controlled the network camera's angle of view appropriately. We confirmed that the robot could follow target subject again even if the target subject moves to outside the angle of view of the camera.

3.5   Noise Level of the New Robot

An experiment was conducted to measure the noise level of the new robot. Figure 7 shows the environment. A noise meter was set 3.5 m away from the robot. Then, the robot moved 5 m at 0.5 m/s in front of the noise meter. As a result, we confirmed that the noise level of the new robot was under 40 dB.

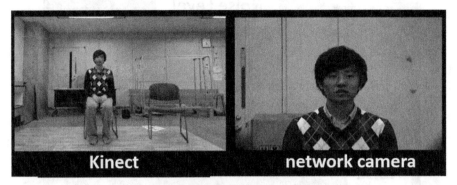

(i) Kinect and network camera captured a target.

(ii) The target moved, and the network camera lost him.

(iii) Network camera followed and acquire appropriate angle of view

Fig. 6. Acquired image of Kinect and network camera

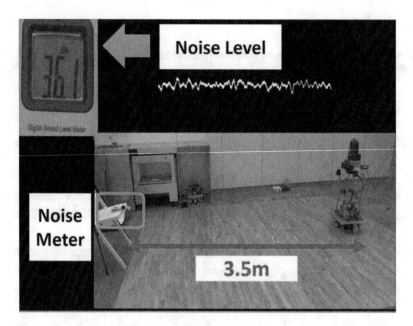

Fig. 7. The environment of noise level experiment

4   Conclusion

In this paper, we introduce Behavior Observation Robot for the substitute of observer in field studies. The problems of prototype robot were discussed, and then new robot was proposed and developed. The behavior algorithm and observation experiment will be presented in the future.

References

1. Wada, K., Sugioka, T., Shimoyama, A.: "Behavior Observation Robot" for field study. In: The 22nd IEEE International Symposium on Robot and Human Interactive Communication (RO-MAN), pp. 674–679, August 2013
2. Wada, K., Shibata, T., Saito, T., Tanie, K.: Effects of robot assisted activity for elderly people and nurses at a day service center. Proc. IEEE **92**(11), 1780–1788 (2004)
3. Nakashima, K., Sato, H.: Personal distance against mobile robot. J. Jpn. Ergon. Res. Soc. **35** (2), 87–95 (1999)
4. Nakashima, K., Sato, H.: Personal distance of older persons against mobile robot. Jpn. J. Ergon. **35** (1999). ISSN 0549-4974
5. Yokoyama, K., Watanabe, A.: A Study on feelings of worriment due to middle-size robot in living space. J. Civil Eng. Archit. **7**(4), 409–413 (2013). Serial No. 65. ISSN 1934-7359
6. Sakai, M., Watanabe, A.: Personal space of small mobile robot moving towards standing or sitting elderly individuals. J. Civil Eng. Archit. **7**(7), 827–832 (2013). Serial No. 68. ISSN 1934-7359

7. Ota, S., Watanabe, A.: Study on individual distance of the Non-Japanese for the small mobile robot. J. Civil Eng. Archit. **7**(3), 323–327 (2013). Serial No. 64. ISSN 1934-7359
8. Aoki, M., Watanabe, A.: A study on the distances of an upright/char-sitting small mobile robot to male adult individuals. J. Archit. Plann. Environ. Eng. **76**(664), 1093–1100 (2011)
9. Yoshikawa, S.: Trends in recent research on face and facial expression processing. Inst. Image Inf. Telev. Eng. **54**(9), 1245–1251 (2000). Special Edition Human Vision Science: Landmarks in the Field of Multimedia Technology over 50 Years and the State of the Art
10. Kyota, A., Souksakhone, B.: Estimation of emotional states based on body movements observed using videos. In: IEICE, vol. 109, no. 470, pp. 13–18 (2010)
11. Konishi, Y.: Real-time estimation of smile intensities. In: IPSJ Symposium, vol. 2008, p. 47 (2008)
12. Nishida, Y., Aizawa, H., Kitamura, K.: Robustly observing human activity by sensor room and its application. IPSJSIG Technical report, HI-106, vol. 2003, no. 111, pp. 37–44 (2003)

A Collaborative Task Experiment by Multiple Robots in a Human Environment Using the Kukanchi System

Kazuma Fujimoto, Takeshi Sasaki, Midori Sugaya, Takashi Yoshimi,
Makoto Mizukawa, and Nobuto Matsuhira[✉]

Shibaura Institute of Technology, 3-7-5, Toyosu, Koto ku, Tokyo 135-8548, Japan
matsuhir@shibaura-it.ac.jp

Abstract. We have developed a collaborative robot system to deliver services to people. The Kukanchi system has been used as an intelligent space that comprises sensors, robots, and management servers. Specifically, we expanded and improved Kukanchi's middleware functionality by adding sensors and robots outside the Kukanchi space. Furthermore, we verified the effectiveness of the system through a guidance service task using the task scheduler in the Kukanchi system. The task scheduler manages the service task by disassembling it and selects and sends requests to robots. In this study, we introduce the system and present the experimental results.

Keywords: Intelligent space · Kukanchi · RT middleware · Robot · Collaboration

1 Introduction

Household affairs and physical and information support, such as elderly care owing to an aging society and reductions in birth rate, has recently been requested in society. To cope with these problems, we expect robot technology to be deployed. However, it is difficult for a robot or a group of similar robots to solve all problems because a variety of services are requested by today's society. Thus, an intelligent space for robot systems is a growing area of research. A mobile robot, sensors, and actuators are distributed over a network in the space, wherein they utilize each other's information to efficiently perform a task [1, 2]. Furthermore, it is necessary to solve these problems for various robots to collaborate work.

To build a robot system comprising many robots from scratch, a significant amount of time and money is required. A solution to this is to use existing robots with only marginal modifications and realize such a framework. Thus, we have used the Kukanchi system [3] and RT middleware [4]. Here, we introduce a collaborative robot service, which utilizes the database that accumulates information regarding human beings, things, and robots in the developed intelligent space using Kukanchi as a platform.

© Springer International Publishing Switzerland 2016
N. Kubota et al. (Eds.): ICIRA 2016, Part II, LNAI 9835, pp. 276–282, 2016.
DOI: 10.1007/978-3-319-43518-3_27

2 Robot Collaboration Through the Database

2.1 Kukanchi Middleware

To realize the intelligent space, sensor and actuator networks, accumulation of the observed information, decision of support contents according to the observed situation, and task generation are important functions that should be integrated as a system. Therefore, Kukanchi middleware using RT middleware has been proposed as the basic system. The Kukanchi middleware has a processing module and a database and provides an abstraction interface with sensors, users, and robots. Thus, we can obtain environmental information using sensors and process and accumulate the gathered information. The existing robots in the space can provide services through commands that are issued when the system receives the service request. A Kukanchi system using robot service technology was developed in a model room of a home living environment to deliver services such carrying objects in the house [5]. Figure 1 shows the developed Kukanchi system. The system is expected to expand and provide other robot services using the middleware and database. Some major applications of this technology could include its introduction to shopping centers and as a guidance service to shops.

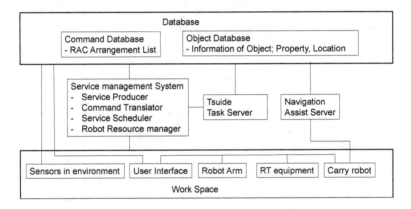

Fig. 1. The Kukanchi system

2.2 Connection of a Robot to the Database

In the Kukanchi middleware, the database that accumulates the structured information of the space is called the Local Data Store (LDS) as shown in the service management system in Fig. 1. LDS can record the information about things, users, and robots existing in the space.

Acquisition and registration of the information to the database is performed by the abstraction interface of the Kukanchi middleware. Therefore, even an existing robot in the intelligent space can collaborate with other robots through the database after adding the Kukanchi middleware and a common interface. First, through sending and acquisition of the sensor information of the robot, it is possible to share the environmental information with other robots. Therefore, a mobile robot is considered as a mobile sensor in the intelligent space. The robot information of LDS shows the function and location of the robot in the space. The service is managed as a combination of tasks. The task scheduler implemented in Kukanchi judges the feasibility of the service and requests the assigned robot to perform the task. As a result, a service is realized through the collaboration of robots.

2.3 Task Scheduler

The task scheduler is an element of the Kukanchi system. It disassembles the user's request into tasks that the robots can conduct, selects the robot that has the assigned function for the required task in the space, and orders the robot to perform the task. In this study, we improved the task scheduler to add the position information to each task as shown in Fig. 2. Figure 3 shows the example of the connection between the Kukanchi system and external robots. This connection was used in the experiment.

Fig. 2. Improvement of the LDS component

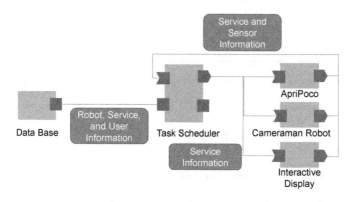

Fig. 3. Example of the connections in the Kukanchi system

3 Algorithm for Robot Selection

3.1 Disassembly of Service Task

To perform the robot collaboration work, we disassemble a series of services for a user to tasks. Then, the required function and position information are added to each task.

The function is described when a task is to be performed and the position is described when the task needs collaboration between robots or between the robot and the user. The contents of the task and the function of the robot are all registered in the LDS in advance. Thus, the service is realized by allocating the task to the suitable robot.

3.2   Algorithm of Robot Selection

Under the condition that a task is conducted by a robot, the task is allocated to a robot using the following procedure:

(1)   Confirm which robot has the required function
(2)   Confirm the status of the robot to which the request is to be sent
(3)   Confirm if the robot position is adequate to satisfy the task
(4)   Add a robot with a movable function as a candidate because of a change in position.

On selecting and exploring the robot using the above procedure, the task can be allocated to the suitable robots. Figure 4 shows the breakdown of the task according to the requested task and the selected sequence. In each task, the suitable robot is selected by considering the location of the person.

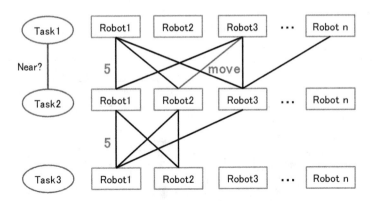

Fig. 4.   Robot selection flow according to the task contents

4   Experiment of Collaborative Service

A demonstration of the guidance service as a collaborative service using various types of robots at multiple locations was conducted in the experiment as shown in Fig. 5. Figure 6 shows the control system of the collaborative service task. Figure 7 shows the experimental scene. However, the exhibition space is not shown. The robots are a reception robot on the desk, an interactive display that can communicate with a user, and a mobile robot acting as a guidance robot, as shown in Fig. 6. The interactive display shows the menu on the screen and detects the input when a user touches the menu as same as a computer touch panel. In the guidance service, the robot greets a user at the reception desk, the user selects the place where he/she wants to go using the interactive display, and the mobile robot guides the user to the desired place. This service was disassembled into three main tasks: reception, selection, and guidance. Furthermore, these tasks were allocated to the robot to perform the service to meet the user by considering the location relationship between robots and between the robots and the user. Using this system, the person was guided to the desired place through the collaboration of multiple robots.

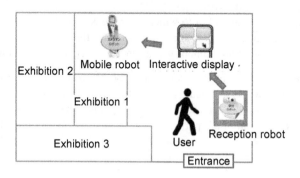

Fig. 5. Example of guidance task by robot collaboration

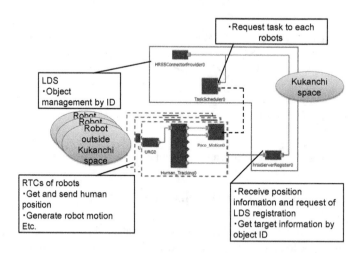

Fig. 6. Connection diagram between mobile robots and LDS of the Kukanchi system in the experiment

Fig. 7. Demonstration of collaborative service by multiple robots: guidance service

5   Discussion

In the demonstration task, directory enquiries were conducted at the reception robot. The service comprises the following tasks in detail. When a person arrives at the reception desk, a reception robot detects the position of the person (person detection task) and registers the position with the LDS of the Kukanchi system. Further, a task scheduler of the Kukanchi system chooses a robot that can talk, and the chosen robot (among other robots in the space) leaves quickly for the reception desk and directs the guide (robot choice task). A mobile robot expands its coordinate system and leaves for the meeting (wide area traveling task). On the other hand, the reception robot guides the interactive display to be an input to the demands of the places that the user wants to visit. Meanwhile, the mobile robot arrives (service choice task) and registers the place with the LDS from the interactive display (interaction task). When the guidance robot arrives at the reception desk, the robot guides the person to the input destination (guidance task). In the experiment, the exhibition space and robot capabilities were limited and the task was not flexible. For instance, changing the destination of the guidance robot and communication with it in real-time was not possible.

6   Conclusion

We have developed a multiple robot collaboration system using the database and task scheduler of the Kukanchi system. Furthermore, the position information between the user and robots has been added to the task, making the service possible according to the user's request. We have confirmed the collaboration of robots through a basic experiment providing a guidance service. As future work, we plan to develop the system to cater for multiple users and to realize a more flexible service.

We express our special thanks to Mr. Taiki Endo and Mr. Daiki Nomiyama who developed the system and performed the experiment.

References

1. Sato, M., Kamei, K., Nishio, S., Hagita, N.: The ubiquitous network robot platform: common platform for continuous daily robotic service. In: IEEE/SICE International Symposium on System Integration, pp. 318–323 (2011)

2. Kato, Y., Izui, T., Tsuchiya, Y., Narita, M., Ueki, M., Murakawa, Y., Okabayashi, K.: RSi-cloud for integrating robot services with internet services. In: IECON, pp. 2158–2163 (2011)
3. Ishiguro, Y., Maeda, Y., Trung, N.L., Sakamoto, T., Yoshimi, T., Ando, Y., Mizukawa, M.: Architecture of Kukanchi middleware. In: The 8th International Conference on Ubiquitous Robots and Ambient Intelligence, pp. 23–26 (2011)
4. RT-Middleware (OpenRTM-aist). http://www.openrtm.org/OpenRTM-aist/html-en/
5. Maeda, Y., Yoshimi, T., Ando, Y., Mizukawa, M.: Task management of object delivery service in Kukanchi. In: IEEE/SICE International Symposium on System Integration (SII), pp. 283–288 (2012)

Mel-Frequency Cepstral Coefficient (MFCC) for Music Feature Extraction for the Dancing Robot Movement Decision

Indra Adji Sulistijono[1(✉)], Renita Chulafa Urrosyda[2], and Zaqiatud Darojah[2]

[1] Graduate School of Engineering Technology,
Politeknik Elektronika Negeri Surabaya, Surabaya, Indonesia
indra@pens.ac.id
[2] Mechatronics Engineering Division,
Politeknik Elektronika Negeri Surabaya Kampus PENS,
Jalan Raya ITS, Sukolilo, Surabaya 60111, Indonesia
renita@me.student.pens.ac.id, zaqiah@pens.ac.id

Abstract. In this research, we built a system that has the capability to recognize some of music patterns which each of the music pattern is obtained by cutting the whole music into some sections. The Music that used has the title of Kicir-Kicir, one of the famous traditional music belongs to Indonesia. We used the Mel Frequency Cepstral Coefficients (MFCCs) method to extract the feature of the music. The neural network (NN) is proposed as the music recognition algorithm. In this study, we used Backpropagation Neural Network (BNN) algorithm, the most popular NN and is used worldwide in many different types of applications. Each section of the music will be recorded many times to obtain the train and test set data. Those set of data will be extracted using MFCC and then classified using BNN. In the examination, one of the section of Kicir-Kicir music is recorded, then it is extracted using MFCC and recognized using BNN. Result of BNN will decide which one of movement the dancing robot should be done. Experimental results show that MFCC and BNN are capable to recognize the music with 76 % of success rate.

Keywords: MFCC · Mel-Frequency cepstral coefficient · Music feature extraction · Backpropagation neural network · Frequency

1 Introduction

Research on voice recognition begins from signal processing research which was not a new thing, starting at nearly two and a half centuries ago, in 1773 by Christian Kratzenstein, a professor from Russia, managed to create a tool that can produce the vowel sound. Several years later, research appears be an inspiration in developing voice recognition. The study produced by Harvey Fletcher where Fletcher documenting the relationship between noise spectrum in which

© Springer International Publishing Switzerland 2016
N. Kubota et al. (Eds.): ICIRA 2016, Part II, LNAI 9835, pp. 283–294, 2016.
DOI: 10.1007/978-3-319-43518-3_28

the distribution of the power of the voice using the frequency. The study resulted in the characteristic sound that can be heard and easily understood by the human listener. In 1881, Chichester Bell and Charles Summer Tainter invented the sound recording device that lead the researchers continue to develop the research of voice [1]. Early emergence of the voice recognition system is based on the discovery of pattern recognition (feature extraction) using Linear Predictive Coding (LPC) to represent the spectral shape introduced by Atal and Itakura [1]. Research in voice recognition will continue to develop ranging from small-scale word recognition (10–100 words) to large scale (1000 - an infinite number of words). Until now, has been a lot of methods to enhance the LPC, such as the emergence of Hidden Makarov Model (HMM) and Mel-Frequency Cepstral Coefficient (MFCC). Methods of feature extraction for speech recognition has been widely utilized its use in this modern era suppose the bird voice recognition [2], characteristics of music identification [3], automatic recognition of the insects [4], and sound detection system in the cradle [5]. Nowadays, voice recognition technology that has developed rapidly, can be implemented in many ways. In this research, voice recognition will be implemented in the field of robotics where at the current era, the field of robotics engineering technology had also been growing rapidly, even where robots are now used not only in industry, but also has been used in various fields such as education, agriculture, health, competition and in terms of entertainment and leisure. At present, the development of robotics in the entertainment and leisure began in the spotlight. Based on statistical data conducted by the International Federation of Robotics (IFR) that is uploaded in www.ifr.org site states that the development of the robot in the entertainment and leisure experienced a significant increase in a year, from 2012 to 2013 the increasing number of demand reach 12 % and it is forecasted that in the next 4 years that number still increases reach 84 %. Considering the trend of interest in entertainment and leisure robot, in this research, we created a robot that has elements of Indonesian culture that is a traditional dancer robot. This robot has the ability to recognize patterns of music, so it can move in harmony with the rhythm of the music. The feature extraction of musical patterns using Mel-Frequency Cepstral Coefficient (MFCC) which is based on previous research, it is a good and efficient method compared with other methods because it is able to achieve 97.55 % accuracy [6]. After the music is extracted, it will be recognized using BNN and the result of BNN is used to decide which the movement of the robot should be taken.

The main purpose of this research is to implement engineering robotics technology in the field of culture entertainment and leisure as well as apply the MFCC for extracting certain musical patterns that will make the robot dancer are able to move harmoniously in accordance with musical accompaniment. In this research, we just focused in the MFCC method for extracting the characteristic of the music to be recognized by BNN. The movement of robot based on the music is not discussed.

This paper is organize as follow. Section 2 explains Pre-Processing. In Sect. 3, applied Mel Frequency Cepstral Coefficients (MFCC). After that, Sect. 4

discusses the Backpropagation Neural Network (BNN). In Sect. 5, we show the experimental result for proving the effectiveness of the proposed model.

2   Pre-processing

At this research, input from MFCC is Kicir-Kicir musical, one of famous traditional music in Indonesia. Kicir-kicir has the overall duration of approximately 2.5 min and the music will be cut into 5 pieces contain of 30 s of music where each piece is represented by the first 5 s. How inputs are obtained is illustrated in Fig. 1.

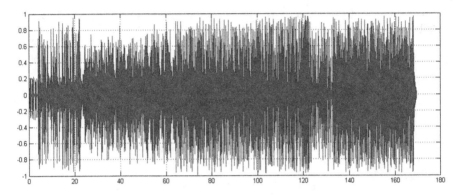

Fig. 1. Song of "Kicir-Kicir" signal. One kernel at x_s (*dotted kernel*) or two kernels at x_i and x_j (*left and right*) lead to the same summed estimate at x_s. This shows a figure consisting of different types of lines. Elements of the figure described in the caption should be set in italics, in parentheses, as shown in this sample caption.

From Fig. 1, the inputs of MFCC are 5 s of the first second from pieces of music which would represent a 30 s of music. 30 s of the music represents one cycle of movement of the robot dancer. So in 2.5 min, Kicir-Kicir has 5 kinds of dance in which each dance will be triggered by the beginning of the 5 samples that have been taken, extracted and identified. Results of MFCC is coefficients that represents each signal. The number of coefficients used in this research was 13 and 26 coefficients to be compared which one is better. Those coefficients will become the input to BNN to be recognized.

3   Mel Frequency Cepstral Coefficients (MFCC)

MFCC are coefficient that make up an MFC, a representation of the short-term power spectrum of a sound based on a log power spectrums linear cosine transform on a nonlinear mel scale of frequency. MFCC process is depicted in Fig. 2.

3.1   Pre-emphasis

Pre-emphasis has an aim to improve signal from noise interference, so as to improve the accuracy of the voice characteristics search. The music signal is pre-emphasized using Eq. (1).

$$y(n) = x(n) - 0.97x(n - 1) \qquad (1)$$

where which, $y(n)$ is emphasized output signal, $x(n)$ is input signal $x(n - 1)$ is previous input signal.

Figure 3 shows the different of signal that before and after emphasized. After emphasized, some of the amplitude is raised decreases the chance of information loss.

Equations should be punctuated in the same way as ordinary text but with a small space before the end punctuation mark.

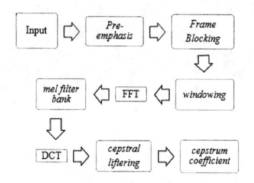

Fig. 2. MFCC process.

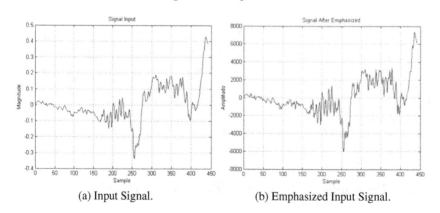

(a) Input Signal.   (b) Emphasized Input Signal.

Fig. 3. Pre-emphasis process.

3.2   Frame Blocking

Frame blocking is a process that aims to divide the music samples into multiple frames or slots with a certain length. The number of frames that will be formed is determined on the value of the time duration (Tw) and time shift (Ts) that follows the Eqs. 2 and 3. Time duration will determine the length of data in each frame. Time shift determines the length of the sample overlap in every frame. Formation of the frame illustrated in Fig. 4.

$$Nw = 10^{-3}(fs.Tw) \tag{2}$$

$$Ns = 10^{-3}(fs.Ts) \tag{3}$$

where Nw is number of samples in each frame and Ns is number of shifted samples.

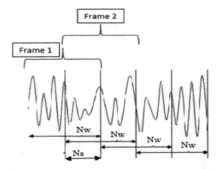

Fig. 4. Frames forming.

This research use Tw 40 ms and Ts 20 ms and based on Eqs. (2) and (3). Number of frame in every input signal is 237 frames. Figure 5 shows the result of frame blocking in input signal.

From the figure above, the continue signal is blocked become frames which each frame is able to process independently.

3.3   Windowing

In basic principle, the use of windowing functions is convolute the input signal with any particular window function so as to reduce signals belonging to leak prior to the transformation process. Here, we used hamming window that is described by Eq. (4).

$$w(n) = 0.54 - (0.46 \; cos(2\pi n/N - 1)) \; ; \quad \text{for } n = 0, 1, 2, .., N - 1 \tag{4}$$

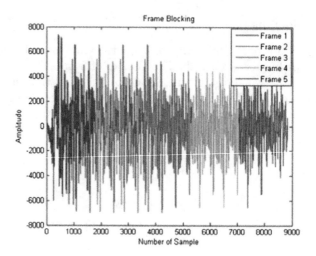

Fig. 5. Frame blocking process in first 5 samples.

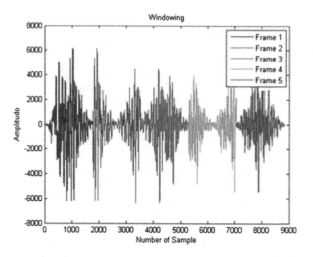

Fig. 6. Windowed signal.

where $w(n)$ is windowed signal and N is length of filter.

Figure 6 shows the result of windowing after signal in blocked into frames.

After the signal is windowed, samples that are shifted after frame blocking process, is filtered thus will make the amplitude of overlapped samples are minimized.

3.4   Fast Fourier Transform (FFT)

An analysis based on Fourier transform is tantamount to spectrum analysis, as fourier transform change digital from the time domain to the frequency domain.

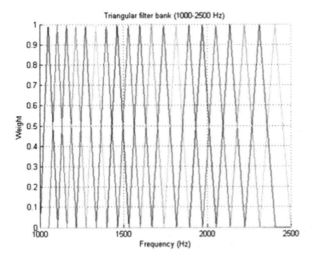

Fig. 7. Triangle filter bank.

FFT is a Fourier transform algorithm developed from algorithms Discrete Fourier Transform (DFT). With the FFT method, computing the rate of transformation fouirer calculation can be improved. After the domain of signal is changed into frequency domain, the magnitude of each frequency is gotten.

3.5 Mel Filter Bank

Magnitude result of the FFT process further through the phase filter bank. This phase was conducted to determine the size of the energy of each frequency band. Before becoming a mel filter bank, the magnitude of the FFT results are changed in the Mel frequency scale. Mel frequency scale is linear frequency below 1 kHz and logarithmic above 1 kHz. Mel scale can be obtained by Eq. (5) approaches. Mel filter bank that is used in this research is illustrated in Fig. 7.

$$Mel(f) = 1125 \, log_{10}(1 + \frac{f}{700})$$ (5)

where $Mel(f)$ is Mel scale frequency and f is linier frequency.

3.6 Discrete Cosine Transform (DCT)

DCT return the domain analysis from frequency to time domain. Through this process the mel cepstral is obtained.

3.7 Liftering

This process is performed to smooth the spectrum of the signal generated from the DCT process. Cepstral liftering can be done by implementing the window

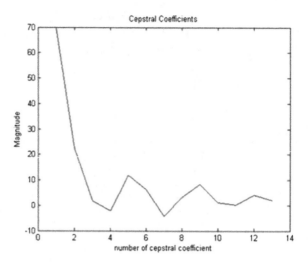

Fig. 8. Coefficient of MFCC.

function to the cepstral features. Result of liftering is 13 coefficients that known as MFCC coefficient. The value of coefficient is showed in Fig. 8.

4 Backpropagation Neural Network (BNN)

Artificial Neural Network (ANN) is an information processing system that has characteristics similar to biological neural networks in human [7]. BNN is feed-forward multilayer which sets the weighting network in the opposite direction by using the change in instantaneous error. BNN has been widely used for many problem solving, one of them recorded that it is capable to identify the characteristic of music successfully.

In this research, the architecture of network is depicted in Fig. 9.

BNN architecture consist of three layers, the input layer, one hidden layer, and output layer. The input layer consist of the number of the MFCCs, 13 or 26 coeffecients, so there are 13 or 26 nodes in the input layer. The hidden layer consists of non-linear activation using 12 nodes. The output layer consist of 3 nodes, which the values are in the binary form. The combination of binary numbers on the output of BNN will determine the robot movement decision. Determination of the combination of binary digits shown in Table 1.

If the result of a combination of digits BNN is 1 0 0, then the robot will do the first movement, if the results of BNN in the form of a combination of digits 0 1 0, then the robot will do the 2nd movements and so on. In the process neural network, the input is obtained by recording each of 5 samples music 20 times. We divide it into two data, 60 % for training data and 40 % for testing data.

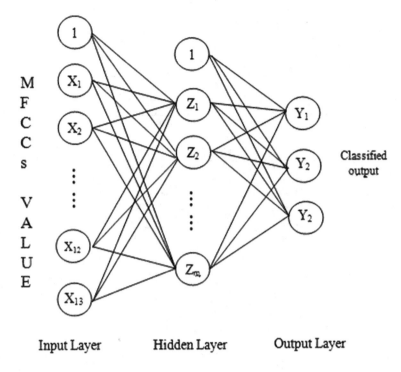

Fig. 9. Architecture of BNN.

Table 1. Digits combination of BNN and movement.

Digit combination	Represent	Movement of robot
100	1st Cycle	1st Movement
010	2st Cycle	2st Movement
110	3st Cycle	3st Movement
101	4st Cycle	4st Movement
111	5st Cycle	5st Movement

5   Experimental Results

The experimental process is begun by getting some data that is used for training and testing data. The data is obtained by recording the piece of Kicir-kicir music with duration of 5 s, which has been determined, as many as 10 times, as has been mentioned in previously section. The data will then be extracted with MFCC and identified by BNN. BNN results in the form of a combination of three binary digits are stored in the database as a reference value. After the training and testing data obtained, to test the reliability of MFCC and BNN some of processes should be done. The processes are illustrated in Fig. 10.

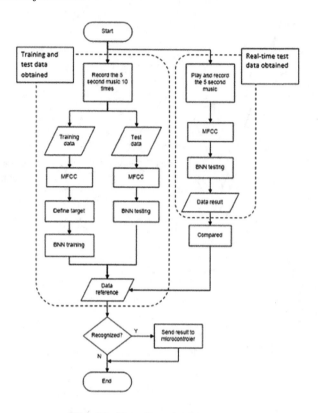

Fig. 10. Experiment test process.

Based on Fig. 10, the experiment begins by playing one of a piece of Kicir-kicir music with duration of 5 s and then recorded. The tape is extracted and the extraction will be identified by BNN whether including 1st or 2nd movement, and so on. If the identification results have been obtained either 1st movement or the other movement, the value of the identification of a combination of three binary digits are transmitted to microcontroller contained in the robot dancers so that the robot can specify what action should be done. The experiment against one of the inputs is done 20 times to see the success rate of the system.

The experimental results show that the average success in recognizing the input by using the comparison training and test data by 6 to 4 with number of coefficients 13 is 69 % and 76 % successfully recognized for the 26 coefficients. The results of the experiment are shown in Table 2.

Table 2 shows that testing with comparison of training and test data by 6 to 4 using 26 coefficients generate better value. In some cases, using 13 coefficient is better such as when recognize the 5th input. It happen because 5th input has the highest differences than other inputs in the first 13 coefficient. So it will easily recognized using 13 coefficient. But in other hand, in case of 3rd input recognition, using 13 coefficient produce the unsatisfied result because

Table 2. Experimental results of the average success in recognizing the input.

Input	Number of MFCC's coefficient	
	13	26
1	70	70
2	85	85
3	50	85
4	60	80
5	80	60
Average(%)	69	76

2nd input and 3rd input have similar characteristic but 2nd input has more prominent magnitude thus almost of 3rd identification was mistaken recognized with 2nd input. By using 26 coefficient, more information are able to be gained so when use 26 coefficient the differences of 2nd input and 3rd input are clearly visible and both 2nd input and 3rd input are succeed to be recognized. The difference result of 13 and 26 coefficient for 2nd input and 3rd input is depicted in Fig. 11.

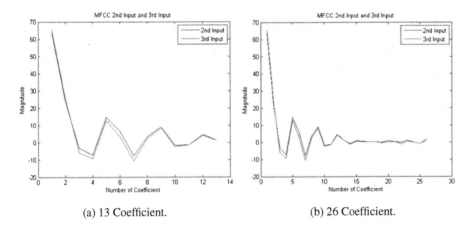

(a) 13 Coefficient.   (b) 26 Coefficient.

Fig. 11. MFCC of 2nd and 3rd input.

From Fig. 11 we can see that the 20th until 26th coefficient have a main role to make 2nd input and 3rd input are recognized to be different.

6   Conclusion and Future Work

In this research we have proposed a system that is capable to recognized piece of music for dancing robot using MFCC and BNN. Research of MFCC in voice recognition mainly discuss about speech and speaker recognition nevertheless, MFCC is not harmful to be implemented on complex voice such as music extraction as its recognition result above 50 % succeed. From this research we found that 26 coefficient in MFCC is not always better than 13 coefficient to be chosen for extraction, some of them may cause worse result. 26 coefficient is necessary used if there are some input that have a familiar characteristic. Noise analysis in this research is ignored. So when record the music in case of data collection, the room must sealed from influential noise as noise is able to make information of data changed. To make data more robust from noise, a noise reduction system must be built and our ongoing work will focus on this problem.

References

1. Juang, B.H., Rabiner, L.R.: Automatic speech recognition a brief history of the technology development. In: Georgia Institute of Technology, Atlanta, Rutgers University and the University of California, Santa Barbara (2004)
2. Vilches, E., Escobar, I.A., Vallejo, E.E., Taylor, C.E.: Targeting input data for acoustic bird species recognition using data mining and HMMs. In: 7th IEEE Conferenceon ICDMW 2007, E-ISBN: 978-0-7695-3033-8, Print ISBN: 978-0-7695-3019-2. Omaha, NE (2007)
3. Ahmad, R.: Identifikasi Ciri Musik dengan Menggunakan Mel-Frequency Cepstral Coefficien (MFCC), Scription of Elektronika PENS-ITS (2009)
4. LeQing, Z., Zhen, Z.: Automatic recognition of insect sounds using MFCC adn GMM. J. Acta Entomologica Sinica $455(4)$, 466–471 (2012)
5. Zahro, F., Ningrum, S., Endah, S., Darojah, Z.: Sistem Deteksi Suara Pada Ayunan Bayi Menggunakan Filter Digital, Scription of Mekatronika PENS-ITS (2014)
6. Universite Pierre & Marie Curie, LA Science A Paris (2004)
7. Fausett, L.: Fundamentals of Neural Network. Prentice Hall, Singapore (1994)

Sensing and Monitoring in Environment and Agricultural Sciences

Random Forests Hydrodynamic Flow Classification in a Vertical Slot Fishway Using a Bioinspired Artificial Lateral Line Probe

Shinji Fukuda[1(✉)], Jeffrey A. Tuhtan[2,3],
Juan Francisco Fuentes-Perez[3], Martin Schletterer[4],
and Maarja Kruusmaa[3]

[1] Tokyo University of Agriculture and Technology, Tokyo, Japan
`shinji-f@cc.tuat.ac.jp`
[2] SJE Ecohydraulic Engineering GmbH, Stuttgart, Germany
`tuhtan@sjeweb.de`
[3] Centre for Biorobotics, Tallinn, Estonia
`{Juan.Fuentes,Maarja.Kruusmaa}@ttu.ee`
[4] TIWAG - Tiroler Wasserkraft AG, Innsbruck, Austria
`martin.schletterer@tiwag.at`

Abstract. Ecohydraulic studies rely on observations of fish behavior with hydrodynamic measurements. Most commonly, observed fish locations are compared with maps of the bulk flow velocity and depth. Fish use their lateral line to sense hydrodynamic interactions mediated by body-oriented spatial gradients. To improve studies on fish an artificial lateral line probe (LLP) is tested on its ability to classify either the "slot" or "pool" regions within 28 basins of a vertical slot fishway. Random forests classification is applied using four models based on high-frequency (200 Hz) recordings using 11 collocated pressure sensors and two triaxial accelerometers. It was found that the assigned classification task proved to be reliable, with 100 % correct classification of all four models, across all 28 basins. Preliminary results from the first field study of this new sensing platform show the LLP-random forests workflow can provide robust, highly accurate classification of turbulent flows experienced by fish innatura.

Keywords: Random forests · Lateral line probe · Vertical slot fishway · Turbulent flow · Hydrodynamic classification

1 Introduction

The study of complex innatura flows has been focused on the analysis of a collection of point measurements, usually using Doppler velocimetry or more recently time-averaged maps based on acoustic Doppler current profiling [1]. Such maps show the Eulerian distribution of the velocity field. Considering fish, which sense the flow field via their lateral line, the biologically relevant stimulus is not the velocity or its

© Springer International Publishing Switzerland 2016
N. Kubota et al. (Eds.): ICIRA 2016, Part II, LNAI 9835, pp. 297–307, 2016.
DOI: 10.1007/978-3-319-43518-3_29

fluctuations *per-se*, but rather the fractional derivative of the near body velocity and spatial distribution of the near body pressure field, driven by local flow accelerations and filtered through the boundary layer [2]. Thus the sampling of a "naked" flow field, and its subsequent representation as a time-averaged velocity distribution makes it difficult to physically relate such mappings recorded in the field to observed fish hydraulic preference [3].

The physics of the lateral line sensory organs and the differentiation of their roles in flow sensing comprise a broad range of active research topics, spanning several decades [4]. It was first fisheries researchers who began to experiment with basic electromechanical lateral line probes (LLPs) in the 1990s [5], and concurrently, a second generation fish-sized, pressure and inertial sampling device was developed and has been successfully applied internationally in studies of the often extreme conditions fish experience as they pass through hydroelectric turbines [6]. Current artificial lateral line research remains focused on new developments in materials and devices, especially the testing of near-laminar and ideal flows. The majority of investigations are focused on the careful recreation of biological sensing capabilities, most often under laboratory conditions. Although the field is small in comparison to other flow sensing technologies, there exist multiple LLP technologies which have been and continue to be actively developed and tested; examples are thermal hotwire anemometry [7], illuminated silicon bars embedded in machined PVC canals [8], and most recently multimodal system including both visual and inertial-base sensory inputs [9]. An excellent overview of the background on bioinspired lateral line flow sensors is provided by Shizhe [10].

Primarily in order to expand the usage of LLPs outside of the laboratory and to sense turbulent flows in hydraulic structures relevant to fish passage, we have built a bioinspired flow sensing and classification platform consisting of a fish-shaped body outfitted with a high-frequency synchronous array of body-oriented pressure sensors, referring to our particular design as simply the artificial lateral line probe. Here we take the device out of the lab and into the field, and present preliminary analysis of two sets of measurements taken in an operational vertical slot fishway, measuring a total of 28 different basins. Our primary goal was to determine if the lateral line probe could successfully classify if a given measurement was made either within the "pool", or in a "slot". We developed a series of novel hydrodynamic features and four different models were tested using random forests classification. Random forests [11] has been widely applied as a predictive tool for classification and regression for its advantages including its very high accuracy, its ability to determine variable importance, its ability to model complex interactions between predictor variables, the flexibility to perform several types of statistical data analysis including regression, classification and unsupervised learning, and its ability to handle missing values [12]. Its high predictive capability has been supported by previous comparative studies with other machine learning techniques [13].

It was found that due to the large amount of data collected by the LLP as a high-frequency collocated sensor array, the classification task proved to be reliable, with 100 % correct classification of all four models, across all 28 basins.

2   Materials and Methods

2.1   Lateral Line Probe

The LLP used in this work consists of 11 piezoresistive pressure sensors (SM5420C-030-A-P-S, Silicon Microstructures) and two triaxial accelerometers mounted within an ABS plastic fish-shaped body (Fig. 1). In contrast to our plastic body, the biological lateral line also includes mechanical filtering due to viscous dampening of the neuromasts in the boundary layer (especially the superficial) and within the canals. However, the extent of any effect of difference in surface roughness and its impacts on the LLP has not yet been investigated. The LLP geometry is taken from a 3D scan of an adult rainbow trout (*Oncorhynchus mykiss*) having a body length of 45 cm. The pressure transducers have a sensitivity of 8 Pa/LSB over a 0-207 kPa span, of which the resistor used for current measurements has 470 O resistance with 0.1 % tolerance and is in 0402 SMD package. The signals undergo first stage amplification (AD8656 ARMZ, Analog Devices) digitized using a 16-bit analog to digital converter (AD768 BSPZ, Analog Devices). The accelerometers used were (ADXL325BCPZ, Analog Devices) with a sensitivity of 191.4 mV per g along each of the three axes. The signals were digitized using the same 16-bit conversion, providing 2.63 E-04 g per least significant bit resolution. Temperature is estimated via current consumption with a shunt resistor. The output signals are first obtained at 2.5 kHz and then 10× over-sampled and transmitted with a serial connection at 200 Hz. Detailed descriptions of the device with signal processing workflows for time-averaged velocity estimation can be found in [14–16].

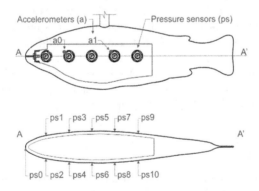

Fig. 1. LLP used in this work and sensor distribution.

2.2   In-situ Flow Measurement

The natural flow experiments were conducted in the new fishpass Wenns installed along the Pitze River in Tirol, Austria. The Pitze River is a tributary of the Inn, and has a total length of about 40 km and a total catchment area of 309 km^2 (Fig. 2). The installation of a fish pass at the weir Wenns (a water intake of the HPP Imst) was stipulated by the Water Framework Directive as an objective in the 1st River Basin

Management Plan (RBMP). Construction work of the fishpass Wenns started in late 2013 and it was successfully completed in April 2014. The 140 m long fishpass is a combination of a vertical slot and a near-natural bypass channel, which re-establishes the river continuum, was built according to the Austrian guidance document on the construction of fishpasses [17], i.e. represents a state-of-the art fishpass in the metar-hithral (lower brown trout region).

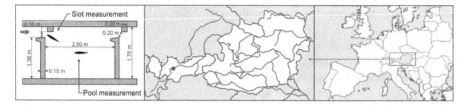

Fig. 2. Study site and sampling points

The LLP was mounted on a metal rod such that it had a fixed submerged height of 77 cm above the river bed for all measurements, depending on local substrate conditions. Sampling locations were chosen based on "slot" or "pool" at each basin (Fig. 2). The field surveys were carried out in all 28 basins of the vertical slot, with one set of slot and pool measurements within each basin. This sampling methodology was selected in order to test the practicality and robustness of signal acquisition and classification for further field studies in multiple fishways. In this experiment, the LLS unit was connected to a field laptop and external power supply. Data was recorded at 200 Hz for two triaxial accelerometers and the 11 collocated pressure sensors, one at the "nose" and five on each side, located bilaterally across the sensor body. Precise geographic measurement positions were not taken, as the ultimate goal of the LLP is to provide a technology for the robust field classification of complex flows which can be achieved using simple maps to locate the classification sites. All measurements had a recording interval of 60 s duration within each of slot and pool, where the first 10 s were removed due to the signal response of the LLP being submerged.

2.3   Data Preparation

In this study, a set of sensors (two triaxial accelerometers (i.e., a0 and a1) and seven pressure sensors (i.e., ps0, ps1, ps2, ps5, ps6, ps9 and ps10) were selected for flow pattern classification. Of these pressure sensors, ps0 is located at the stagnation point, while ps1-ps2, ps5-ps6, and ps9-ps10 were mounted in parallel to the plastic body (Fig. 1). All the sensor values were first normalized by subtracting the median value of each sensor, and four datasets, namely accelerometer-only data, pressure sensor-only data, accelerometer plus pressure sensor data, and pressure difference between the stagnation point (ps0) and a second point (other sensors), were prepared for modelling the flow patterns of "slot" or "pool". Each data set consists of 28 measurements each at "slot" and "pool". From each measurement, descriptive statistics of minimum, 1st and 3rd

quantile, and maximum values were calculated and used as input variables for flow pattern classification in this study. Median value was additionally included for the pressure difference data set.

2.4   Random Forests for Flow Pattern Classification

We used the randomForest package [18] of the R software [19], in which default settings were applied. The variability of the resulting models was evaluated using 50 different sets of initial conditions in model building. Performance of RF models was evaluated with four measures, namely mean squared error (MSE), Nash-Sutcliffe Efficiency (NSE [20]), correctly classified instances (CCI) and area under the receiver operating characteristics (ROC) curve (AUC [21]). MSE and NSE are defined as follows.

$$MSE = \frac{1}{n} \sum_{i=1}^{n} (FP_i - RF_i)^2 \tag{1}$$

$$NSE = 1 - \sum_{i=1}^{n} (FP_i - RF_i)^2 \left/ \sum_{i=1}^{n} (FP_i - \overline{FP})^2 \right. \tag{2}$$

where FP_i is the flow pattern of ith sampling point, RF_i is the probability of class "slot" computed by RF, $\overline{FP}$ is the mean of FP_i and n is the size of data set. NSE takes a range $[-\infty, 1]$ with 1 being a perfect fit. CCI is the ratio of correctly classified cases to the entire cases. AUC was computed using the package pROC [22]. The best fit for CCI and AUC takes a value of one.

The importance of a habitat variable was measured using the mean decrease in accuracy, which was evaluated as the normalized mean difference in classification accuracy on the out-of-bag data when permuting the values of the variable in question [12]. To illustrate important conditions for classifying a flow pattern as "slot", a response curve for each input variable was obtained from the partial dependence plot computed with the function 'partialPlot' of RF. The two kinds of information, namely variable importance and response curves, were used as tools for knowledge extraction and used for comparing characteristics of the four models obtained in this study.

3   Results and Discussion

All four RF models achieved ideal classification of the recorded flow patterns (Table 1). The model performance was almost the same irrespective of the differences in the number of input variables. Variability in model structure originating from the 50 different sets of initial conditions was small as shown in the standard deviation of performance measures. This high performance is one of the benefits of RF application in natural science. Our preliminary analysis on varying numbers of sensors did not show a decrease in model performance, which suggests appropriateness of the use of descriptive statistics for classifying flow patterns of "slot" and "pool" in this study.

Table 1. Model performance of the four models, namely accelerometer (AC), pressure sensor (PS), Accelerometer plus pressure sensor (AC+PS) and pressure differences (ΔPS), based on the four performance measures of mean squared error (MSE), Nash-Sutcliffe Efficiency (NSE), area under the ROC curve (AUC), and correctly classified instances (CCI). Mean ± standard deviation are presented for each performance measure.

Model	MSE	NSE	AUC	CCI
AC	0.0098 ± 0.0005	0.9608 ± 0.0020	1 ± 0	1 ± 0
PS	0.0098 ± 0.0005	0.9610 ± 0.0019	1 ± 0	1 ± 0
AC + PS	0.0083 ± 0.0004	0.9667 ± 0.0015	1 ± 0	1 ± 0
ΔPS	0.0094 ± 0.0005	0.9625 ± 0.0021	1 ± 0	1 ± 0

Variable importance (Figs. 3-4) indicates that 1st and 3rd quantiles are important in most cases irrespective of the four models developed in this study. Regarding the accelerometer model, quantile values of x-axis (a0 and a1) and y- and z-axes (a1) were found to be important, whereas minimum and maximum values were not (Fig. 3a). Sensors on the x-axis were the most important, followed by y- and z-axes. This may

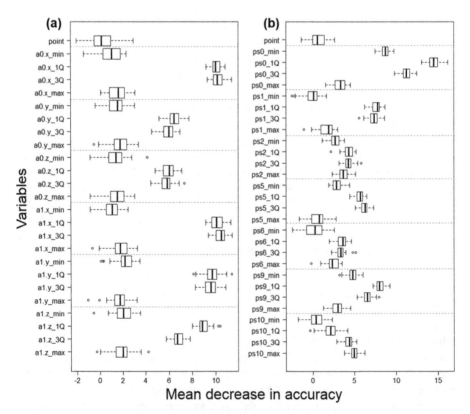

Fig. 3. Variable importance computed by random forests: (a) accelerometer and (b) pressure sensor

imply that the flow patterns of "slot" and "pool" in this study were characterized largely by horizontal flow dynamics, which is in line with the findings by [23]. In the pressure sensor model, the pressure sensor at the stagnation point (ps0) was the most important, followed by 1st and 3rd quantiles of other sensors (Fig. 3b). The importance of 1st and 3rd quantiles implies that the temporal variation of the sensor values characterized flow patterns (i.e., "slot" and "pool") observed in this study. The use of quantile values for flow pattern classification can improve practicality in terms of time needed for sensing and computation. When accelerometer and pressure sensors were both considered as input data sources in the analysis, it was found that the 1st and 3rd quantiles of the pressure sensor at the stagnation point (nose sensor only) were the most important variables, whereas the accelerometers alone were more important when compared to the lateral pressure sensors (ps1–ps10) (Fig. 4a). The findings were almost the same with the accelerometer model and the pressure sensor models. This can be partially because of high predictive capability of RF as reported in previous papers [24, 25]. It was found from Fig. 4(b) that median, minimum and maximum values were not important for flow pattern classification based on pressure differences. This result suggests that 1st and 3rd quantiles are the major descriptors of flow patterns observed

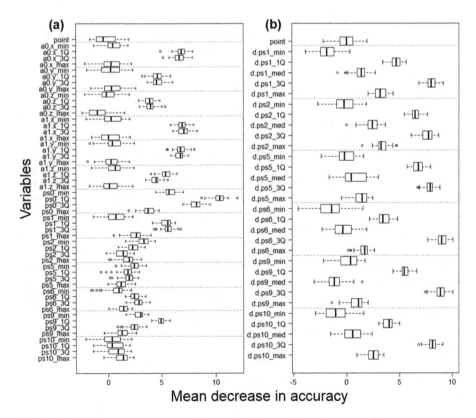

Fig. 4. Variable importance computed by random forests: (a) accelerometer plus pressure sensor and (b) pressure difference

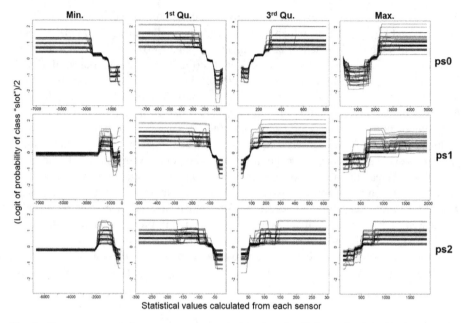

Fig. 5. Response curves for pressure sensors (ps0, ps1 and ps2) for each of descriptive statistic variables: Minimum, 1st and 3rd quantile, and maximum values.

using the LLP in this study. Future applications of the LLP may focus on the use of 1st and 3rd quantile values for flow pattern classification. However, examining the LLPs for more complex and dynamic flow may be needed to test their applicability and practicality for innatura flows.

Response curves indicate specific conditions that are important for classifying the flow pattern "slot" in this study. Figure 5 shows an example of response curves of ps0, ps1 and ps2 from the pressure sensor model. One of the clear trends observed in the response curves of all the models is the high probability in lower range of 1st quantile and higher range of 3rd quantile of the sensor values. That is, these ranges are the key features of the flow pattern "slot". As partially presented in the variable importance (Figs. 3-4), response curves for the position of basins were inconsistent across the models (result not shown). While it is expected that flow patterns in a fishway can vary according to its position (e.g., upstream or downstream end), the position did not matter when classifying flow patterns. This is a benefit of the use of LLP for flow pattern classification, in which descriptive statistics from high frequency measurement within 60 s and predictive classification method such as RF are the only requirement for accurate flow pattern classification. The measurement time can further be reduced as partially shown by [26], which will further strengthen the applicability of LLP in practice.

4 Conclusions

This paper demonstrated how a lateral line probe sensing with predictive machine learning method can be integrated for flow pattern classification in a fishway. Descriptive statistics such as 1st and 3rd quantiles from high frequency measurement of acceleration and pressure were successfully used for modelling complex innatura flow patterns at different sampling positions in the fishway (i.e., 28 sets of "slot" and "pool" from upstream to downstream). Variable importance and response curves computed in random forests could provide useful information to better understand flow patterns observed. We show using results from a first field study that the LLP in conjunction with random forests can be used as a tool for robust hydrodynamic classification in a turbulent flow system. Although preliminary, these results show that the type and volume of information recorded using this new measurement technology in conjunction with well-chosen machine learning algorithms can provide robust, highly accurate classifications of complex and turbulent flows experienced by fish innatura. Further studies considering the measurement time and more complex innatura flow patterns are needed for an improved assessment of flow dynamics, contributing to improving the necessary field monitoring studies required by sustainable use and management of water resources for human and ecosystems.

Acknowledgement. This work was supported in part by the Grant-in-aid for Young Scientists A (25712026) and Grant-in-aid for Challenging Exploratory Research (26660190) of the Ministry of Education, Culture, Sports, Science and Technology (MEXT), Japan. The research leading to these results has received funding from BONUS, the joint Baltic research and development programme, co-financed by the European Union's Seventh Framework Programme (2007–2013) under the BONUS Implementation Agreement. National funding for this work has been provided by the German Federal Ministry for Education and Research (BMBF FKZ:03F0687A) and the Estonian Environmental Investment Centre (KIK P.7254 C.3255). The work has also been partly financed by the EU FP7 project ROBOCADEMY (No. 608096) and the TIWAG research project "Robofish", Austria.

References

1. Yorke, T.H., Oberg, K.A.: Measuring river velocity and discharge with acoustic Doppler profilers. Flow Meas. Instrum. **13**, 191–195 (2002)
2. Kalmijn, A.J.: Hydrodynamic and acoustic field detection. In: Atema, J., Fay, R.R., Popper, A.N., Tavolga, W.N. (eds.) Sensory Biology of Aquatic Animals, pp. 83–130. Springer, New York (1988)
3. Tuhtan, J.A., Fuentes-Pérez, J.F., Strokina, N., Toming, G., Musall, M., Noack, M., Kämäräinen, J.K., Kruusmaa, M.: Design and application of a fish-shaped lateral line probe for flow measurement. Rev. Sci. Instrum. **87**(4), 45110 (2016)
4. Dijkgraaf, S.: The functioning and significance of the lateral-line organs. Biol. Rev. **38**(1), 51–105 (1963)
5. Nestler, J. M., Pickens, J. L., Evans, J., Haskins, R. W.: Multiple sensor fish surrogate for acoustic and hydraulic data collection. US5517465 A (1996)

6. Deng, Z.D., Lu, J., Myjak, M.J., Martinez, J.J., Tian, C., Morris, S.J., Carlson, T.J., Zhou, D., Hou, H.: Design and implementation of a new autonomous sensor fish to support advanced hydropower development. Rev. Sci. Instrum. **85**(11), 115001 (2014)

7. Yang, Y., Chen, J., Engel, J., Pandya, S., Chen, N., Tucker, C., Coombs, S., Jones, D.L., Liu, C.: Distant touch hydrodynamic imaging with an artificial lateral line. Proc. Natl. Acad. Sci. **103**(50), 18891–18895 (2006)

8. Klein, A., Bleckmann, H.: Determination of object position, vortex shed-ding frequency and flow velocity using artificial lateral line canals. Beilstein J. Nanotechnol. **2**, 276–283 (2011)

9. Wang, W., Xie, G.: Online high-precision probabilistic localization of robotic fish using visual and inertial cues. IEEE Trans. Ind. Electron. **62**(2), 1113–1124 (2015)

10. Shizhe, T.: Underwater artificial lateral line flow sensors. Microsyst. Technol. **20**, 2123–2136 (2014)

11. Breiman, L.: Random forests. Mach. Learn. **45**, 5–32 (2001)

12. Cutler, R.D., Edwards, T.C., Beard, K.H., Cutler, K.T., Gibson, H.J., Lawler, J.J.: Random forests for classification in ecology. Ecology **88**, 2783–2792 (2007)

13. Fukuda, S., De Baets, B., Waegeman, W., Verwaeren, J., Mouton, A.M.: Habitat prediction and knowledge extraction for spawning European grayling (*Thymallus thymallus* L.) using a broad range of species distribution models. Environ. Model Softw. **47**, 1–6 (2013)

14. Strokina, N., Kämäräinen, J.K., Tuhtan, J.A., Fuentes-Pérez, J.F., Kruusmaa, M.: Joint estimation of bulk flow velocity and angle using a lateral line probe. IEEE Trans. Instrum. Meas. **65**(3), 601–613 (2016)

15. Fuentes-Pérez, J.F., Tuhtan, J.A., Carbonell-Baeza, R., Musall, M., Toming, G., Muhammad, N., Kruusmaa, M.: Current velocity estimation using a lateral line probe. Ecol. Eng. **85**, 296–300 (2015)

16. Tuhtan, J.A., Strokina, N., Fuentes-Pérez, J.F., Muhammad, N., Musall, M., Noack, M., Toming, G., Kämäräinen, J.-K., Kruusmaa, M., Schletterer, M.: Ecohydraulic flow sensing and classification using a lateral line probe. In: Proceedings of 11th International Symposium on Ecohydraulics, Melbourne, Australia (2016)

17. Bundesministerium für Land- und Forstwirt-schaft, Umwelt und Wasserwirtschaft (Hrsg.): Leitfaden zum Bau von Fischaufstiegshilfen. Wien (2012)

18. Liaw, A., Wiener, M.: Classification and regression by random forest. R News **2**(3), 18–22 (2002)

19. R Core Team: R: A language and environment for statistical computing. R Foundation for Statistical Computing, Vienna, Austria (2015). http://www.R-project.org/

20. Nash, J.E., Sutcliffe, J.V.: River flow forecasting through conceptual models. part I: a discussion of principles. J. Hydrol. **10**, 282–290 (1970)

21. Fawcett, T.: An introduction to ROC analysis. Pattern Recogn. Lett. **27**(8), 861–874 (2006)

22. Robin, X., Turck, N., Hainard, A., Tiberti, N., Lisacek, F., Sanchez, J.C., Müller, M.: pROC: an open-source package for R and S + to analyze and compare ROC curves. BMC Bioinform. **12**, 77 (2011)

23. Cea, L., Pena, L., Puertas, J., Vázquez-Cendón, M.E., Peña, E.: Application of several depth-averaged turbulence models to simulate flow in vertical slot fishways. J. Hydraul. Eng. **133**, 160–172 (2007)

24. Fukuda, S., Spreer, W., Yasunaga, E., Yuge, K., Sardsud, V., Muller, J.: Random Forests modelling for the estimation of mango (Mangifera indica L. cv. Chok Anan) fruit yields under different irrigation regimes. Agric. Water Manage. **116**, 142–150 (2013)

25. Fukuda, S., Yasunaga, E., Nagle, M., Yuge, K., Sardsud, V., Spreer, W., Muller, J.: Modelling the relationship between peel colour and the quality of fresh mango fruit using random forests. J. Food Eng. **131**, 7–17 (2014)
26. Tuhtan, A., Strokina, N., Toming, G., Muhammad, N., Kruusmaa, M., Kämäräinen J.: Hydrodynamic classification of natural flows using an artificial lateral line and frequency domain feature. In: 36th IAHR World Congress (2015)

Assessment of Depth Measurement Using an Acoustic Doppler Current Profiler and a CTD Profiler in a Small River in Japan

Shinji Fukuda[1]([✉]), Kazuaki Hiramatsu[2], and Masayoshi Harada[2]

[1] Tokyo University of Agriculture and Technology, Tokyo, Japan
shinji-f@cc.tuat.ac.jp
[2] Kyushu University, Fukuoka, Japan
{Hiramatsu,mharada}@bpes.kyushu-u.ac.jp

Abstract. Recent advances in acoustic Doppler current profiler (ADCP) allow for measuring spatially continuous and high resolution hydraulic data such as water depth and flow velocity that are the basis for ecohydraulic analysis and modelling. In this paper, an ADCP and a conductivity-temperature-depth (CTD) profiler were used for depth measurement in Zuibaji river in Fukuoka, Japan, and accuracies of these measurements were assessed and compared based on tape-measured water depth with a GPS coordinate. As a result, water depth measured with a CTD profiler and an ADCP showed a good agreement with observation data, which supports the applicability of the ADCP and CTD for bathymetric survey. A major source of error seems to be a positioning error for obtaining GPS coordinates on site. Further study considering measurement errors, systematic errors and positioning errors, may be needed for a deeper understanding of error characteristics, leading to a better application of these innovative technologies for ecohydraulic surveys.

Keywords: Acoustic Doppler current profiler · Accuracy · Bathymetry · Ecohydraulics · CTD profiler

1 Introduction

Underwater mapping has been a key issue in field sciences such as hydraulics and ecology. In hydraulics, bathymetric data is the basis for discharge measurement and hydraulic modelling specifically in two-dimensional (2D) and three dimensional (3D) computations. In aquatic ecology, habitat mapping is one of the major topics, for which spatial maps of water depth, flow velocity and substrates are used to relate spatial distributions of a target species with physical habitat conditions. Conventionally, bathymetric data were collected based on transect surveys or point sampling across target reaches. Such surveys assume that the sampled transect or points are representative of the reaches in focus, and thus topographic accuracies depend on survey design including the density of transects and sampling points. Intensive measurement is labor intensive and time consuming, but can produce detailed and realistic maps. On the other hand, sparse measurement may result in rough and unreliable maps for an intended purpose. In order to establish a cost-efficient survey method in both time and labor,

© Springer International Publishing Switzerland 2016
N. Kubota et al. (Eds.): ICIRA 2016, Part II, LNAI 9835, pp. 308–316, 2016.
DOI: 10.1007/978-3-319-43518-3_30

new techniques using photogrammetry and acoustic sensing have been proposed. Acoustic Doppler current profiler (ADCP) is a method that is now being widely used by hydraulic and ecological engineers. ADCPs have been validated for the use in hydro-dynamics measurements (for instance, see [1] and references therein). However, accu-racies of such measurement techniques depend on local conditions during a survey and the methods used for positioning and mapping.

This study aims to investigate applicability of an ADCP and a conductivity-temperature-depth (CTD) profiler for water depth measurement in a small river in Japan. All data obtained were compiled with a GIS software, based on which perfor-mance was assessed with field measured water depth. Based on results, possible sources of errors are discussed for a better application of these measurement methods in practice.

2 Methods

2.1 Equipment

CastAway-CTD (Sontek, USA [2]) is a handheld CTD profiler which can be used to measure conductivity, temperature and pressure in a 5-Hz frequency with accuracies of 0.1 PSU and 0.05 °C. The three measured components are used to calculate salinity, sound speed, density, and water depth. It has an internal GPS component to record every measurement coordinate. The main usage of CastAway is to provide a vertical profile of temperature and conductivity in a water column, whose applicability has been supported by [3]. Regarding water depth, CastAway can measure water depth from 0 to 100 m with a resolution of 0.01 m. As briefly mentioned below, CastAway data can be used with ADCP measurement, which is one of the main features of this equipment.

In this study, an ADCP equipped with RTK GPS (M9, Sontek, USA [4]) was used. M9 has several advanced features such as multi-band beams with an intelligent algo-rithm to automatically define cell size based on water depth, flow velocity and tur-bulence. It has nine transducers dual 4-beam 3.0 MHz/1.0 MHz Janus at 25° slant angle and a 0.5 MHz vertical beam echosounder, which has capability to measure water depth from 0.2 to 80 m, in a resolution of 0.0001 m, and with an accuracy of 1 %. With the HydroSurveyor software (Sontek, USA), M9 measurement is converted into a three dimensional map of flow velocity. CastAway measurement can be inte-grated to correct sound speed based on water temperature and salinity profiles which can further improve the accuracy of M9 measurement.

2.2 Study Site and Field Measurement

A 430-m-long reach located between the Yashoi and Tsujii weirs (33°34′N, 130°14′E) in Zuibaiji river, Fukuoka, Japan (Fig. 1) was selected as our bathymetric survey site for its stable water level specifically during non-irrigation period. Zuibaiji river originates in Mt. Mihara-yama, having a catchment area of 52.6 km^2 with a total length of 13.2 km, and flowing into Imazu bay in Fukuoka. This area serves as a crop production area for surrounding cities and thus Zuibaiji river is a major source of water for irrigation.

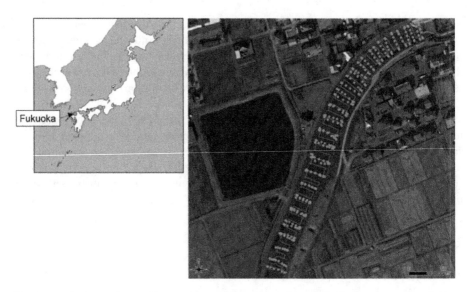

Fig. 1. Study site: yellow marks are the sampling points where water depth was measured using CastAway and a measuring tape, while red line is the boat track surveyed with M9. Water flows upward from the bottom of the figure. (Color figure online)

For this reason, a number of irrigation weirs have been built, which may potentially hamper migration of aquatic species in the area. Bathymetric survey is important to better understand the current ecohydraulic status of the river.

Surveys with CastAway were conducted on December 4 and 5, 2013. A series of cross sections having seven sampling points each were located to cover the entire reach and totalled 308 points were surveyed (44 transects; yellow marks in Fig. 1). At each measuring point, CastAway was casted into water until it reached the bottom; after recording water depth with a measuring tape, it was rolled up to water surface to record the coordinate of the position. The tape measured water depth was used as reference data in this study. All the records were summarized in a CSV format and used for further analyses.

A survey with the ADCP was conducted on March 26, 2014; water level at a fixed point was similar on these days (on average 0.878 m on December 4 and 5, 2013 and 0.885 m on March 26, 2014). M9 was mounted on the Hydroboat and towed with a rope from both sides of the river to cover the entire study reach. The movement track was recorded with an RTK-GPS module as the red line shown in Fig. 1. The M9 system was connected via a Bluetooth module and all data were compiled in the HydroSurveyor software which spatially interpolate observation data in a real-time manner. After the survey, the surveyed area was defined as a polygon based on which bathymetric data were interpolated. Then, data sets containing interpolated bathymetry and boat track were exported as a CSV format for subsequent analyses to calculate water depth distribution in the study reach.

2.3   Data Analysis

All records were analyzed using QGIS [5] for mapping water depth in the target reach. Because GPS coordinates of CastAway data were scattered around the target reach, it was aligned based on the cross-section predefined for the survey. The tape measured water depths were compared with CastAway water depths at each sampling point.

For M9, data sets of bathymetry and water surface elevation were extracted from measurement data. Water depth was calculated by subtracting bathymetric elevation from water surface elevation, and then used to create a two-dimensional water depth map. To validate M9 data, water depth at the CastAway sampling points were extracted using a buffer of approximately 1.4 m radius, mean of which was used as M9 water depths at the sampling points. Water depths obtained from CastAway and M9 were assessed using mean error (ME), mean squared error (MSE), root mean squared error (RMSE) and Nash-Sutcliffe Efficiency (NSE) [6]:

$$ME = \frac{1}{n}\sum_{i=1}^{n}\left(D_{\text{ref},i} - D_{\text{m},i}\right) \tag{1}$$

$$MSE = \frac{1}{n}\sum_{i=1}^{n}\left(D_{\text{ref},i} - D_{\text{m},i}\right)^2 \tag{2}$$

$$RMSE = \sqrt{\frac{1}{n}\sum_{i=1}^{n}\left(D_{\text{ref},i} - D_{\text{m},i_i}\right)^2} \tag{3}$$

$$NSE = 1 - \sum_{i=1}^{n}\left(D_{\text{ref},i} - D_{\text{m},i}\right)^2 \Big/ \sum_{i=1}^{n}\left(D_{\text{ref},i} - \overline{D_{\text{ref}}}\right)^2 \tag{4}$$

where $D_{\text{ref},i}$ is the reference water depth at i^{th} sampling point, $D_{\text{m},i}$ is the water depth measured with either CastAway or M9, $\overline{D_{\text{ref}}}$ is the mean of $D_{\text{ref},i}$, and n is the number of sampling points. NSE takes a range [$-\infty$, 1] with 1 being a perfect fit.

3   Results and Discussion

With M9 measurement, a spatial map of water depth was created as shown in Fig. 2c. This spatially continuous map is one of the major features of the use of M9 and HydroSurveyor, in which other flow components such as flow velocity (e.g., depth-averaged, surface, middle and bottom) can also be mapped. In Fig. 2c, we can observe that thalweg changes its side: left in most of upstream area, right in midstream area, and left in the downstream area. Spatially continuous, high resolution mapping as in this system is very difficult to achieve with the conventional point sampling methods, thereby contributing to better hydraulic modelling [7] and habitat mapping for aquatic species [8]. The applicability and reliability of the map is discussed later.

It is found from Fig. 3 that both CastAway and M9 could capture the trends of tape-measured water depth in the target reach. Color differences between a mark and

312   S. Fukuda et al.

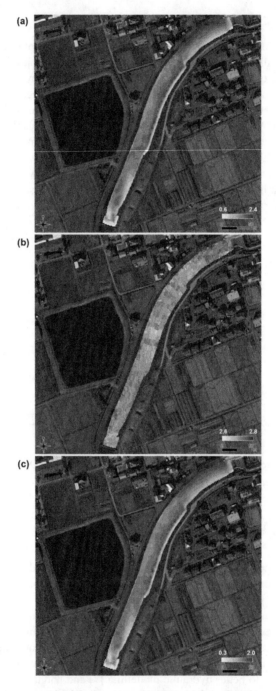

Fig. 2. ADCP measurements: (a) bathymetry, (b) water surface elevation and (c) water depth.

background indicate errors in the measurement. Most distinct color difference was found around water's edge where tape-measured depth was small but M9 measurement indicate larger water depth. This may be ascribed to positioning errors originated from the GPS of CastAway as well as the use of buffer to calculate ADCP-measured water depth. In practice, care should be taken when CastAway is used for a small river as in this study because a small positioning error can result in an unrealistic bathymetric map. Positioning error may be reduced by improved measurements using a longer time when obtaining a GPS coordinate.

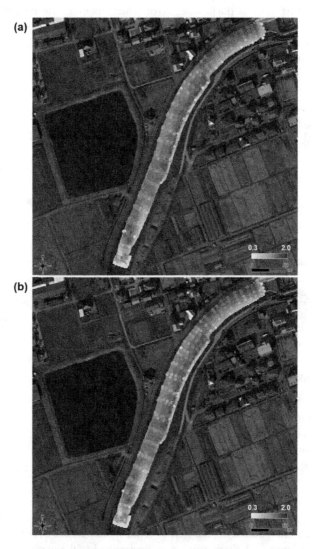

Fig. 3. Maps comparing water depth measured by ADCP with (a) filed measurement and (b) CastAway. Color scale indicate water depth from shallow (0.3 m; white) to deep (2 m; blue). (Color figure online)

Scatter plot between tape-measured water depth and CastAway-measured one showed a good agreement, but distinct errors were found at a depth of 0 m measured by CastAway (Fig. 4a). This may be due to the sensitivity of pressure sensor of CastAway right after being casted into water. That is, CastAway could not sense water depth less than 0.6 m. It may be possible to keep CastAway underwater for a longer time for better measurement. Removal of such data points greatly improved measurement accuracy of CastAway (Fig. 4b). ME decreased from 9.2 cm to 7.3 cm, while RMSE improved from 14.3 cm to 9.6 cm (Table 1). NSE values of both cases support the applicability of CastAway for depth measurement. As discussed earlier, there may be a problem originating from a positioning error when mapping the measurement result spatially. However, such a positioning error matters only when it is used in a small water body. Both positioning and depth measurement errors can be reduced by an improved measurement practices in field.

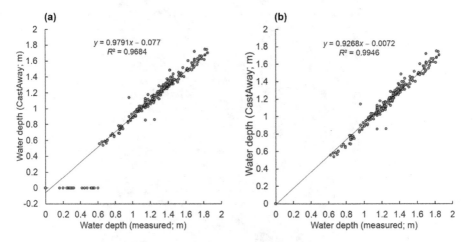

Fig. 4. Scatter plot of water depth measured by hand versus that measured by CastAway: (a) all data points and (b) data points in a CastAway-feasible range.

Table 1. Performance of CastAway-based water depth measurement

Measure	All data points	CastAway feasible range
Mean error	0.092	0.073
Mean squared error	0.021	0.009
Root mean squared error	0.143	0.096
Nash-Sutcliffe Efficiency	0.947	0.977

Compared to the CastAway case, M9 results showed larger errors (Fig. 5) and lower performance values (Table 2). Distinct errors were observed at a tape-measured depth of 0 m. This means interpolated depths based on M9 measurement overestimated the tape measured water depth of 0. Such errors may be ascribed to the method to extract water depth from M9 measurement points, in which a buffer size of approximately 1.4 m was used. Specifically, several data points can be included in a buffer which may contain both shallow and deep points. This is possible near the thalweg (Fig. 2c) where river banks were lined with concrete and form a steep slope. Removing zero-depth data points decreased measurement errors (Table 2). For instance, ME decreased from 34.2 cm to 15.5 cm, while RMSE decreased from 55.2 cm to 24.8 cm. NSE values greatly improved from 0.1 to 0.5, which also support applicability of M9 measurement. Whereas it seems that M9 measurement resulted in poorer performance, the errors are, as discussed earlier, ascribed to positioning errors and the data extraction method for validating the measurement results. Because M9 survey can be done within a few hours, the applicability is high compared to conventional methods based on transect that can take days to cover entire reach of a target site. Future study may integrate an ADCP measurement with an unmanned aerial vehicle (UAV) measurement to cover both terrestrial and aquatic regions for complete landscape mapping of a target site.

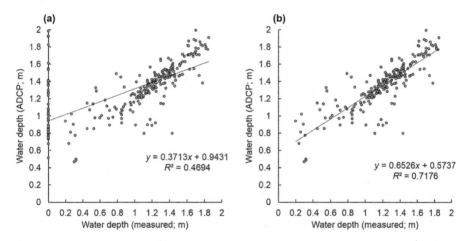

Fig. 5. Scatter plot of water depth measured by hand versus that measured by ADCP: (a) all data points and (b) zero-depth omitted data.

Table 2. Performance of ADCP-based water depth measurement.

Measure	All data points	Zero-depth omitted data
Mean error	−0.342	−0.155
Mean squared error	0.305	0.061
Root mean squared error	0.552	0.248
Nash-Sutcliffe Efficiency	0.106	0.526

4 Conclusions

This paper illustrated the applicability of an acoustic Doppler current profiler with an RTK GPS module and a CTD profiler for bathymetric survey in a small river in Japan. Accuracy of depth measurement was acceptable, whereas positioning errors were problematic when surveying a small water body. With a spatial interpolation software,

ADCP can provide spatially continuous maps of depth, velocity and other related measurement in a reasonably short time compared to conventional point- or transect-based surveys. As such, innovative methods allow for data collection that was not possible in the past, thereby contributing to advances in field sciences and computer simulations based on field measurement data. There may be more opportunities for field measurement by integrating different measurement methods with powerful data analytic tools to obtain useful information for a target system.

Acknowledgement. This work was supported in part by the Grant-in-aid for Young Scientists A (25712026) and Grant-in-Aid for Challenging Exploratory Research (26660190) of the Ministry of Education, Culture, Sports, Science and Technology (MEXT), Japan. The authors thank Mr. Nguyen Viet Anh, Mr. Taichi Tanakura, Ms. Mio Hanada, and Mr. Keisuke Tanaka for their assistance in the field work.

References

1. Oberg, K., Mueller, D.S.: Validation of streamflow measurements made with acoustic Doppler current profilers. J. Hydraul. Eng. **133**(12), 1421–1432 (2007)
2. Sontek Inc.: CastAway®-CTD. http://www.sontek.com/productsdetail.php?CastAway-CTD-11. Accessed 14 Apr 2016
3. Patterson, J., Bushnell, M., Haenggi, M.: A field evaluation of the CastAway CTD at the Deepwater Horizon Oil Spill site in the Gulf of Mexico. http://ysi.actonsoftware.com/acton/attachment/1253/f-018d/1/-/-/-/-/Field%20Evaluation%20of%20the%20CastAway%20at%20the%20Deepwater%20Horizon%20Oill%20Spill.pdf. Accessed 14 Apr 2016
4. Sontek Inc.: HydroSurveyor™ – exclusive 5-beam depth sounding device. http://www.sontek.com/productsdetail.php?HydroSurveyor-13. Accessed 14 Apr 2016
5. QGIS Development Team: QGIS Geographic Information System. Open Source Geospatial Foundation Project (2015). http://qgis.osgeo.org
6. Nash, J.E., Sutcliffe, J.V.: River flow forecasting through conceptual models. Part I: a discussion of principles. J. Hydrol. **10**, 282–290 (1970)
7. Kasvi, E., Alho, P., Lotsari, E., Wang, Y., Kukko, A., Hyyppä, H., Hyyppä, J.: Two-dimensional and three-dimensional computational models in hydrodynamic and morphodynamic reconstructions of a river bend: sensitivity and functionality. Hydrol. Process. **29**, 1604–1629 (2015)
8. Vezza, P., Parasiewicz, P., Spairani, M., Comoglio, C.: Habitat modeling in high-gradient streams: the mesoscale approach and application. Ecol. Applic. **24**, 844–861 (2014)

Potential for Automated Systems to Monitor Drying of Agricultural Products Using Optical Scattering

Marcus Nagle[1(✉)], Giuseppe Romano[1,2], Patchimaporn Udomkun[1,3],
Dimitrios Argyropoulos[1], and Joachim Müller[1]

[1] Institute of Agricultural Engineering (440e), Tropics and Subtropics Group,
Universität Hohenheim, 70599 Stuttgart, Germany
{naglem,dimargy,jmueller}@uni-hohenheim.de
[2] Research Centre for Agricultural and Forestry, 39040 Ora/Auer, Italy
Giuseppe.Romano@provincia.bz.it
[3] International Institute of Tropical Agriculture, Bukavu, Democratic Republic of Congo
P.Udomkun@cgiar.org

Abstract. Drying of agricultural products is a critical and energy-intensive processing step in the production of many foodstuffs. During convective drying, products are highly susceptible to thermal damage. In recent years, novel techniques have been established based on optical scattering due to the interaction of light with organic materials. The presented research investigated this approach using vis/NIR wavelengths to observe changes of quality parameters during drying of foodstuffs. The method was proven useful to monitor changes in moisture, color, and texture in a variety of products such as apple, mango, papaya, litchi, and bell pepper. Although many applications have been confirmed, additional hardware and software aspects still need to be refined. Optical scattering shows strong potential for implementation as a non-destructive method for in-line control of product qualities during industrial drying processes. A robotic prototype should be developed that is capable of automated measurement of agricultural products during drying. Optimization of product quality and prevention of energy waste by over-drying are among the potential impacts of the developed technology.

Keywords: Laser diode · Non-destructive evaluation · Image analysis · Processing

1 Background

Convective hot air drying of high-value agricultural products is a typical process applied to reduce a typically high initial moisture content to a final state which is suitable to guarantee optimal conservation parameters and to satisfy consumer acceptance [1]. According to the drying conditions, excessive temperature or prolonged processing time can lead to reduced product quality via degradation of chemical compounds and pigment formation (browning) in perishable products [2–4] as well as physical changes in texture [5]. Essential oil content of aromatic plants is a critical quality parameter for the pharmaceutical and food industries because of the desired medicinal and gastronomic

© Springer International Publishing Switzerland 2016
N. Kubota et al. (Eds.): ICIRA 2016, Part II, LNAI 9835, pp. 317–327, 2016.
DOI: 10.1007/978-3-319-43518-3_31

properties. In order to ensure good shelf life and to preserve the heat-sensitive compounds in dried products, deterioration of quality parameters and losses of nutrients and essential oils should be minimized by optimizing drying processes [6]. Much research has already been done to document drying behavior and to optimize drying procedures of fruits [7] and medicinal plants [8]. Moreover, effects of different pretreatments to preserve quality of agricultural products during drying is also well established [5]. Meanwhile, control of influential food qualities is of utmost importance for the industrial production of foodstuffs. Routine moisture content analysis, electromechanical testers to assess product texture and quantification techniques for pigment and essential oils such us chromatography and distillation do not allow continuous regulation during drying because they are damaging, expensive and cause adulteration.

Over the last two decades, near-infrared spectroscopy (NIRS) has been applied as a non-invasive method for extensive measurement of foodstuff composition [9, 10]. Specifically related to drying, NIRS has been used to predict water content of various products such as mushrooms [11], rice kernels [12] semolina pasta [13], and blueberries [14]. Despite its high speed and precision and, with more recent instruments, the elimination of sample preparation requirements, unfavorable factors of NIRS technology remain, most of all high investments costs and calibration constraints. More recently, spectroscopy has been fused with imaging techniques to develop hyperspectral cameras which can be used for spatial investigation of quality parameters. This technology has been tested to detect major plant pigments, for example in anthocyanin contents of apples and carotenoid contents of apples, pears and kiwifruits [15]. So far, hyperspectral images are mainly applied for grading and detection of surface defects (such as bruising) in agricultural products [16]. A few studies are available about the use of hyperspectral technology to continuously monitor quality changes of horticultural products during various drying processes.

As an alternative approach to NIRS, another optical method has been developed based on the interaction of light with organic materials. Electromagnetic energy in vis/NIR spectrum delivered to medium will interact in certain manners, such as reflection, transmittance, or absorbance. This is based on resonance frequencies of the molecules and their respective bonds. The behavior of the light with the material can then be characteristic of the properties of the material. In actuality, only about 5 % of the light is directly reflected in fruits and vegetables and the rest is either absorbed, transmitted, or propagated back through the interface [17]. The composite structure of many plant tissues causes high optical density which restricts light penetration. In many cases, a majority of the light enters the material and only travels a small distance and disseminates close to the incident point. In cases where the material is translucent or slightly opaque, light propagates deeper (up to several millimeters) into the tissues and is more affected by interactions with molecular constituents before exiting. The latter is common with many biological tissues and is a phenomenon which is termed 'diffuse reflection' or 'scattering'. It is the combination of the direct reflectance together with this scattered light, measured by particular optical setups, which can be used to extract useful physicochemical information.

Laser diodes in the vis/NIR range have been tested together with digital cameras to estimate various quality properties of fruits and vegetables based on photon migration, absorption of electromagnetic radiation and optical scattering [18–20]. Here, the scattering profile is determined by absorption and reflectance reactions as influenced by composite physicochemical properties of the crop. For example, direct reflection may give information about surface properties such as porosity and texture, which absorption could be affected by chemical concentrations. Previous studies have shown that absorption of the light at 670 nm wavelength on tomato tissue was considered to be higher due to chlorophyll and lycopene content [21] resulting in less scattering and diffuse reflectance in the optical profile [22].

Optical scattering systems present several benefits with regard to other non-destructive techniques are high precision, possibility for use under practical conditions, smaller size, increased portability and lower cost. Laser diodes offer a highly proficient light transport system that is stable and strong enough to interact with biological materials. In contrast, near infrared radiation has limited penetration and consequently minimal quantitative information from the sample properties. Furthermore, NIRS does not make use of quantitative information on optical scattering in materials [23, 24]. In this case, the scattering effect that contains useful information is usually treated as interference in conventional NIR techniques, and is therefore eliminated or corrected in analysis [15, 25]. Furthermore, in comparison with hyperspectral imaging, optical scattering does not require complex spectroscopic analyses or chemometric approaches [26, 27].

With respect to drying, scattering is influenced by physicochemical properties of the tissue, as well as cell structure [25]. Softer fruit tissue tends to have a larger scattering profile than firmer tissue [26, 27]. During drying, biological materials become denser because of reduced volume and porosity as result of dehydration [28–30]. Shrinkage associated with drying results in a higher density of cells and their components [31]. These structural changes greatly impact the interaction of light with the material as well as an influence on the measurement of scattering by established methods. Furthermore, the behavior of light is affected by surface properties of the samples. If the surface of incidence becomes rougher, the light will be scattered in particular directions causing distortion of the profiles.

2   Approach

Initial investigations were conducted in laboratory under controlled and stationary settings (Fig. 1). The optical scattering system was composed of a high-resolution CCD camera (PAX Cam P1-CMO, Villa Park IL, USA) outfitted with a macro lens (F2.5 and focal lengths of 18–108 mm, TV Zoom 7000, Navitar, Japan) with the focal length set at 22 cm. The incident angles of the laser diodes were positioned at 30°. Camera settings were controlled by Pax-It software (MIS, Villa Park IL, USA), which was also used for image capture. Measurements were taken in a closed system to mitigate environmental light. Settings such as exposure, contrast and saturation were set to optimize the visualization of scattering while avoiding noise thresholds. For subsequent work, the optical system was mounted to a test dryer for in-line measurement made on samples during experiments without interrupting the drying process (Fig. 2). Drying materials were placed on a microbalance to monitor mass reduction as related to water removal. The dryer was described in a previous study [32].

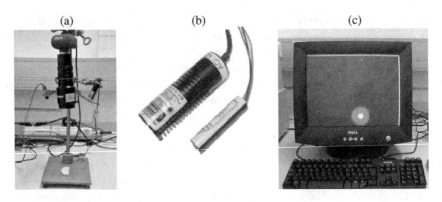

Fig. 1. Images showing components of the optical system: (a) camera mounted with lasers, (b) laser diode modules, and (c) PC with frame-grabbing software.

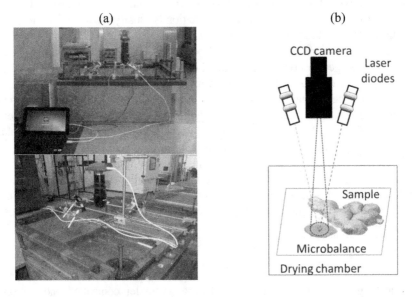

Fig. 2. Installation of the optical system on the test dryer for in-line measurement; (a) images of the setup and (b) schematic diagram.

Example images taken during sample processing are presented in Fig. 3a, where the effect of drying on the scattering image is evident. Different approaches have been taken with regard to image analysis. Total area, scattering area, light intensity, and Lorentzian distribution function were used. Scattering area was calculated determining the pixel number within a routinely defined illumination area. Light intensity (I_L) was calculated using (Eq. 1):

$$I_L = 0.3red + 0.6green + 0.1blue \qquad (1)$$

(a)

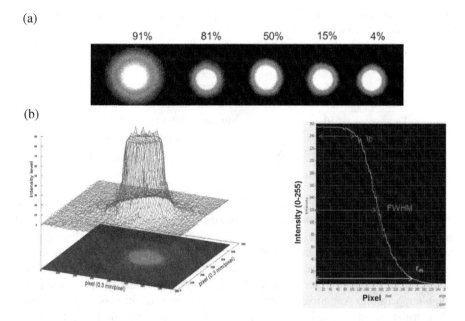

(b)

Fig. 3. Scattering analysis: (a) examples of images at different moisture content and (b) visualization and defining features [Ip = incident point, FWHM = full width at half maximum, rm = maximum radius] extracted for calibrations. (Color figure online)

where *red*, *green* and *blue* are defined by mean RGB denominations of selected pixels. I_L values ranged from 0 (black) to 255 (white). Calculation of Lorentzian distribution (*LD*) was considered to represent the mean intensity for each optimized circular scattering band, as given in Eq. 2:

$$LD = \frac{b}{1 + \left(\frac{z}{c}\right)} \qquad (2)$$

where b is the maximum light intensity at the incident light point at $z = 0$, z estimates the maximum radius of the scattering profile, and c is the full width at half maximum height value (FWMH). The image processing approach to analyze the scattering profile was previously described [33]. Examples of analyzed parameters are shown in Fig. 3b.

3   Results

The research is well-established regarding the optical scattering technique for prediction of product quality parameters during drying processes. Specific results are highlighted in Table 1.

Table 1. Overview non-destructive measurement of product quality during drying by application of optical scattering technique

Product	Laser diode (nm)	Measurable parameter	Reference
Banana	670	Moisture	[34]
Apple	635	Moisture, SSC	[35]
Bell pepper	532, 635	Moisture, color	[36]
Papaya	650	Moisture, shrinkage	[37]
Mango	473, 532, 785	Moisture, color	[38]
Litchi	473, 532, 785	Moisture, hardness	[38]

SSC = soluble solids content

Scattering images of banana slices subjected to different air drying temperatures showed scattering of measurable photons from the fruit surface at 670 nm was reduced when samples lost moisture. However, light distribution was also influenced by changing absorption of the banana flesh caused by browning at higher drying temperatures. In order to reduce surface browning and to analyze the influence of discoloration on the scattering images, experiments were repeated after pre-treating the banana slices at high temperature. An untreated sample for each pre-treatment was used as control. High correlation was found between relative laser area and moisture content by averaging the values at each interval during drying for the different treatments of banana. ANOVA demonstrated that scattering parameters were only significantly affected by moisture content ($p < 0.001$), whereas pretreatment did not have significant influence.

Additional experiments were carried to evaluate optical scattering analysis for synchronized observation of physicochemical properties during apple drying such as moisture content, soluble solids content (SSC) and hardness. Linear regression revealed a strong relationship between luminescence and moisture content ($R^2 = 0.89$), while a weaker association was found for SSC ($R^2 = 0.74$). Nonlinear models were tested to evaluate regressions between scattering area and both moisture content and SSC, with fits of $R^2 = 0.84$ and 0.80, respectively. Decreases in moisture and increases in SSC were found to cause a reduction in photon scattering. Convincing results were not found for prediction of hardness.

To evaluate the effect of product color on scattering measurements, green and red lasers were applied to red, yellow and green bell pepper slices [36]. Results showed that 37 % of the variance in scattering was caused by surface color. However, combined variance increased to 63 % when both wavelengths were considered. Validation results indicated that moisture content of yellow samples could be predicted with both red (SEP = 8.90) and green (SEP = 7.30) light. Similar results were observed for green samples at the green wavelength and red samples at the red wavelength, with SEPs of 9.97 and 8.80 respectively. Greater SEP (> 22) indicated that laser wavelengths were not accurate in predicting moisture content for materials with contrasting colors. Although results for yellow peppers were convincing, the performance of green and red light in samples with complimentary colors raises concerns about general applicability. Presumably, product color should be considered for applying suitable light wavelengths. For bell pepper slices, regression results were found to be affected by measurements on the outer (cuticle) and inner (flesh) surfaces. More detailed study should be conducted to determine whether tissue morphology affects scattering during drying.

Another study aimed to assess the feasibility of using optical scattering for predicting shrinkage of papaya cubes during drying [37]. The illuminated area (A_I) and light intensity (I_L) were used to monitor the photon migration profiles into the fruit tissue. The relationship between moisture content and shrinkage in terms of volume and area reduction during drying was satisfactorily explained by a quadratic model. Scattering parameters A_I and I_L were found to be acceptable for describing shrinkage, especially I_L values. Multivariate correlations of measured digital image parameters and scattering properties showed a superior fit for prediction of papaya shrinkage.

It was reported by Romano et al. (2011) that scattering measurements at 635 nm were not suitable for the assessment of product hardness during drying [35]. Consequently, the practicality of NIR lasers to predict textural changes during drying was examined in litchi and mango [38]. Linear models showed that laser at 473 nm wavelength was the most suitable to monitor browning ($R^2 = 0.81$) and moisture content ($R^2 = 0.81$) of litchi. For mango NIR, scattering at 785 nm exhibited good regression results with product hardness ($R^2 = 0.70$; $p < 0.001$), while moisture content had the highest influence on regression at 532 ($p < 0.001$) and 785 nm wavelengths ($p < 0.05$).

4 Concepts for Design and Automation of In-line Technology

It is envisioned that a robotic prototype could be developed that is capable of automated measurement of agricultural products during drying using optical scattering techniques. This system would have both mechanical automation coupled with artificial intelligence and would be able to react and compensate not only for variability of different products, but also for measurement inconsistency, both spatially and temporally. Although many applications have been confirmed in laboratory, additional aspects still need to be refined, mainly with respect to hardware and software issues.

4.1 Hardware Issues

One technical aspect is that the distance of the camera to the product is imperative and must be maintained. This becomes challenging when the product height changes during drying or spatially varies within the dryer. A self-adjusting system should be designed where the camera is mounted on a track which can regulate vertical distance to the product. An additional time-of-flight camera could be used to determine the distance to the product to establish the required compensation. The system would then be able to self-regulate the camera distance. The camera should also be adjusted to perform in environments with different illumination. To minimize light interference, exposure and contrast of the settings could be automatically compensated to optimize the image acquisition and to reduce the saturated pixel due to the direct reflected laser light on the product surface. Another issue is the circularity of the scattering area, which is the reason a low incident angle of the laser diodes in the laboratory setup is critical. In cases where the circularity deviates, image analysis becomes difficult. Thus, a system to compensate the positioning of the diodes in order to maintain circularity would make industrial measurement more feasible, especially on products where surface properties vary. As well, fiber optics should be explored as a system for the delivery of light to the material and transport of the scattering signal back to the camera. This would mainly counteract the need to place sensitive equipment like the camera sensor and optics in a harsh environment such as a dryer.

4.2   Software and Data Analysis Issues

Development of an integrated software program for adjusting the camera will allow for proper visualization of the scattering image. As well, this program should be integrated with the existing image analysis software to calculate parameters from the scattering profile and apply the previously developed calibration models. This would allow one-step monitoring of product qualities during drying into a user-friendly interface, possibly using LabView software which is intended for embedded control and monitoring systems. For development of a functional prototype for automated monitoring of drying processes, a robust classification system is required. This should be done via a multi-modal integration approach. Filter algorithms must be developed to clean sensor signals from noise, and then sensor data can be aligned. Data mining and data fusion can be applied for finding cause and relationship of sensor and reference measurements. Finally transducer markup language could be formulated combining data from multiple lasers into a single signal in a general format so that data handling can be realized between distinct measurements.

A number of machine learning methods have shown potential for the classification of food products by computer vision, with varying degrees of success [39]. Such approaches include Bayesian or artificial neural networks [40, 41], discriminant analysis [42, 43], nearest neighbor [44], random forests [45], and support vector machine [46]. With the abundance of classification techniques, best methods cannot simply be defined, but requires investigation. For analysis of optical scattering images, pattern recognition techniques [47] in combination with classification algorithms may prove to be optimal for further application.

5   Conclusion and Outlook

In previous studies, the working group has investigated optical scattering in the vis/NIR spectrum as a valid and novel technique to predict physicochemical changes during drying of horticultural products. Although the work is still in pilot phase and additional aspects need to be clarified, this innovative method can provide the foundation for forthcoming design of a robust in-line control of quality during industrial drying processes. Quality parameters of biological materials are affected by the applied drying conditions and are also product-specific. During convective drying, the prevailing industrial practice, materials and equipment are susceptible to a high level of thermal damage. Therefore, a robust automated system is required for real-time inline application of optical scattering to monitor agricultural products during drying.

A prototype system should be developed which is able to endue harsh conditions found in industrial drying and which can adapt to different situations and product changes during the process. Optimization of final products and prevention of over-drying (energy efficiency) are among the potential impacts of the proposed technology. The in-line optical sensor would have the following advantages: (i) rapid, real-time product information via non-destructive measurements during drying, (ii) reduction of laborious and destructive analytical methods, together with respective costs, (iii) continuous monitoring of drying to avoid over-drying and quality losses, and (iv) possibility for integration into existing drying equipment for process control and automation.

References

1. McMinn, W.A.M., Magee, T.R.A.: Principles, methods and applications of the convective drying of foodstuffs. Food Bioprod. Process. **77**, 175–193 (1999)
2. Kim, S., Park, J., Hwang, I.: Composition of main carotenoids in Korean red pepper (Capsicum annuum, L) and changes of pigment stability during the drying and storage process. J. Food Sci. **69**, FCT39–FCT44 (2004)
3. Mahayothee, B., Udomkun, P., Nagle, M., Haewsungcharoen, M., Janjai, S., Müller, J.: Effects of pretreatments on colour alterations of litchi during drying and storage. Eur. Food Res. Technol. **229**, 329–337 (2009)
4. Vega-Gálvez, A., Di Scala, K., Rodríguez, K., Lemus-Mondaca, R., Miranda, M., López, J., Perez-Won, M.: Effect of air-drying temperature on physico-chemical properties, antioxidant capacity, colour and total phenolic content of red pepper (Capsicum annuum, L. var. Hungarian). Food Chem. **117**, 647–653 (2009)
5. Lewicki, P.P.: Effect of pre-drying treatment, drying and rehydration on plant tissue properties: a review. Int. J. Food Prop. **1**, 1–22 (1998)
6. Arabhosseini, A., Huisman, W., Van Boxtel, A., Müller, J.: Modeling of thin layer drying of tarragon (Artemisia dracunculus L.). Ind. Crops Prod. **29**, 53–59 (2009)
7. Fernandes, F.A., Rodrigues, S., Law, C.L., Mujumdar, A.S.: Drying of exotic tropical fruits: a comprehensive review. Food Bioprocess Technol. **4**, 163–185 (2011)
8. Müller, J.: Convective drying of medicinal, aromatic and spice plants: a review. Stewart Postharvest Rev. **3**, 1–6 (2007)
9. Connolly, C.: NIR spectroscopy for foodstuff monitoring. Sens. Rev. **25**, 192–194 (2005)
10. Nicolai, B.M., Beullens, K., Bobelyn, E., Peirs, A., Saeys, W., Theron, K.I., Lammertyn, J.: Nondestructive measurement of fruit and vegetable quality by means of NIR spectroscopy: a review. Postharvest Biol. Technol. **46**, 99–118 (2007)
11. Roy, S., Anantheswaran, R.C., Shenk, J.S., Westerhaus, M.O., Beelman, R.B.: Determination of moisture content of mushrooms by Vis—NIR spectroscopy. J. Sci. Food Agric. **63**, 355–360 (1993)
12. Kawamura, S., Natsuga, M., Takekura, K., Itoh, K.: Development of an automatic rice-quality inspection system. Comput. Electron. Agric. **40**, 115–126 (2003)
13. De Temmerman, J., Saeys, W., Nicolaï, B., Ramon, H.: Near infrared reflectance spectroscopy as a tool for the in-line determination of the moisture concentration in extruded semolina pasta. Biosyst. Eng. **97**, 313–321 (2007)
14. Sinelli, N., Casiraghi, E., Barzaghi, S., Brambilla, A., Giovanelli, G.: Near infrared (NIR) spectroscopy as a tool for monitoring blueberry osmo–air dehydration process. Food Res. Int. **44**, 1427–1433 (2011)

15. Qin, J., Lu, R.: Measurement of the optical properties of fruits and vegetables using spatially resolved hyperspectral diffuse reflectance imaging technique. Postharvest Biol. Technol. **49**, 355–365 (2008)
16. Lorente, D., Aleixos, N., Gómez-Sanchis, J., Cubero, S., García-Navarrete, O.L., Blasco, J.: Recent advances and applications of hyperspectral imaging for fruit and vegetable quality assessment. Food Bioprocess Technol. **5**, 1121–1142 (2012)
17. Birth, G.S.: The light scattering properties of foods. J. Food Sci. **43**, 916–925 (1978)
18. Adebayo, S.E., Hashim, N., Abdan, K., Hanafi, M.: Application and potential of backscattering imaging techniques in agricultural and food processing–a review. J. Food Eng. **169**, 155–164 (2016)
19. Qing, Z., Ji, B., Zude, M.: Predicting soluble solid content and firmness in apple fruit by means of laser light backscattering image analysis. J. Food Eng. **82**, 58–67 (2007)
20. Qing, Z., Ji, B., Zude, M.: Non-destructive analyses of apple quality parameters by means of laser-induced light backscattering imaging. Postharvest Biol. Technol. **48**, 215–222 (2008)
21. Tu, K., Jancsók, P., Nicolaï, B., De Baerdemaeker, J.: Use of laser-scattering imaging to study tomato-fruit quality in relation to acoustic and compression measurements. Int. J. Food Sci. Technol. **35**, 503–510 (2000)
22. De Belie, N., Tu, K., Jancsok, P., De Baerdemaeker, J.: Preliminary study on the influence of turgor pressure on body reflectance of red laser light as a ripeness indicator for apples. Postharvest Biol. Technol. **16**, 279–284 (1999)
23. Lu, R.: Multispectral imaging for predicting firmness and soluble solids content of apple fruit. Postharvest Biol. Technol. **31**, 147–157 (2004)
24. Lu, R., Ariana, D.: A near-infrared sensing technique for measuring internal quality of apple fruit. Appl. Eng. Agric. **18**, 585 (2002)
25. Qin, J., Lu, R.: Monte Carlo simulation for quantification of light transport features in apples. Comput. Electron. Agric. **68**, 44–51 (2009)
26. Peng, Y., Lu, R.: Prediction of apple fruit firmness and soluble solids content using characteristics of multispectral scattering images. J. Food Eng. **82**, 142–152 (2007)
27. Peng, Y., Lu, R.: Analysis of spatially resolved hyperspectral scattering images for assessing apple fruit firmness and soluble solids content. Postharvest Biol. Technol. **48**, 52–62 (2008)
28. Janjai, S., Mahayothee, B., Lamlert, N., Bala, B.K., Precoppe, M.F., Nagle, M., Müller, J.: Diffusivity, shrinkage and simulated drying of litchi fruit (Litchi chinensis Sonn.). J. Food Eng. **96**, 214–221 (2010)
29. Talla, A., Puiggali, J.-R., Jomaa, W., Jannot, Y.: Shrinkage and density evolution during drying of tropical fruits: application to banana. J. Food Eng. **64**, 103–109 (2004)
30. Zogzas, N., Maroulis, Z., Marinos-Kouris, D.: Densities, shrinkage and porosity of some vegetables during air drying. Drying Technol. **12**, 1653–1666 (1994)
31. Torricelli, A., Spinelli, L., Contini, D., Vanoli, M., Rizzolo, A., Zerbini, P.E.: Time-resolved reflectance spectroscopy for non-destructive assessment of food quality. Sens. Instrum. Food Qual. Saf. **2**, 82–89 (2008)
32. Argyropoulos, D., Heindl, A., Müller, J.: Assessment of convection, hot-air combined with microwave-vacuum and freeze-drying methods for mushrooms with regard to product quality. Int. J. Food Sci. Technol. **46**, 333–342 (2011)
33. Baranyai, L., Zude, M.: Analysis of laser light propagation in kiwifruit using backscattering imaging and Monte Carlo simulation. Comput. Electron. Agric. **69**, 33–39 (2009)
34. Romano, G., Argyropoulos, D., Gottschalk, K., Cerruto, E., Müller, J.: Influence of colour changes and moisture content during banana drying on laser backscattering. Int. J. Agric. Biol. Eng. **3**, 46–51 (2010)

35. Romano, G., Nagle, M., Argyropoulos, D., Müller, J.: Laser light backscattering to monitor moisture content, soluble solid content and hardness of apple tissue during drying. J. Food Eng. **104**, 657–662 (2011)

36. Romano, G., Argyropoulos, D., Nagle, M., Khan, M.T., Müller, J.: Combination of digital images and laser light to predict moisture content and color of bell pepper simultaneously during drying. J. Food Eng. **109**, 438–448 (2012)

37. Udomkun, P., Nagle, M., Mahayothee, B., Müller, J.: Laser-based imaging system for non-invasive monitoring of quality changes of papaya during drying. Food Control **42**, 225–233 (2014)

38. Romano, G., Nagle, M., Müller, J.: Two-parameter Lorentzian distribution for monitoring physical parameters of golden colored fruits during drying by application of laser light in the Vis/NIR spectrum. Innov. Food Sci. Emerg. Technol. **33**, 498–505 (2016)

39. Du, C.-J., Sun, D.-W.: Learning techniques used in computer vision for food quality evaluation: a review. J. Food Eng. **72**, 39–55 (2006)

40. Al Ohali, Y.: Computer vision based date fruit grading system: design and implementation. J. King Saud Univ. Comput. Info. Sci. **23**, 29–36 (2011)

41. Schulze, K., Nagle, M., Spreer, W., Mahayothee, B., Müller, J.: Development and assessment of different modeling approaches for size-mass estimation of mango fruits (Mangifera indica L., cv. 'Nam Dokmai'). Comput. Electron. Agric. **114**, 269–276 (2015)

42. Mendoza, F., Aguilera, J.M.: Application of image analysis for classification of ripening bananas. J. Food Sci. **69**, E471–E477 (2004)

43. Blasco, J., Cubero, S., Gómez-Sanchís, J., Mira, P., Moltó, E.: Development of a machine for the automatic sorting of pomegranate (Punica granatum) arils based on computer vision. J. Food Eng. **90**, 27–34 (2009)

44. Vélez Rivera, N., Gómez-Sanchis, J., Chanona-Pérez, J., Carrasco, J.J., Millán-Giraldo, M., Lorente, D., Cubero, S., Blasco, J.: Early detection of mechanical damage in mango using NIR hyperspectral images and machine learning. Biosystems Eng. **122**, 91–98 (2014)

45. Fukuda, S., Yasunaga, E., Nagle, M., Yuge, K., Sardsud, V., Spreer, W., Müller, J.: Modelling the relationship between peel colour and the quality of fresh mango fruit using Random Forests. J. Food Eng. **131**, 7–17 (2014)

46. Du, C.-J., Sun, D.-W.: Multi-classification of pizza using computer vision and support vector machine. J. Food Eng. **86**, 234–242 (2008)

47. Costa, C., Antonucci, F., Pallottino, F., Aguzzi, J., Sun, D.-W., Menesatti, P.: Shape analysis of agricultural products: a review of recent research advances and potential application to computer vision. Food Bioprocess Technol. **4**, 673–692 (2011)

Online Monitoring System on Controlled Irrigation Experiment for Export Quality Mango in Thailand

Eriko Yasunaga[1]([⊠]), Shinji Fukuda[2], Wolfram Spreer[3],
and Daisuke Takata[1]

[1] The University of Tokyo, Tokyo, Japan
{erikoy, takata}@isas.a.u-tokyo.ac.jp
[2] Tokyo University of Agriculture and Technology, Tokyo, Japan
shinji-f@cc.tuat.ac.jp
[3] Chiang Mai University, Chiang Mai, Thailand
wolfram.spreer@gmx.net

Abstract. Export quality fresh mango fruit requires a careful management from production to distribution in order to meet a high standard specifically for a long supply chain. In this paper, we describe our online monitoring system to observe responses of mango trees to different levels of irrigation intensities, based on which an optimal irrigation regime can be established. The system consists of different sensors for soil moisture content, sap-flow, photographs, and so on. To fully control the amount of water irrigated, rain-out shelter was installed for every tree for the experiment. Field observation data are used to compute water consumption and water stress of target trees. Preliminary results supported the practicality of this system that can further be used for quantitative assessment and numerical simulation for optimizing a production system aiming at a long supply chain.

Keywords: Real-time monitoring · ICT · Mango · Irrigation management · Yield · Fruit quality

1 Introduction

Japanese, in general, concerns food safety, thereby setting strict import/export criteria compared to other countries. In practice, excessive amount of agrochemicals and fertilizers have been used for instance in Southeast Asian countries. Also, inappropriate farm management such as extensive water management has been commonly practiced specifically in rural and remote areas. This hampers production of quality crops and fruits that can be sent to a big market such as Japan and Europe. However, such an inappropriate production management is ascribed not only to famers' inexperience and lower motivation but also to the availability of resources including water for irrigation. Therefore, development projects often focus on infrastructures to improve agricultural productivity by improving access to such fundamental natural resources for agriculture. Production systems can be improved on the basis of precise measurement and assessment of soil-plant-atmosphere status. Recent advances of sensing technologies

© Springer International Publishing Switzerland 2016
N. Kubota et al. (Eds.): ICIRA 2016, Part II, LNAI 9835, pp. 328–334, 2016.
DOI: 10.1007/978-3-319-43518-3_32

further allow for continuous monitoring and careful management based on such a monitoring result.

Transportation of fresh agricultural produces during a long supply chain requires careful management considering ecophysiological characteristics such as chilling sensitivity of fresh fruits. Respiration rate and postharvest ripening should also be considered for climacteric fruits. It is however difficult to realize such a system in tropical/subtropical countries due to local weather conditions and infrastructures needed for a well-controlled transportation and distribution system. Therefore, interdisciplinary research covering from production to distribution of high-value fresh produces can contribute to the establishment of an improved supply chain.

Mango (*Mangifera indica* L.) is one of the major fresh fruit for export [1]. According to [1], there are several issues when exporting fresh mango fruit. The most critical issue is quality losses during transportation, which is caused mainly by harvesting fruit at improper maturity, mechanical damage caused during harvesting, improper field handling, sap burn, spongy tissue, lenticels discoloration, fruit softening, decay, chilling injury, disease and pest damage [1]. First of all, in order to ensure high yield, proper irrigation management is needed [2, 3]. It has been shown that the amount of irrigated and/or precipitated water can be used for estimating the yield of fresh mango fruit [4]. One of the key issues in production is to design a good irrigation practice such as timing and amount for a specific cultivar. A process-based model can support such a decision, but detailed information required for modelling the dynamics of a mango tree is still limited. Harvest timing is also a critical issue after production as partly discussed in [1]. This is very much difficult to quantitatively assess because flowering does not occur at the same time and fruit maturity is different even within a tree. This raises a need for well-controlled experiment for better designing harvest timing. Most importantly, technological innovation is needed to reduce quality losses during transportation specifically for high-value fresh fruit subjected to a long supply chain. In this regard, machine learning-based approach can be used as a non-destructive tool to estimate the quality of fresh mango fruit based on peel color [5]. However, for a long supply chain, a process-based model is needed to predict how fast or slowly ripening of a mango fruit proceeds. For this purpose, ecophysiological processes including ripening and respiration should be clarified through a well-controlled experiment.

To advance mango research from production to distribution, we established an experimental field in which production management such as application of water, chemicals and fertilizers can be controlled. Online monitoring system was installed to track soil moisture content at different depths and sap-flow of selected trees. This paper illustrates our real-time monitoring system for mango, based on which management practices such as irrigation can be optimized for producing quality fresh mango fruit.

2 Methods

2.1 Online Monitoring System

In our study, three-year-old trees of mango without bearing were used for monitoring, because bearing has strong effects on the dynamics of water in plants. For precise management of agricultural production, precise assessment based on continuous and detailed monitoring of a target crop or plant are necessary. We therefore established a solar-powered monitoring system for soil moisture content, soil temperature, electrical conductivity, sap-flow and photographs (Fig. 1) in an experimental field at Chiang Mai University (CMU), Chiang Mai, Thailand. Soil moisture content, soil temperature and electrical conductivity were measured with the 5TE sensors (Decagon Devices, Inc., USA) placed at the depths of 5 cm and 60 cm. The 5TE sensors are with accuracies of soil volumetric water content being ±3 %; electrical conductivity being ±10 % from 0 to 7 dS/m; and soil temperature being ±1 °C. Observation data were recorded with the Em50 data logger (Decagon Devices, Inc., USA). The FieldRouter is equipped with a fixed camera with 8 MP camera resolution. Measurement interval was set at 10 min except for the camera that takes a photograph of appearance of mango field and trees at noon every day (Fig. 2). All sensors were connected via the FieldRouter® system (X-ability, Japan; http://x-ability.co.jp/en/xasensor.html) for collecting and storing data from the field, of which observation data are sent to a cloud server every day. The FieldRouter was connected to another data logger (CR-10) via Bluetooth connection. Sap-flow amount in plant stem was measured by a sap-flow sensor which is based on the stem heat balance method [6], and obtained data were recorded with the CR-10 data logger (Decagon Devices, Inc., USA). In addition, we obtain weather information such as air temperature, humidity, wind speed and solar radiation at the nearby weather station of CMU (about 500 m away).

Fig. 1. Online field monitoring system.

2.2 Irrigation Management

In our experiment, five different irrigation treatments with ten trees each can be designed. Rain-out shelter is installed to eliminate the effects of rainfall on the irrigation management. Figure 3 shows a conceptual diagram of our irrigation experiment, in which two targets of water stress and harvest timing are set. Irrigation regimes are designed based on the daily potential evapotranspiration calculated using the weather station data.

Fig. 2. An example photograph taken by the FieldRouter

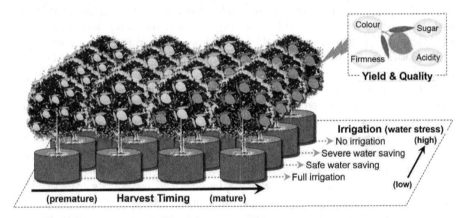

Fig. 3. Conceptual figure for experiments considering harvest timing and irrigation treatment.

The basic equation for water balance in a pot is as follows:

$$SW_t = SW_0 + \sum_{t=0}^{t} \left(PR_i + IR_i - ET_i - Q_{out,i} \right)$$

where SW_t is the soil moisture content at time t, SW_0 is the initial soil moisture content, PR_i is the precipitation at time i, IR_i is the irrigation at time i, ET_i is the evapotranspiration at time i, and $Q_{out,i}$ is the outflow at time i. In case a rain-out shelter is used, the amount of precipitation can be omitted. The amount of irrigated water was recorded with flow meter at a water tank every day. As such, we can monitor water balance in a selected pot. With a regular monitoring of soil moisture content at all pots using a handheld soil moisture content meter, we can estimate water balance for the entire trees with possible variability between individual trees.

3 Results and Discussion

Figure 4 shows our field observation data in December 2015. Due to the local condition, we could upload observation data once a day, which is a clear limitation of our system. However, this basic information is valuable for water balance analysis for each individual mango tree, thereby we can evaluate variability within the same irrigation treatment. With sap-flow sensors installed in each irrigation treatment (result not shown), important information such as evapotranspiration and/or translocation of nutrients may be quantitatively obtained. Specifically soil moisture content at different depths exhibited different dynamics. With a better understanding of such dynamics, improved water saving and deficit irrigation schemes such as partial rootzone drying [7] and sensor-based irrigation scheduling [8] can be established. Specifically, different amount of water can be irrigated according to the different stages of a mango tree such as flowering, fruiting and fruit growing stages. Moreover, in future, we might determine and monitor the status of fertilizer application (i.e., timing and amount) in soil from the values of electrical conductivity [9]. As such, both irrigation and fertilizer application can be optimized for yield and quality of mango fruit for a long supply chain. Innovation in irrigation shall contribute to a better and sustainable water resources management specifically in a water scarce region.

Recently, there have been many innovative monitoring systems proposed and marketed for agricultural production. An example is the "e-kakashi" system (PS Solutions Corp., Tokyo, Japan; https://www.e-kakashi.com/en). It has a good user interface and user-friendly display of data and information. Unfortunately this monitoring systems can only be used in Japan because of limited compatibility in internet. The system can be used for monitoring weather information as well as CO_2 concentration. Such information is very useful for farmers who control, for instance, CO_2 concentration in a greenhouse to improve fruit quality. This online monitoring system is one of the future directions to be considered for better production-distribution system. As such, not only scientists but also farmers can use this system to check status of their own field, which is needed for a precision agriculture in practice.

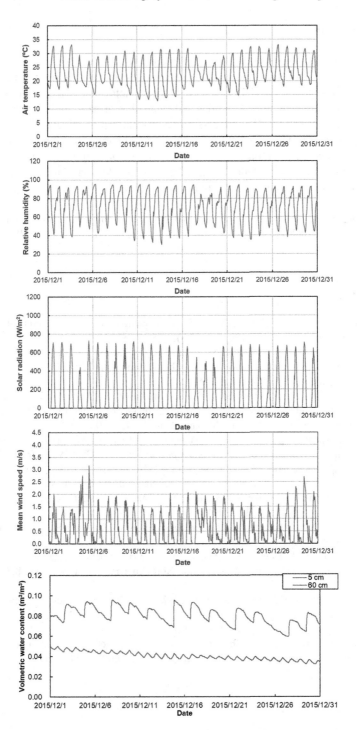

Fig. 4. An observation data set from the CMU experimental field (December 2015).

4   Conclusions

In this paper we demonstrated an online monitoring system for mango production under controlled irrigation treatment. The system is now under preliminary test for subsequent experiments on yield assessment. Sensor information will be supplemented by a direct measurement and continuous and detailed information for all the pots, based on which data-driven as well as process-based models will be developed for predictive modelling of yield and qualities of fresh mango fruit at harvest. Postharvest fruit quality changes will then be monitored at varying transportation conditions such as temperature, humidity and gas content. Results are expected to contribute to an improved production-distribution system for a long supply chain.

Acknowledgement. This study was supported in part by the Grants-in-Aid for Scientific Research (B) (16H05800) of the Japan Society for the Promotion of Science (JSPS).

References

1. Sivakumar, D., Jiang, Y., Yahia, E.M.: Maintaining mango (*Mangifera indica* L.) fruit quality during the export chain. Food Res. Int. **44**(5), 1254–1263 (2011)
2. Spreer, W., Ongprasert, S., Hegele, M., Wünsche, J.N., Müller, J.: Yield and fruit development in mango (*Mangifera indica* L., cv. Chok Anan) under different irrigation regimes. Agric. Water Manag. **96**, 574–584 (2009)
3. Spreer, W., Nagle, M., Neidhart, S., Carle, R., Ongprasert, S., Müller, J.: Effect of regulated deficit irrigation and partial rootzone drying on the quality of mango fruits (*Mangifera indica* L., cv. 'Chok Anan'). Agric. Water Manag. **88**(1–3), 173–180 (2007)
4. Fukuda, S., Spreer, W., Yasunaga, E., Yuge, K., Sardsud, V., Muller, J.: Random Forests modelling for the estimation of mango (*Mangifera indica* L. cv. Chok Anan) fruit yields under different irrigation regimes. Agric. Water Manag. **116**, 142–150 (2013)
5. Fukuda, S., Yasunaga, E., Nagle, M., Yuge, K., Sardsud, V., Spreer, W., Muller, J.: Modelling the relationship between peel colour and the quality of fresh mango fruit using Random Forests. J. Food Eng. **131**, 7–17 (2014)
6. Sakuratani, T.: A heat balance method for measuring water flux in the stem of intact plants. J. Agric. Meteorol. **37**, 9–17 (1981)
7. Spreer, W., Müller, J., Hegele, M., Ongprasert, S.: Effect of deficit irrigation on fruit growth and yield of mango (*Mangifera indica*, L.) in Northern Thailand. Acta Hortic. **820**, 357–364 (2009)
8. Migliaccio, K.W., Schaffer, B., Crane, J.H., Davies, F.S.: Plant response to evapotranspiration and soil water sensor irrigation scheduling methods for papaya production in South Florida. Agric. Water Manag. **97**, 1452–1460 (2010)
9. Miyamoto, T., Kameyama, K., Shinogi, Y.: Electrical conductivity and nitrate concentration in an Andisol field using time domain reflectometry. In: 19th World Congress of Soil Science 2010, pp. 792–795 (2010)

Intensity Histogram Based Segmentation of 3D Point Cloud Using Growing Neural Gas

Shin Miyake[✉], Yuichiro Toda, Naoyuki Kubota, Naoyuki Takesue,
and Kazuyoshi Wada

System Design Graduate School, Tokyo Metropolitan University, Hino, Tokyo, Japan
{miyake-shin,toda-yuuichirou}@ed.tmu.ac.jp,
{kubota,ntakesue,k_wada}@tmu.ac.jp

Abstract. This paper proposes a 3D point cloud segmentation method using a reflection intensity of Laser Range Finder (LRF). In this paper, we use LRF and tilt unit for acquiring a 3D point cloud. First of all, we apply Growing Neural Gas (GNG) to the point cloud for learning a topological structure of the point cloud. Next, we proposed a segmentation method based on an intensity histogram that is composed of the nearest data of each node. Finally, we show experimental results of the proposed method and discuss the effectiveness of the proposed method.

Keywords: Robot sensing · LRF intensity · Clustering

1 Introduction

Recently, many large-scale disasters such as an earthquake have been occurred all over the world, and a response to the large-scale disasters has been discussed. Especially, it is important to restore main highways and ensure a lifeline as soon as possible. Under the circumstances, intelligent robots are expected for a rescue in the disasters because the robots can measure the environment safely and quickly for determining the extent of the disaster damage such as materials and volume of wreckage. In this paper, we focus on a measurement method using a 3D point cloud for providing the efficient and effective information of the disaster damage. Specifically, we propose a material segmentation method of the point cloud because detecting the material features in the point cloud enables the robot to utilize some objectives such as an environmental perception and recognition.

Many segmentation methods of the 3D point cloud have been proposed in resent years. The researches of segmentation can be divided into two main streams from the viewpoint of the sensor devise. One is to use only a 3D distance sensor such as a distance and surface feature based segmentation method [1,2]. Actually, these researches is very useful to divide the area in the 3D space. However, these kinds of methods cannot cluster the material because of lack of the key features. Another stream is to combine the 3D distance sensor and a camera devise [3–6]. These researches can effectively cluster the material using

© Springer International Publishing Switzerland 2016
N. Kubota et al. (Eds.): ICIRA 2016, Part II, LNAI 9835, pp. 335–345, 2016.
DOI: 10.1007/978-3-319-43518-3_33

color and luminance features and synthesizing the 3D coordinate system and camera coordinate system. However, it is unstable to cluster in outdoor environment because these researches strongly depend on a lighting environment. Therefore, we use a reflection intensity of LRF for the material segmentation because the reflection intensity is robust to the lighting environment and acquire the distance data and reflection intensity simultaneously. In addition, it is known that different materials has different reflection intensities. For instance, Y. Hara, et al. proposed a localization and mapping method using this property, and showed the effective and efficient results [7]. Our proposed method also use this property for the material segmentation.

In this paper, our objective is to cluster the materials from the 3D point clouds using LRF. At first, we explain our 3D sensor devise composed of LRF and tilt unit for acquiring the 3D point clouds. Next, we explain Growing Neural Gas (GNG) [8] for performing a distance-based segmentation. GNG is one of the competitive learning methods, and can learning a topological structure of the 3D point cloud. In the distance based segmentation, each point cloud data is assigned the nearest node of the topological structure. Next, we propose a intensity histogram based segmentation method. For calculating the similarity of the intensity and removing the noise, we use the intensity histogram which is composed of the point cloud data assigned each node in the distance based segmentation. Furthermore, we propose the segmentation algorithm which utilizes the topological structure of GNG and growing region method. Finally, we show experimental results of the material segmentation by the proposed methods, and discusses the effectiveness of the proposed methods.

2 Measurement System

In this paper, we use a Laser Range Finder, URG-04LX-UG01, developed by Hokuyo Automatic Co. Ltd, [11]. Table 1 shows the specification of the LRF. Furthermore, the LRF is equipped on a tilt mechanism composed of a servo motor for measuring the 3D distance (Fig. 1). Next, we explain how to construct a 3D point cloud using the distance information by the LRF. The position (x', y', z') of the measurement point on a global coordinate system is calculated by the following equation,

$$
\begin{bmatrix} x'_{i,j} \\ y'_{i,j} \\ z'_{i,j} \\ 1 \end{bmatrix} = \begin{bmatrix} 1 & 0 & 0 & x_p \\ 0 & 1 & 0 & 0 \\ 0 & 0 & 0 & z_p \\ 0 & 0 & 0 & 1 \end{bmatrix} \begin{bmatrix} 1 & 0 & 0 & 0 \\ 0 & \cos\theta_j & -\sin\theta_j & 0 \\ 0 & \sin\theta_j & \cos\theta_j & 0 \\ 0 & 0 & 0 & 1 \end{bmatrix} \begin{bmatrix} x_i \\ y_i \\ 0 \\ 1 \end{bmatrix} \tag{1}
$$

where $(x_p, 0, z_p)$ is the position of the LRF; (x_i, y_i) is the ith distance data of the LRF with the tilt angle θ_j. In this way, the measured distance information from the LRF can be reflected on the global coordinate system. Table 2 shows the specification of the measuring system.

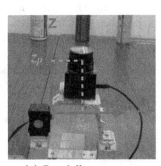

(a) Paralell movement

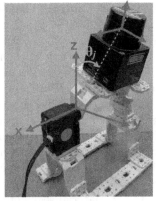

(b) Tilt angle

Fig. 1. Configuration of 3D measurement system

Table 1. Specification of the LRF [11]

Model	URG-04LX-UG01
Power source	5 V DC ± 5 %
Detection distance	20 mm–4,000 mm
Resolution	1 mm
Scan angle	240°
Scan time	100 ms/scan

Table 2. Operating range and resolution of tilt feature

Operation range	[−90°−90°]
Resolution	[0.36°]

3 Clustering of Measurement Object with Reflection Intensity

3.1 Relation of Distance and Reflection Intensity

In this paper, we assume the outdoor environment. For clustering the 3D point clouds, many researchers use a color information integrated with the 3D point clouds. However, using the color information strongly depends on a lighting environment. Therefore, we use reflection intensity of the LRF for clustering because the reflection intensity does not depend on a lighting environment.

First of all, we show a primary experiment for observing a relationship about the distance and reflection intensity. This experiment uses a concrete and marble stone as the materials. Our purpose of this experiment is to observe the relationship. Therefore, we measure the reflection intensity between 0.5 m and 4.0 m in 2D coordinate system for collecting the measurement data. In addition, we assume that the measurement data includes some noise and other materials. Therefore, we classify the measurement data by calculating the distance between the nearest measurement data in each scan for removing the noise and the other materials. After the classification, we calculate a histogram of the reflect intensity (intensity histogram) in each distance. Figure 2 represents examples of the intensity histogram. In Fig. 2, the each material has the different shape of the intensity histogram which has the different peak clearly. In addition, Fig. 3 shows the comparison result of the relationship between distance and reflection intensity. In Fig. 3, the reflection intensities which has the maximum number of bins are plotted in each distance. It is difficult to cluster each material between 500 and 1500 [mm] (Near distance) because the measured data has the almost same degree of the reflection intensity. However, as mentioned before, the shape of each intensity histogram is different (Fig. 2). On the other hand, the measurement data between 2000 and 4000 [mm] (Far distance) has the different degree of intensity in each material. In this range, we can easily cluster each material. From these properties of the reflection intensity of LRF, we apply the intensity histogram to the material segmentation method.

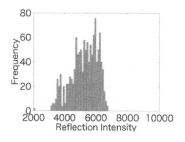 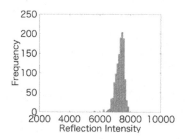

Fig. 2. Examples of the intensity histogram

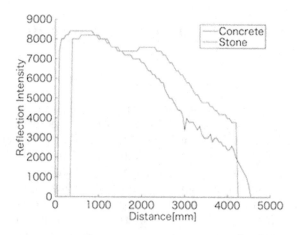

Fig. 3. Relation of distance and reflection intensity

3.2   Growing Neural Gas

We use a Growing Neural Gas (GNG) for learning a topological structure of the 3D point clouds. GNG proposed by Fritzke [8] is one of the unsupervised learning methods. Unsupervised learning is performed by using only data without any teaching signals [13]. Self-organized map (SOM), neural gas (NG), growing cell structures (GCS), and GNG are well known as unsupervised learning methods. Basically, these methods use the competitive learning. The number of nodes and the topological structure of the network in SOM are designed beforehand [13]. In NG, the number of nodes is fixed beforehand, but the topological structure is updated according to the distribution of sample data [14]. On the other hand, GCS and GNG can dynamically change the topological structure based on the adjacent relation (edge) referring to the ignition frequency of the adjacent node according to the error index. However, GCS does not delete nodes and edges, while GNG can delete nodes and edges based on the concept of ages [8,12]. Furthermore, GCS must consist of k-dimensional simplexes whereby k is a positive integer chosen in advance. The initial configuration of each network is a k-dimensional simplex, e.g., a line is used for $k = 1$, a triangle for $k = 2$, and a tetrahedron for $k = 3$ [8,15]. GCS has applied to construct 3D surface models by triangulation based on 2-dimensional simplex. However, because the GCS does not delete nodes and edges, the number of nodes and edges is over increasing. Furthermore, GCS cannot divide the sample data into several segments. Therefore, we use growing neural gas for learning the topological structure of the 3D environmental space.

At first, we explain the learning algorithm of GNG as follows;

w_i: the nth dimensional vector of a node
A: a set of nodes
N_i: a set of nodes connected to the ith node
c: a set of edges
a_{ij}: Age of the edge between the ith and jth nodes

Step 0. Generate two units at random position, w_1, w_2 in $\boldsymbol{R}_n$ where n is the dimension of input data. Initialize the connection set.

Step 1. Generate at random an input data v. In this paper, we use the position information (x, y, z) of a point measured by the LRF and tilt unit system.

Step 2. Select the nearest unit (winner) s_1 and the second-nearest unit s_2 from the set of nodes by

$$s_1 = \arg\min_{i \in A} \|v - w_i\| \tag{2}$$

$$s_2 = \arg\min_{i \in A \backslash \{s_1\}} \|v - w_i\| \tag{3}$$

Step 3. If a connection between s_1 and s_2 does not yet exist, create the connection ($c_{s1,s2} = 1$). Set the age of the connection between s_1 and s_2 at zero;

$$a_{s_1,s_2} = 0 \tag{4}$$

Step 4. Add the squared distance between the input data and the winner to a local error variable;

$$E_{s_1} \leftarrow E_{s_1} + ||v - w_{s_1}||^2 \tag{5}$$

Step 5. Update the reference vectors of the winner and its direct topological neighbors by the learning rate η_1 and η_2 respectively, of the total distance to the input data.

$$w_{s_1} \leftarrow w_{s_1} + \eta_1 \cdot (v - w_{s_1}), \tag{6}$$

$$w_j \leftarrow w_j + \eta_2 \cdot (v - w_j) \quad if \ c_{s_1,j} = 1 \tag{7}$$

Step 6. Increment the age of all edges emanating from s_1

$$a_{s_1,j} \leftarrow a_{s_1,j} + 1 \quad if \ c_{s_1,j} = 1 \tag{8}$$

Step 7. Remove edges with an age larger than a_{max}. If this results in units having no more connecting edges, remove those units as well.

Step 8. If the number of input data generated so far is an integer multiple of a parameter λ, insert a new unit as follows.

  i. Select the unit f with the maximal accumulated error.

$$q = \arg\max_{i \in A} E_i \tag{9}$$

 ii. Select the unit f with the maximal accumulated error among the neighbors of q.

iii. Add a new unit r to the network and interpolate its reference vector form q and f.

$$w_r = 0.5 \cdot (w_q + w_f) \tag{10}$$

 iv. Insert edges connecting the new unit r with units q and f, and remove the original edge between q and f.

  v. Decrease the error variables of q and f by a temporal discounting rate α.

$$E_q \leftarrow E_q - \alpha E_q \tag{11}$$

$$E_f \leftarrow E_f - \alpha E_f \tag{12}$$

 vi. Interpolate the local error variable of r from q and f.

$$E_r = 0.5 \cdot (E_q + E_f) \tag{13}$$

Step 9. Decrease the local error variables of all units by a temporal discounting rate β.

$$E_i \leftarrow E_i - \beta E_i \quad (\forall i \in A) \tag{14}$$

Step 10. Continue with Step 2 if a stopping criterion (e.g., the number of nodes or some performance measure) is not yet fulfilled.

Figure 4 shows an example of learning the topological structure from the 3D point clouds. The parameters are $\lambda = 200$; $\mu_1 = 0.04$; $mu_2 = 0.001$; $\alpha = 1.0$; $\beta = 0.0005$ in GNG. In Fig. 4, GNG efficiently cover the data distribution without redundant nodes and edges because of the edge and node removal algorithm.

(a) Original Environment

(b) A result of GNG. Green dot indicates the 3D point cloud and blue dot and line indicate the topological structure of GNG.

Fig. 4. Examples of the intensity histogram (Color figure online)

3.3 Intensity Histogram Based Segmentation

This subsection proposes an intensity histogram based 3D point clouds segmentation algorithm. Our segmentation algorithm is divided into two parts. One is a distance based segmentation using GNG as mentioned before. Another is the intensity histogram based segmentation. Figure 5 shows the total flowchart of our proposed method.

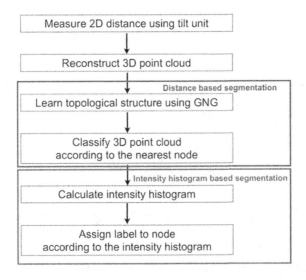

Fig. 5. Algorithm of our proposed method

At first, we explain the distance based segmentation. After learning the topological structure by using GNG, each 3D point cloud's data belongs to the nearest node as follows;

$$S_i \leftarrow S_i + \{k\} \tag{15}$$

$$k = \arg\min_{j \in A} ||v - w_j|| \tag{16}$$

where S_i is the distance segmentation set of the ith node. In this way, GNG can segment the 3D point cloud into the number of nodes according to the distance between the input data and node.

After the distance based segmentation, the intensity histogram of the ith node is calculated as follows;

$$b_{ij} = \sum_{x \in S_i} \rho_j(x) \tag{17}$$

$$\rho_j(x) = \{ \begin{matrix} 1 & if\ j = \lfloor v_x^{int}/width + 0.5 \rfloor \\ 0 & otherwise \end{matrix} \tag{18}$$

where b_{ij} is the jth bin's histogram of the ith node; v_x^{int} is the reflection intensity value of the xth point cloud's data in set; $width$ indicates the range of each bin. In this way, the intensity histogram of each node is created by using the result of distance segmentation.

Next, we explain the intensity histogram segmentation algorithm. The idea of the segmentation algorithm is similar to a region growing method. Our proposed method utilizes the topological structure of GNG for growing the region. The total algorithm is as follows;

Step 0. All labels l_i of the node set to -1 ($l_i = -1, \forall i \in A$), and label number n set to 0 ($n = 0$). The temporary set $\Pi = \emptyset$

Step 1. Select the ith node whose label is -1 from the set A, and the jth label sets to n ($l_i = n$).

Step 2. Select the all nearest nodes from the ith node ($c_{ij} = 1$), and if the difference d_{ij} of intensity values between the ith and jth nodes less than threshold value $sigma$, the jth label sets to n. In addition, the jth node set to the temporary set ($\Pi \leftarrow \Pi + \{j\}$);

$$l_j = n (if \quad d_{ij} < \sigma) \tag{19}$$

$$d_{ij} = |v_i^{max} - v_j^{max}| \tag{20}$$

Step 3. Select and remove the ith node from the set Π ($\Pi \leftarrow \Pi - \{i\}$) and go to Step 2 if the temporary set is not empty. Otherwise, go to Step 4.

Step 4. Increment the label n and go to Step 1 if unlabeled node ($l_i = -1$) exists. Otherwise, Stop the segmentation.

where v_i^{max} is the intensity value of the ith node and the intensity value indicates the bin which has the maximum frequency of the histogram. In this way, our proposed method utilizes the topological structure of 3D point clouds and the region growing method for clustering the materials according to the intensity histogram.

4   Experiment

This section shows experimental results of the proposed method for the material segmentation. The parameters used for GNG and intensity histogram are shown in the following; $\lambda = 200$; $\mu_1 = 0.04$; $\mu_2 = 0.001$; $\alpha = 1.0$; $\beta = 0.0005$; the threshold for the segmentation is 3, respectively. In this experiment, Fig. 6 shows the environment of this experiment. This experiments divided into two experimental conditions. In condition 1, the distance from the LRF to the target objects is 1 [m] for evaluating the near distance (Fig. 6(a)). In condition 2, the distance is 3 [m] for evaluating the far distance (Fig. 6(b)).

(a) Near distance (Condition 1)   (b) Far distance (Condition 2)

Fig. 6. Experimental environment

Figures 7 and 8 show the experimental results of learning the topological structure and the material segmentation by our proposed method. In Figs. 7(b) and 8(b), each topological structure can cover the data distribution of each 3D point cloud because of the rule of the edge and node removal in GNG. Therefore, the distance based segmentation can efficiently divide the point cloud data into each region according to the distance between node and the point cloud data. In Figs. 7(c) and 8(c), our proposed method can cluster the each material because each material has the different histogram (Fig. 3). In Fig. 8(c), each objects can also be clustered as different material. In this way, our proposed method can segment the region of the 3D point cloud by using the topological structure of GNG and the intensity histogram. However, some parts of the chair are merged with each object because these materials are almost the same reflection intensity (Fig. 2).

However, the floors and walls cannot be clustered as the same objects despite of the same materials because the reflection intensity decreases with distance from the object. We consider that the decreasing of the reflection intensity can be approximated by a linear function. For realizing the material segmentation independent of the distance, we should utilize the linear function and the distance between each node of GNG to determine the suitable range of the intensity histogram's bin.

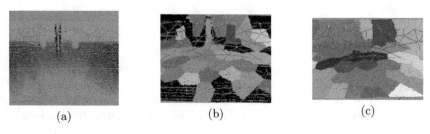

Fig. 7. Experimental result of condition 1. In (a), green dot indicates 3D point cloud and White dot and blue line are the topological structure of GNG. In (b), each color represent each segment. In (c), each color represent the region of same materials. (Color figure online)

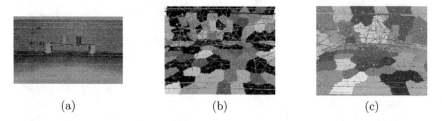

Fig. 8. Experimental results of condition 2. In (a), green dot indicates 3D point cloud and White dot and blue line are the topological structure of GNG. In (b), each color represent each segment. In (c), each color represent the region of same materials. (Color figure online)

5   Conclusion

In this paper, we proposed the material segmentation method using the topological structure of GNG and reflection intensity. First, we proposed the intensity histogram to assign the reflection intensity to each node of GNG because the reflection intensity of LRF is unstable. Next, we proposed the segmentation algorithm using the region growing like method. To verify the effectiveness of our proposed method, we showed the experimental results in two conditions. In the experiment, Our proposed method could cluster each material in the measured environment.

However, some parts of the node have the different materials because of using the same threshold value regardless of the distance. As future work, it is necessary to learn the appropriate threshold value in proportion to distance for coping with the attenuation of reflection intensity.

Acknowledgment. This work was funded by ImPACT Program of the Council for Science, Technology and Innovation.

References

1. Sithole, G., Mapurisa, W.T.: 3D object segmentation of point clouds using profiling techniques. S. Afr. J. Geomatics **1**(1), 60–76 (2012)
2. Khan, A.A., Riaz, S., Iqbal, J.: Surface estimation of a pedestrian walk for outdoor use of power wheelchair based robot. Life Sci. J. **10**, 1697–1704 (2013)
3. To, A.W.K., Paul, G., Liu, D.: Surface-type classification using RGB-D. IEEE Trans. Autom. Sci. Eng. **11**, 359–366 (2014)
4. Kim, B., Xu, S., Savarese, S.: Accurate localization of 3D objects from RGB-D data using segmentation hypotheses. In: IEEE Conference on Computer Vision and Pattern Recognition, pp. 3182–3189 (2013)
5. Diebold, J., Demmel, N., Hazırbaş, C., Moeller, M., Cremers, D.: Interactive multi-label segmentation of RGB-D images. In: Aujol, J.-F., Nikolova, M., Papadakis, N. (eds.) SSVM 2015. LNCS, vol. 9087, pp. 294–306. Springer, Heidelberg (2015)
6. Richtsfeld, A., Mrwald, T., Prankl, J., Zillich, M., Vincze, M.: Learning of perceptual grouping for object segmentation on RGB-D data. J. Vis. Commun. Image Repersentation **25**(1), 64–73 (2014)
7. Hara, Y., Kawata, H., Ohya, A., Yuta, S.: Mobile robot localization and mapping by scan matching using laser reflection intensity of the SOKUIKI sensor. In: 32nd Annual Conference on IEEE Industrial Electronics, IECON 2006, pp. 3018–3023 (2006)
8. Fritzke, B.: A growing neural gas network learns topologies. In: Advances in Neural Information Processing Systems, vol. 7, pp. 625–632 (1995)
9. Choi, B., Mericli, C., Biswasa, J., Veloso, M.M.: Fast human detection for indoor mobile robots using depth images. In: Advances in Neural Information Processing Systems, vol. 7, pp. 625–632 (1995)
10. Oishi, S., Kurazume, R., Iwashita, Y., Hasegawa, T.: Colorization of 3D geometric model utilizing laser reflectivity. In: IEEE International Conference on Robotics and Automation (ICRA), pp. 2319–2326 (2013)
11. HOKUYO AUTOMATICCO, LTD. http://www.hokuyo-aut.co.jp/
12. Fritzke, B.: Growing cell structures - a self-organising network for unsupervised and supervised learning. Neural Netw. **7**(9), 1441–1460 (1994)
13. Kohonen, T.: Self-Organizing Maps. Springer, Heidelberg (2000)
14. Martinetz, T.M., Schulten, K.J.: "A neural-gas" network learns topologies. In: Artificial Neural Networks, vol. 1, pp. 397–402 (1991)
15. Fritzke, B.: Growing self-organizing networks why? In: European Symposium on Artificial, Neural Networks, pp. 61–72 (1996)

Influence of Water Supply on CO_2 Concentration in the Rootzone of Split-Root Potted Longan Trees

Winai Wiriya-Alongkorn[1], Wolfram Spreer[2,3(✉)],
Somchai Ongprasert[4], Klaus Spohrer[2], and Joachim Müller[2]

[1] Department of Horticulture, Mae Jo University, Chiang Mai, Thailand
[2] Institute of Agricultural Engineering, University of Hohenheim,
(440e), Stuttgart, Germany
wolfram.spreer@uni-hohenheim.de
[3] Faculty of Agriculture, Chiang Mai University, Chiang Mai, Thailand
[4] Department of Soil Resources and Environment,
Mae Jo University, Chiang Mai, Thailand

Abstract. Longan trees are irrigated in Thailand as fruit growth takes place during the dry season. Due to the scarcity of water resources, ways for water saving irrigation are investigated. As deficit irrigation was found to have a high water saving potential, the focus was on the investigation of plant stress responses to drought, which can be used for optimizing the deficit irrigation regime. Five split-root potted longan trees in sand culture were subjected to partial rootzone drying (PRD), and during six months the CO_2 concentration in the rootzone was measured by rootzone probes and compared to a well-watered control. The CO_2 efflux from the rootzone was found to be well correlated to the moisture regime in the substrate. However, it was necessary to correct the measured values by the values from the control to obtain a significant correlation between CO_2 concentration and soil moisture. The main observed external factor influencing CO_2 was the ambient temperature. It was shown that the CO_2 efflux from the soil can be used as a non-destructive method for drought stress monitoring, but continuous measurement will be necessary to externalize disturbing environmental effects.

Keywords: Drought · Photosynthesis · Water stress · Partial rootzone drying

1 Introduction

As longan trees (*Dimocarpus longan* Lour.) are particularly sensitive to drought during flowering and early fruit development, in growing regions which have a distinctly summer rainfall pattern, high fruit yields can only be obtained under irrigation [7]. This is the case in Thailand, where nowadays limited water resources are an obstacle to increase longan production. Thus, deficit irrigation strategies have lately been investigated. Studies on partial rootzone drying (PRD) had encouraging results. It was shown that water saving is possible without substantial reduction in yield [19]. However, in pot experiments it was found that sustained PRD negatively affects the

© Springer International Publishing Switzerland 2016
N. Kubota et al. (Eds.): ICIRA 2016, Part II, LNAI 9835, pp. 346–356, 2016.
DOI: 10.1007/978-3-319-43518-3_34

growth of the trees [22]. It is a desired effect that vegetative growth is reduced during the deficit irrigation period. On the other hand the knowledge about the right degree of water stress applied helps to avoid damage to the trees. It was hypothesized that for a good stress management, besides soil moisture data, monitoring of plant responses is important. The monitoring of canopy temperature by thermal imaging was found to have a certain potential, as it can be correlated to stomatal aperture [23]. However, changes in stomatal aperture or photosynthesis happen at a rather advanced stage of plant stress.

In contrast, root respiration reacts more immediately to changes in the environment and may be used as an indicator for early occurrence of stress. Root respiration is an important indicator for growth and health of a plant as dependent on biomass growth and nutritional status more than half of the daily photosynthesis products can be respired by roots [17]. The most important abiotic influences are temperature, nutrient supply and water. Soil water content influences availability of carbohydrates in the soil and thereby root borne CO_2 production [18]. Reactions of roots on water stress are closely related to adaptation mechanisms on changing temperature regimes. A study showed a high temperature dependency CO_2 efflux from the rootzone of citrus trees at good water supply, while there was no influence of temperature on water-stressed trees [4]. In grape wine it was observed that at a higher temperature root respiration in the drying soil decreases faster and more than at a lower temperature, which means that roots are more sensitive to drought if exposed to heat stress [14]. At the often occurring combination of water and heat stress the trees have little chance to adapt to the changing conditions. Further, the daily course of photosynthesis influences the temporal intensity of root respiration [16]. Therefore, typical diurnal variations in CO_2 concentration in the rootzone occur, which are more pronounced after watering and caused by root respiration and microbial activity, which both are influenced by temperature [20].

In nutrient solution, root respiration can be determined exactly [2] and the fact that – under controlled conditions – it is not only possible to establish a high correlation between matric potential and CO_2 efflux, but also to use this information for irrigation control, was first shown by [6]. But in the field additional factors influence the measurement. Typically root respiration is considered the sum of all processes in the soil: Respiration of living roots, respiration of mycorrhizae and microbial activity [1]. Quantification of root respiration is not possible by sheer determination of total CO_2 efflux, as the share of root respiration varies considerably [13]. However, CO_2 from root respiration was found to be the dominating share in the total CO_2 efflux of a cropped area [15]. Moreover, the problem of exact distinction between root born CO_2 and CO_2 from other sources in the soil, is secondary when observing the influence of abiotic stressors on root activity. It can be assumed that CO_2 from other sources decreases in the same proportion as CO_2 from the roots. This correlation, however, needs to be investigated in more detail.

Another factor, which is important to observe, is the movement of CO_2 within the rootzone, which may produce differences in CO_2 concentrations, even though the actual CO_2 production by roots has not changed. Determination of the CO_2 efflux in the field leads to an underestimation of root and soil respiration after an irrigation event. This effect is longer lasting at a finer texture [3]. This is possibly an effect of irrigation water closing a high share of fine pores and, thereby, slowing down the flux. It was shown that CO_2 efflux from the soil does not quickly react on small scale changes in soil moisture, but rather on long term drought events in the rootzone [3, 4]. Obviously, there is the need for monitoring CO_2 in the rootzone, rather than CO_2 efflux, when attempting to establish a correlation between root respiration and water supply. As for reliable information about the degree of water stress continuous measurements are recommended [5], Spohrer et al. [21] introduced a continuous monitoring of the CO_2 concentration in a probe introduced into the rootzone. Gas exchange took place at the lower part of the probe only, so that the alteration of CO_2 movement in response to an irrigation event were documented. In the present study, rather than only measuring the CO_2 concentration in the rootzone, it was attempted to monitor the actual CO_2 production by the roots, by introducing a perforated probe into the rootzone and measuring the CO_2 increase over time. The aim of this study was to investigate the potential of CO_2 monitoring in the rootzone for stress management in split-root longan based on the assumption that rootzone CO_2 monitoring not only enables the consideration of different parts of the root, but that stress reactions can be most immediately detected by alterations in root respiration.

2   Materials and Methods

The experiments were carried out at Mae Jo University, Chiang Mai Province, Thailand (18°53′40″ N, 99°00′58″ E) in the time between 01st of November 2011 and 30th of April 2012. Five year old longan trees with a split root system grown in two separate cement containers (diameter 0.8 m, height 0.5 m) were kept under a transparent plastic shelter with free air circulation. Washed sand was used as growing media. A perforated pipe was introduced at the bottom of the containers to prevent water logging. A modified Hoagland nutrient solution was applied with the irrigation water once per week. The tree height was approx. 2.0 m and the average canopy diameter was 1.8 m.

The experiment was designed during the cold and dry season. To ensure a constant and sufficient water supply during that time trees were irrigated with increasing application rates until steady substrate moisture without leaching of excess water was achieved. By this method the initial daily water consumption was determined as 10 l tree^{-1}. For the duration of the experiment irrigation was applied daily; 5 trees, which served as control (CO), received 5 l of irrigation water in each of the two pots. Another five trees received 5 l of irrigation water in one pot only, either to the left side of the rootzone (L) or its right side (R). The irrigated and the dry side were changed in a two-weeks-interval.

Soil moisture was determined by time domain reflectometry (TDR) using a 1502C cable tester (Tektronix, USA) one hour before and after each irrigation event. Photosynthesis was measured on three leaves on either side of the tree by use of a LCA-4 Leaf Chamber Analyser (ADC BioScientific Ltd., UK) based on gas exchange analyses.

The CO_2 concentration in the rootzone measured by use of a TES 1370 NDIR CO_2 m (TES Electronic Corp., TW), which were placed in a sealed container connected to a 30 cm long 2″ PVC pipe with a perforated area of $(94 \times 20 \text{ mm}^2)$

introduced into the rootzone. In the container there was no gas exchange other than diffusion from the rootzone (Fig. 1).

Fig. 1. Experimental set up for rootzone CO_2 efflux measurement. **Left**: the rootzone probes hung upside down before installation; **Right**: rootzone probes with CO_2 sensor installed in split-root potted longan trees.

The increase in CO^2 concentration as compared to the initial (ambient) concentration in the container was measured after 20 min. This interval was chosen after a series of pre-tests, where the increase in CO_2 concentration in the measuring chamber after irrigation was monitored over a longer interval. It was found that differences were visible after a comparatively short time. As measuring time 10 a.m. was chosen, as based on initial measurements it was observed that during the morning hours there is a time with comparatively low variation and change over time.

3   Results

Figure 2 shows the volumetric substrate water content (swc) during the whole time of experiment. As the water requirement was determined during the cold period with low evaporative demand, initially, the moisture regime is homogeneous, with values which are close to the water holding capacity of the sand. At the end of October, when the treatments were started the wet-dry cycle of PRD is clearly visible. Change between wet and dry side of the rootzone is indicated by the bottom line and coincides with the changes in swc. During March and April swc decreased in all treatments due to lower relative humidity of the air and higher air temperature, which result in a higher evaporative demand. It was observed that swc and CO_2 concentration follow the same pattern on the wet and dry sides (Fig. 3).

However, a mathematical correlation could not be established based on the measurements in the treatments only. This was mainly due to a phase shift between the change in substrate moisture and the subsequent change in root respiration, which was not constant to an extend that the function could have been corrected by time delay. But if the measured values in the treatments were corrected by the control measurements, a

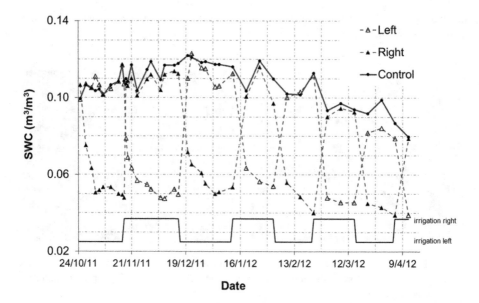

Fig. 2. Volumetric substrate (sand) water content (SWC) of ten split-root potted longan trees. Data points in "control" represent the average of ten full irrigated pots. Data points "left" and "right" are the average of five alternately wet and dry treated pots. The bottom line indicates the alternation between the irrigated sides in the treatment.

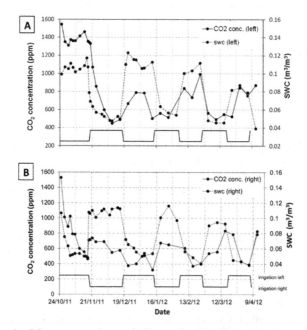

Fig. 3. Increase in CO_2 concentration as indicator for CO_2 production by roots and substrate water content (SWC) on the left side (**A**) and on the right side (**B**) of split-root potted longan trees. The bottom line indicates the alternation between the irrigated sides in the treatment.

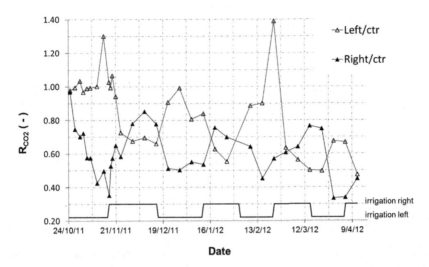

Fig. 4. Ratio of CO_2 concentration as indicator for CO_2 efflux from the rootzone of the left and the right side of the PRD-treated trees divided by control (R_{CO2} [ppm/ppm]). The bottom line indicates the alternation between the irrigated sides in the treatment.

clear picture was obtained, which depicts the reaction of the CO_2 efflux to the soil moisture regime. Figure 4 shows the values of the treatments divided by the control. This ratio follows the same pattern and based on this, a strong correlation was found. The CO_2 efflux ratio between treatment and control was found to be significantly correlated to soil moisture at a confidence level of 0.99 and with a rather high coefficient of determination of $R^2 = 0.51$ (Fig. 5).

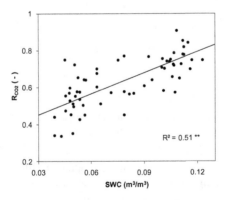

Fig. 5. Correlation between CO_2 concentration ratio of treatment and control (R_{CO2} [ppm/ppm]) and the substrate water content (SWC). ** = Significant at $\alpha = 0.99$

The correction of the treatment values by control was necessary as a strong correlation between the CO_2 and the air temperature was found. Figure 6 shows the

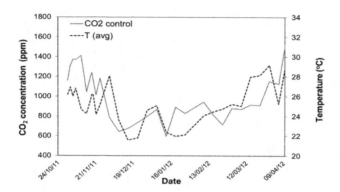

Fig. 6. Average daily ambient temperature and CO_2 concentration as indicator for CO_2 efflux from the rootzone in the control during the time of experiment.

decrease in temperature and the parallel decrease in root respiration in the control treatment. During the cold months (December – February) the average temperature as well as the CO_2 concentration in the rootzone of the control trees was lower. There was no apparent correlation to other external factors, but the correlation between the average daily temperature and the CO_2 efflux was significant, even though the coefficient of determination of $R^2 = 0.47$ was lower than the one found with respect to soil moisture (Fig. 7).

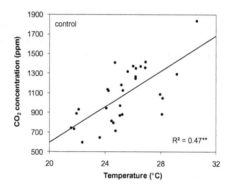

Fig. 7. Correlation between average daily ambient temperature and CO_2 concentration as indicator for CO_2 efflux from the rootzone in the control. ** = Significant at $\alpha = 0.99$

With respect to photosynthesis there was no apparent difference between the PRD irrigated trees and the control trees. This is illustrated in Fig. 8, where the ratio of both sides of the PRD irrigated trees and the control trees fluctuates around 1.0 throughout the experiment. By comparing photosynthesis and swc, no correlation between the irrigation treatment and the photosynthesis was found, so that it can be assumed, that the trees did not suffer severe water stress during the time of experiment (data not shown). Consequently, there was also no correlation between the CO_2 efflux from the rootzone and the photosynthetic rate (Fig. 9).

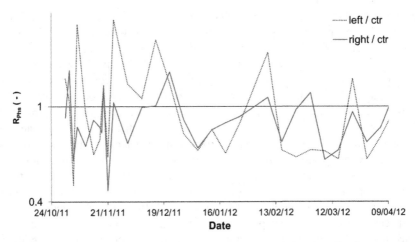

Fig. 8. Ratio of photosynthesis of treatments and control (R_{PHS} [μmol m^{-2} s^{-1}/μmol m^{-2} s^{-1}]) during the time of experiment.

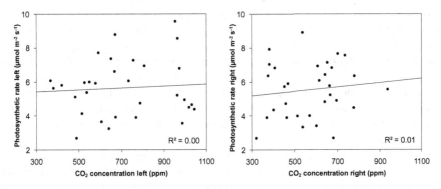

Fig. 9. Correlation between CO2 concentration as indicator for CO2 efflux from the rootzone and photosynthetic rate on the left and the right side of the treatment.

4   Discussion and Conclusions

As shown by the monitoring of substrate moisture during the experiments there was a continuous wet-dry regime in the PRD irrigated trees, with a two weeks alternation interval, as recommended to keep the stress stimulus and at the same time prevent roots from damage due to prolonged drying [8]. The control trees were well watered, as irrigation took place on a daily base due to the low water holding capacity of the substrate. When comparing CO_2 concentration in the rootzone and swc, there was a clear pattern, showing the same development between water supply and CO_2 concentration. Even under conditions where no differences in the photosynthetic rate were observed, the root CO_2 efflux reacted immediately to the change in water supply with a decrease under drought and a subsequent increase after re-wetting. Apparently, during the experiment there was an increasing phase shift between re-watering and recovery of CO_2 efflux, however, no clear pattern was found to quantify the duration of recovery. A continuous measuring of root borne CO_2 may be necessary to clarify this effect.

Two questions need to be clarified in more detail: Firstly, how the variability changes over the days, e.g. with changing temperature in the course of the day. Secondly, it needs to be clarified to what extend the recovery of the root takes place. So far there is little experience with root recovery of fruit trees after partial drying of the rootzone. Fruit trees under drought conserve the existing root system rather than shedding it, and on re-watering recovery depends mainly on existing roots, rather than rapidly building new roots [9]. A rapid recovery is, thus, based on the retention of old roots, which in turn depends on the duration of stress and the degree of adaptation. It was found that citrus trees fully recover after 7 days complete drought [4]. A period, which is shorter than commonly applied in PRD [8]. Recovery also depends on the plant species and its specific adaptation mechanisms. Thus, it cannot be assumed that the cited findings can be applied directly to longan trees which are more susceptible to drought than citrus. On the other hand, partial drought should affect the trees to a lesser extend than complete drought. Usually photosynthesis recovery after a mild water stress is rapid and almost complete, while the recovery of photosynthesis following severe water stress is slow and possibly incomplete [10]. The limitations in recovery of photosynthesis have been found to be in stomatal conductance [12] but photosynthesis was reported to react faster than stomata upon re-watering after drought [11]. This indicates an important role of root-borne hormonal signaling. For irrigation management this implies that measuring the root reactions to water supply is more direct and immediate as the measurement of photosynthesis. As in this study photosynthesis did not decrease significantly under partial drying, it can be concluded that the trees did not suffer severe stress, thus, the observed recovery after re-watering is in line with the cited literature.

Ambient temperature was found to be the main external influence on the CO_2 concentration in the rootzone. As previously described [4], higher temperatures enhance CO_2 efflux from the soil. The correlation between CO_2 efflux and temperature was significant. The investigation of the underlying effects merits more research. Even though a direct correlation between soil moisture and CO_2 concentration, it was visible that the pattern of the CO_2 efflux followed the one of the wet-dry regime in the soil. When compared to the values of non-stressed trees there was a high and significant correlation between soil moisture and CO_2 efflux. As shown in previous studies [6, 21]. CO_2 production was found to be a plant based parameter, which can be used to monitor the plant water status. As compared to the photosynthetic rate, CO_2 production reacted much quicker and more pronounced to the slight drought stress, established in the experiment.

For the use in deficit irrigation the determination of CO_2 concentration in the rootzone can be useful in the future. Unlike other methods of measuring plant reactions, the determination of CO_2 efflux is non-destructive and can be performed continuously. At the same time it does not focus on a small part of the plant only, such as the measurement of photosynthesis or stomatal conductance on single leaves, but focuses on a determined portion of the root system. The root system is where water is taken up and where hormonal and chemical signals about drought stress are produced. It needs to be investigated to what extend stress signals can be correlated to root activity and whether this can be measured by CO_2 efflux. A combination of soil moisture determination and CO_2 efflux measurement may give important information for controlling deficit irrigation systems.

Acknowledgements. This work was financed by Deutsche Forschungsgemeinschaft (DFG) within the collaborative research program "Sustainable rural development in mountainous regions of Southeast Asia" (SFB 564) and co-financed by the National Research Council of Thailand (NRCT). Special thanks to Asst. Prof. Dr. Noporn Boonplod, Mae Jo University, for facilitation of research space.

References

1. Baggs, E.M.: Partitioning the components of soil respiration: a research challenge. Plant Soil **284**, 1–5 (2006)
2. Bloom, A.J., Sukrapanna, S.S., Warner, R.L.: Root respiration associated with ammonium and nitrate absorption and assimilation by barley. Plant Physiol. **99**(4), 1294–1301 (1992)
3. Bouma, T.J., Bryla, D.R.: On the assessment of root and soil respiration for soils of different textures: interactions with soil moisture contents and soil CO_2 concentrations. Plant Soil **227** (1–2), 215–221 (2000)
4. Bryla, D.R., Bouma, T.J., Eissenstat, D.M.: Root respiration in citrus acclimates to temperature and slows during drought. Plant Cell Environ. **20**(11), 1411–1420 (1997)
5. Bryla, D.R., Bouma, T.J., Hartmond, U., Eissenstat, D.M.: Influence of temperature and soil drying on respiration of individual roots in citrus: integrating greenhouse observations into a predictive model for the field. Plant Cell Environ. **24**(8), 781–790 (2001)
6. Casadesus, J., Caceres, R., Marfa, O.: Dynamics of CO_2-efflux from the substrate root system of container-grown plants associated with irrigation cycles. Plant Soil **300**(1–2), 71–82 (2007)
7. Diczbalis, Y., Nicholls, B., Lake, K., Groves, I.: Sapindaceae production and research in Australia. Acta Hortic. **863**, 49–57 (2010)
8. Dry, P.R., Loveys, B.R., Stoll, M., Stewart, D., McCarthy, M.G.: Partial rootzone drying – an update. Aust. Grapegrower Winemaker **438**, 35–39 (2000)
9. Eissenstat, D.M., Whaley, E.L., Volder, A., Wells, C.E.: Recovery of citrus surface roots following prolonged exposure to dry soil. J. Exp. Bot. **50**(341), 1845–1854 (1999)
10. Flexas, J., Bota, J., Galmés, J., Medrano, H., Ribas-Carbó, M.: Keeping a positive carbon balance under adverse conditions: responses of photosynthesis and respiration to water stress. Physiol. Plant. **127**, 343–352 (2006)
11. Gallé, A., Haldimann, P., Feller, U.: Photosynthetic performance and water relations in young pubescent oak (*Quercus pubescens*) trees during drought stress and recovery. New Phytol. **174**, 799–810 (2007)
12. Galmés, J., Medrano, H., Flexas, J.: Photosynthetic limitations in response to water stress and recovery in mediterranean plants with different growth forms. New Phytol. **175**, 81–93 (2007)
13. Hanson, P.J., Edwards, N.T., Garten, C.T., Andrews, J.A.: Separating root and soil microbial contributions to soil respiration: a review of methods and observations. Biogeochemistry **48** (1), 115–146 (2000)

14. Huang, X., Lakso, A.N., Eissenstat, D.M.: Interactive effects of soil temperature and moisture on concord grape root respiration. J. Exp. Bot. **56**(420), 2651–2660 (2005)

15. Kuzyakov, Y., Cheng, W.: Photosynthesis controls of rhizosphere respiration and organic matter decomposition. Soil Biol. Biochem. **33**, 1915–1925 (2001)

16. Kuzyakov, Y., Cheng, W.: Photosynthesis controls of CO_2 efflux from maize rhizosphere. Plant Soil **263**, 85–99 (2004)

17. Lambers, H., Atkin, O.K., Millenaar, F.F.: Respiratory Patterns in Roots in Relation to Their Functioning. In: Kafkafi et al. (ed) Plant Roots The Hidden Half. 3rd edn Marcel Decker, Inc., New York, 323–362 (1996)

18. Liu, H.S., Li, F.M.: Effects of shoot excision on in situ soil and root respiration of wheat and soybean under drought stress. Plant Growth Regul. **50**(1), 1–9 (2006)

19. Ongprasert, S., Spreer, W., Wiriya-Alongkorn, W., Ussahatanonta, S., Köller, K.: Alternative techniques for water-saving irrigation and optimised fertigation in fruit production in Northern Thailand. In: Heidhues, F., et al. (eds.) Sustainable land use in mountainous regions of Southeast Asia: Meeting the challenges of ecological, socio economic and cultural diversity. Environmental Science and Engineering, pp. 120–133. Springer, Heidelberg (2007)

20. Rocksch, T., Gruda, N., Schmidt, U.: The influence of irrigation on gas composition in the rhizosphere and growth of three horticultural plants, Cultivated in Different Substrates. Acta Hortic. **801**, 1143–1148 (2008)

21. Spohrer, K., Zia-Khan, S., Spreer, W., Müller, J.: Real-time detection of root zone-CO_2 and its potential for irrigation scheduling. Landtechnik – Agric. Eng. **70**, 150–156 (2015)

22. Srikasetsarakul, U., Sringarm, K., Sruamsiri, P., Ongprasert, S., Wiriya-alongkorn, W., Spreer, W., Müller, J.: Biomass formation and nutrient partitioning in potted longan trees under partial rootzone drying. Acta Hortic. **889**, 587–592 (2011)

23. Wiriya-Alongkorn, W., Spreer, W., Ongprasert, S., Spohrer, K., Pankasemsuk, T., Müller, J.: Detecting drought stress in longan tree using thermal imaging. Maejo Int. J. Sci. Technol. **7**(1), 166–180 (2013)

Human Data Analysis

Development of an Uchi Self-learning System for Mutsumi-ryu-style Shamisen Using VR Environment

Takeshi Shibata[1]([✉]), Kazutaka Mitobe[2], Takeshi Miura[2], Katsuya Fujiwara[2], Masachika Saito[2], and Hideo Tamamoto[3]

[1] College of Information and Systems, Muroran Institute of Technology, Mizumoto 27-1, Muroran, Hokkaido, Japan
shibata@csse.muroran-it.ac.jp
[2] Graduate School of Engineering Science, Akita University, Tegata gakuenmachi 1-1, Akita, Akita, Japan
[3] Tohoku Koeki University, Iimori-yama 3-5-1, Sakata, Yamagata, Japan

Abstract. The *shamisen* is a traditional Japanese folk musical instrument. It is regarded as a valuable cultural asset that might attract people into moving into unpopulated districts of the country. We focus on the Mutsumi-ryu style of *shamisen*, a famous playing style originating in Akita (a prefecture in northern Japan). Mutsumi-ryu has a characteristic action called the *uchi*, where a plectrum is struck down to the body and strings of the *shamisen*. In this paper, we propose a self-learning system for the *uchi* for students of the Mutsumi-ryu playing style by using Virtual Reality (VR). The system captures a student's actions during *uchi* using a motion capture system, evaluates the student's actions, and interactively represents the movements of an expert performing the *uchi* in a VR environment. Experimental results showed that using our system, students can acquire a skill level comparable to that of those who learn directly from human experts.

Keywords: Self-learning · Motion analysis · Motion capture · Musical instrument

1 Introduction

The entire range of folk cultural heritage in Japan is called Minzoku bunkazai. It consists of Minzoku geino (folk performing arts) involving amateur actors depicting events in daily life. These arts are regarded as a valuable asset that has had a significant influence on the cultural background of the region.

However, some Japanese folk performing arts have begun to disappear over time. Falling birthrate and decreasing rural population, which are significant social problems in Japan, may have contributed to the decline of performing arts in Japan. We think that performing arts might attract people into moving into unpopulated districts of the country. If the number of people in the districts

© Springer International Publishing Switzerland 2016
N. Kubota et al. (Eds.): ICIRA 2016, Part II, LNAI 9835, pp. 359–370, 2016.
DOI: 10.1007/978-3-319-43518-3_35

who are interested in these unique folk performing arts increases, it increases the likelihood of the sustenance of these arts, as they can then be passed down to the next generation and help motivate the revitalization of the districts.

In order to address this problem, various approaches have been proposed ranging from social activities to the development of relevant technical methods. As an instance of the former, the Japanese government enforced several laws designed to protect cultural assets and encourage folk performing arts festivals.

As an example of the latter, methods for digital archiving or analysis of performing arts have been proposed [1–3]. These research activities have contributed to popularizing folk performing arts.

We think that appropriately assisting beginners in training for folk performing arts is important for enhancing the popularity of these arts. It is difficult for an expert to spend much time teaching because experts of folk performing arts tend to not play professionally. Moreover, a student who lives far from any expert has few opportunities to learn by instruction.

Toward the end of teaching such students, self-learning systems using Virtual Reality (VR) and motion capturing systems (MoCaps) have been proposed [4–7]. A high-resolution MoCap has the capacity to capture complex finger movements on musical instruments for learning assist systems [8,9].

In this study, we develop a self-learning system for the *shamisen*, a traditional Japanese musical instrument. We focus on the *Mutsumi-ryu playing style* originating from the Akita prefecture, which also has the most rapidly declining population in Japan. The Mutsumi-ryu playing style has a characteristic action called *uchi*, where a plectrum is struck down to the body and strings of the *shamisen*.

Some past studies have proposed analysis or assistance systems for the *shamisen*. Reference [10] proposed an automatic scoring (music dictation) system for the *Tugaru-jamisen* playing style. Reference [11] proposed an automatic score scrolling system for the *Nagauta* playing style.

These studies, based on sound analysis, did not target beginners. Moreover, little research has focused on the actions involved in *uchi* in the Mitsumi-ryu playing style. Hence, in this paper, we propose a self-learning system for the *uchi* using VR and MoCap.

2   Past Work

A *shamisen*, which is shown in Fig. 1, is a plucked string instrument with a *dou* (body), a *sao* (neck) and three strings. A player holds a *shamisen* like a guitar or a banjo when he/she plays it (Fig. 1(b)). The *dou* works as a sound hole, like in a guitar, as well as a drum.

The player grips a *bachi* (plectrum) in the right hand and strikes it down at the strings and the *dou* face. This striking action is called the *uchi*, which is the most basic way of playing the *shamisen*. When the player performs the *uchi*, the sound of the *shamisen* contains both the sound of the strings and that of the drum. There are many other ways to play the *shamisen*, such as the *sukui*

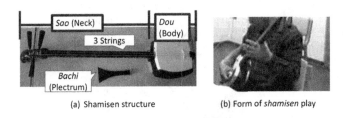

(a) Shamisen structure   (b) Form of *shamisen* play

Fig. 1. A *shamisen*

(upstroke), the *oshi bachi* (pressure technique), the *hajiki* (pluck using the left hand), and so on. The Mutsumi-ryu playing style has the characteristics of the *uchi* [14].

In past work, we discussed the features of the *uchi* in the Mutsumi-ryu playing style by comparing the action of an expert performing the *uchi* with that of a beginner using MoCap [12,13].

Scenes involving the *uchi* in these experiments are shown in Fig. 2. Lines in the computer graphics (CG) describe the trajectories of the actions involved in the *uchi* (*uchi* trajectory). Figure 2(a) shows an expert's *uchi* actions and Fig. 2(b) shows those of a beginner. We can see that the expert moved his *bachi* along a straight line. We can also see that the expert kept the angles between the lines and the *dou* stable. By contrast, the beginner moved his *bachi* along a circle.

This showed that beginners do not strike the *dou*, but only pluck the string. In this case, especially in the Mutsumi-ryu playing style, it is not good execution of the uchi, because the sound of the *shamisen* then consists only of the sound of a string without that of the drum.

In these studies, we proposed two indicators for evaluating the *uchi*, shown in Fig. 3: linearity and the *uchi* angle. We used principal component analysis (PCA) to define linearity and the *uchi* angle.

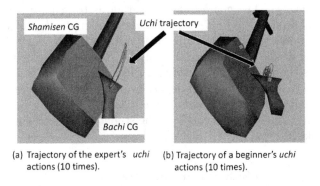

(a) Trajectory of the expert's *uchi*   (b) Trajectory of a beginner's *uchi*
    actions (10 times).   actions (10 times).

Fig. 2. Trajectory of the actions involved in the *uchi* performed an expert and a beginner [13]

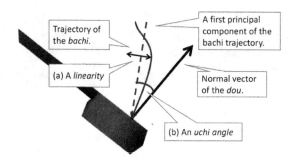

Fig. 3. Definition of linearity and the *uchi* angle [13]

PCA is a statistical procedure which obtains both of principal component vector and contribution ratio using eigenvalue computation of data sets matrix. We defined linearity as a contribution ratio ($0.0 < linearity \leq 1.0$) of the first principal component of a trajectory (Fig. 3(a)). We also defined an *uchi* angle as the angle between the normal vector of the face of the *dou* and the first principal component vector (Fig. 3(b)). We also reported that an expert's linearity was approximately 1.0 and his *uchi* angle was nearly 42°.

In this paper, we employ these indicators for the proposed *uchi* self-learning system. The system encourages a student to play the *shamisen* in a state of high linearity and at a stable *uchi* angle.

3   The *Uchi* Self-learning System

3.1   Educational Theory

Our proposed *uchi* self-learning system is based on the educational theory for Japanese traditional performing arts called *Shu-Ha-Ri*. *Shu* means "keep," and is a beginner's grade. A *Shu*-grade student is expected to imitate and keep up with his/her teacher. The goal of the *Shu* grade is to acquire basic skills. *Ha* means "break," and is the next step after the *Shu* grade. A *Ha*-grade student is expected to break through the superficial imitation of the teacher's skill. The goal of the *Ha* grade is the expression of the student's originality. *Ri* means "leave," and is the final step of learning. A *Ri*-grade student is expected to be independent of his/her teacher and depart from the teacher's style.

The main target of our proposed *uchi* self-learning system is a student in the *Shu* grade. Thus, the system encourages the student to imitate an expert's uchi to learn the skill.

3.2   Overview of *Uchi* Self-learning System

Figure 4 shows a configuration of the *uchi* self-learning system. This system contains a motion capturing system (MoCap: Polhemus Liberty. The spatial

resolution of the system is 0.0038 mm, 0.0012° and accuracy is 0.76 mm, 0.15°),
a computer (PC: Dell Precision M3800), a head-mounted display (HMD: Oculus
DK1), a *shamisen*, and a *bachi*. The MoCap consists of a main unit, a magnetic
transmitter, and three receivers. The system captures the position (three DOF)
and rotation (three DOF) of each MoCap receiver at under 120 data frames per
second (fps) while a student practices.

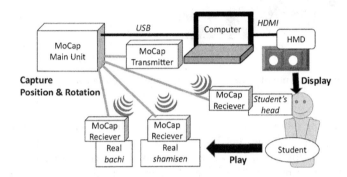

Fig. 4. Configuration of the system

One of the MoCap receivers is set up on the *shamisen*, the second on the
bachi, and the third MoCap receiver is set up on the head of the student using
the system. The main unit of the MoCap is connected to the PC through a USB
cable. A CG image of the VR environment, which reflects the student's motion
in real time, is transmitted to the HMD and displayed to the student.

In order to encourage the student to imitate the expert's *uchi*, this system
provides the two functions: one that evaluates a student's linearity and his/her
uchi angle for each execution of the *uchi* (the evaluating process), and another
that displays an expert's *uchi* as a guide (feedback process).

Figure 5(a) shows a scene where a student is practicing *uchi* actions with the
uchi self-learning system. Figure 5(b) shows CG images of the VR environment
displayed to the student. These images are side-by-side barrel distorted images
to adapt to the HMD.

3.3   Evaluating Process

In order to evaluate a student's execution of the *uchi*, the system needs to recognize the relevant *uchi* action interactively. We define a moment of *uchi* as the
time when the student's *bachi* is close to the face of the *dou*. We define an uchi
action as the action performed between two consecutive moments of *uchi*. We
set up a threshold distance between the *bachi* and the *dou* manually for the
recognition of a moment of *uchi*.

The system runs the evaluating process each time a data frame is captured
by the MoCap. In the first step of the evaluating process, the system stores the

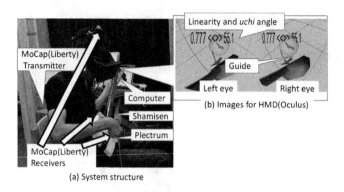

Fig. 5. Student practicing *uchi* using the self-learning system

captured data in memory. In the second step, the system obtains a distance between the *bachi* and the *dou* by using the captured data. If the distance is shorter than the threshold, the system obtains a linearity and an *uchi* angle by applying PCA to time-series position data in the memory. Following this, the system clears the memory. Otherwhise, when the distance is longer than the threshold, the system returns to first step.

3.4   Feedback Process

In the feedback process, the system provides a VR environment using captured data frames and the HMD. The student can see a *shamisen* CG in the same layout as the real *shamisen*. In this VR environment, the results of the evaluation process are shown, as is a guide to the expert's uchi action.

Figure 6 shows a scene from the VR environment. In order to display the results of the evaluating process, the system uses text messages as well as a transparent CG sphere that always follows the *bachi* CG. We call the sphere a "score ball." The system shows an expert's *uchi* action using a transparent *bachi* CG called a "ghost."

Figure 7 shows the way to describe the results of the evaluation process by a score ball. The radius of the score ball describes a linearity (Fig. 7(a)). If the linearity is high, the radius of the score ball is small. The radius R ($0.08 \leq R \leq 0.2$) of the score ball is obtained by Eq. 1, where L_{stu} represents a given student's *uchi* linearity. When the linearity is 1.0, the radius is 0.08 m in the VR environment; when the linearity is 0.0, the radius is 0.2 m in the VR environment.

$$R = (0.2 - 0.08)(1 - L_{stu}) + 0.08 \qquad (1)$$

The color of the score ball describes the *uchi* angle (Fig. 7(b)). Red ($0 \leq r \leq 1$), green ($0 \leq g \leq 1$), and blue ($0 \leq b \leq 1$) components are obtained by Eq. 2, where A_{stu} represents a given student's *uchi* angle and A_{exp} represents the expert's *uchi* angle, which was 42° in this case.

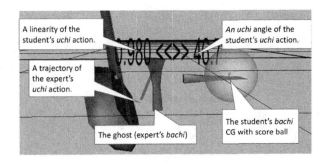

Fig. 6. VR environment

$$(r, g, b) = \begin{cases} (\frac{A_{stu}}{A_{exp}}, 1, \frac{A_{stu}}{A_{exp}}) & if \ A_{stu} < A_{exp} \\ (1, 1, 1) & if \ A_{stu} = A_{exp} \\ (1, \frac{90 - A_{stu}}{90 - A_{exp}}, \frac{90 - A_{stu}}{90 - A_{exp}}) & if \ A_{stu} > A_{exp} \end{cases} \qquad (2)$$

When the student's *uchi* angle is greater than 42°, the score ball turns red; when the *uchi* angle is less than 42°, the score ball turns blue; when the *uchi* angle is exactly 42° (the same as the expert's *uchi* angle), the score ball turns white.

If the score ball is small and white, this means that the student's *uchi* action is similar to that of the expert.

When the student moves the *bachi* close to the ghost (Fig. 8(b)), the system moves the ghost to the point where the expert raises the *bachi* up (Fig. 8(c)). The student raises the *bachi* at the ghost, the system moves down the *bachi* again, and returns to its initial state.

The student practices *uchi* actions by following the ghost until the score ball becomes small and white in order to imitate an expert's *uchi* action.

4   Evaluating the System

4.1   Experimental Instructions

In order to evaluate our proposed system, we compared the *uchi* actions of two groups. One of these, called GpF2F, consisted of students lectured by experts in person (Subjects A to E). The other group, called GpVR, consisted of students who had studied the action using the proposed system (Subjects F to J). All students were beginner *shamisen* students who had never used the instrument before. We did not target the group consisted of students who had studied the action using only the textbook, because there is no textbook for specialized Mutsumi-ryu playing style.

The instruction sequence of this experiments is shown in Table 1. Each member of the GpF2F group was instructed on a basic form of the *shamisen* for approximately five minutes by an expert who was a master of the Mustumi-ryu

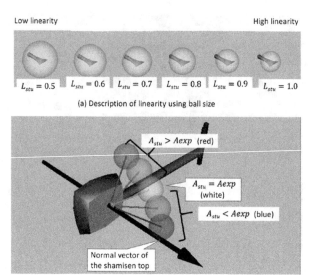

(a) Description of linearity using ball size

$L_{stu} = 0.5$ $L_{stu} = 0.6$ $L_{stu} = 0.7$ $L_{stu} = 0.8$ $L_{stu} = 0.9$ $L_{stu} = 1.0$

Low linearity   High linearity

$A_{stu} > Aexp$ (red)

$A_{stu} = Aexp$ (white)

$A_{stu} < Aexp$ (blue)

Normal vector of the shamisen top

(b) Description of *uchi* angle using colors

Fig. 7. Score ball

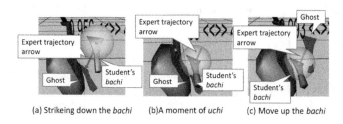

Expert trajectory arrow

Expert trajectory arrow

Expert trajectory arrow

Ghost

Ghost

Ghost

Student's *bachi*

Student's *bachi*

Student's *bachi*

(a) Strikeing down the *bachi*   (b)A moment of *uchi*   (c) Move up the *bachi*

Fig. 8. Guided learning using ghost

playing style. The student then tried the *uchi* action 12 times (PreF2F phase). The members received group lessons for the *uchi* from the expert. After the lesson, they tried the *uchi* action 12 times (PostF2F phase) again.

Each member of the GpVR group read a textbook in the first five minutes. In this phase, the students practiced the basic form of the *shamisen* and learned how to grip the *bachi*. In the second step, the students tried the *uchi* action 27 times (PreVR phase). The students then put on the HMD and began practicing *uchi* actions for five minutes (VR1 phase). Then, each student took a few minutes' rest, and began practicing again (VR2 phase). In the final step, each student tried the *uchi* 27 times (PostVR phase) once again.

The motions of each phase were captured by the MoCap. We captured GpF2F by 30 fps and GpVR by 60 fps. An *uchi* pattern in phases PreF2F and PostF2F is shown in Fig. 9(a), and an *uchi* pattern in phases PreVR and PostVR is shown in Fig. 9(b). These patterns were orally sounded out by staff. We designed these

Table 1. Instruction sequence

	(a) GpF2F		(b) GpVR	
Form practice	5 min	With expert	5 min	Using textbook
Measurement (pre)	12 times (PreF2F)	Without guide	27 times (PreVR)	Without VR feedback
Uchi practice	20 min	Group lesson	5 min (VR1) rest	By the system
			5 min (VR2)	By the system
Measurement (post)	12 times (PostF2F)	Without guide	27 times (PostVR)	Without VR feedback

|♩ ♩ ♩ ♩| ×3

(a) An *uchi* pattern of preF2F and postF2F.

|♩ ♩ ♩ 𝄾|♩ ♩ ♩ 𝄾|♩ ♩ ♩ 𝄾| ×3

(b) An *uchi* pattern of preVR and postVR.

Fig. 9. *Uchi* pattern

sequences and system configurations by keeping in mind the strain experienced by the subjects.

4.2 Result

Figure 10 shows the linearities and the *uchi* angles of Subject I belonging to group GpVR. The vertical line shows the score of evaluation ((a) linearity (ratio), (b) *uchi* angle (degrees)) and the horizontal line shows the *uchi* actions. Phases VR1 and VR2 contained large numbers of *uchi* actions. We chose the first 24 *uchi* data items in VR1 and the last 24 in VR2.

The scores from phases VR1 and VR2 were closer to the expert scoreline than those of the PreVR phase. Deviations in these phases were small. This result shows that the subjects could imitate experts' uchi action using the proposed system.

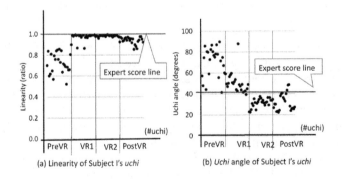

(a) Linearity of Subject I's *uchi*   (b) *Uchi* angle of Subject I's *uchi*

Fig. 10. Linearity and *uchi* angle of Subject I

5   Discussion

The first *uchi* data item of each clause in the *uchi* pattern contained a preparatory motion. Thus, we excluded the first *uchi* data item when discussing the results. The results of interactive *uchi* action recognition provided by the system contained some errors. We revised these errors manually for discussion.

Figure 11 shows the linearities and *uchi* angles of all subjects using boxplot description. The vertical line shows the score of evaluation ((a) linearity (ratio), (b) *uchi* angle (degrees)) and the horizontal one shows those of the *uchi* actions of expert and the subjects. Comparing the pre-phase and the post-phase, we see that both groups GpVR and GpF2F had a shorter range than prior to practice. The median points of the linearities and *uchi* angles were closer to those of the expert after practice as well. We also see that Subject C and Subject D (GpF2F) accorded linearities priority over *uchi* angles. On the other hand, all students in GpVR improved both of linearities and *uchi* angles.

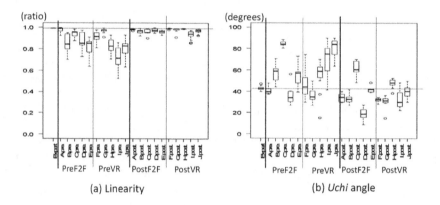

(a) Linearity   (b) *Uchi* angle

Fig. 11. Distribution of linearity and *uchi* angle using boxplot description

To gauge the effectiveness of the proposed system, we used hierarchical cluster analysis with the group-average method. Figure 12 shows a dendrogram of the obtained clusters. The vertical line shows the distance between clusters, and the horizontal one represents the *uchi* actions of expert and the subjects. The expert is highlighted by a filled rectangle, and members of group PostVR are represented by empty rectangles. A branch at a certain distance from an adjacent branch in the dendrogram represents one cluster.

We drew a cutting line in such way that a cluster including an expert was independent of the others. This results shows that the PostVR members and most PostF2F member were organized in the same cluster as the expert (the broken-line rectangle in the figure). This means that subjects who had practiced using our proposed system learned *uchi* actions as well as those taught by the expert. Thus, our proposed system has sufficient functionality for *uchi* self-learning.

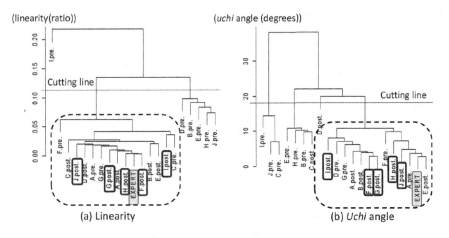

Fig. 12. Dendrogram of cluster analysis

6   Conclusion

In this paper, we proposed an *uchi* self-learning system using VR. The system captures *uchi* actions by MoCap, and simultaneously evaluates linearities and *uchi* angles. The system also send feedback results of an evaluating process, and shows the feature of an expert's *uchi* action as a guide for students.

In an experiment, we compared a group of subjects taught *uchi* actions by an expert in person with one that had practiced *uchi* actions using the proposed system. The results showed that our system was able to teach students comparable skill to that taught by the expert. This means that a student can practice *uchi* using the proposed system.

Some experts have pointed out that form is important while playing a *shamisen*. We need a system that can allow student to practice both *uchi* actions and playing form. A force a player takes for a *bachi* is also important. We need to discuss the contact force between a *bachi* and a body of a *shamisen*. We should also consider methods to evaluate the music produced by a *shamisen* student. These issues will form the subject of our future research.

Acknowledgment. We are grateful to Toshimitsu Ono, a teacher of the Mutsumi-ryu playing style, for his generous support. Hiroaki Katsura, emeritus professor at Akita University, gave insightful comments and suggestions on earlier drafts of this paper. We also thank Shin Kosaka for insightful discussions. This work was supported by the JSPS KAKENHI Grant Number 15K16103.

References

1. Soga, A., Shiba, M., Niwa, Y., Okada, Y.: Archives and exhibits of Buddhist ceremonial processions for museum. TVRSJ **19**(3), 405–421 (2014)
2. Tamamoto, H., Yukawa, T., Kaiga, T., Mitobe, K., Miura, T., Yoshimura, N.: Development of digital contents production techniques for bequeathing folkloric performing arts in cooperation among industry, academia and local government. J. Inst. Electron. Inf. Commun. Eng. **91**(4), 303–308 (2008). Japanese
3. Miura, T., Kaiga, T., Katsura, H., Tajima, K., Sihbata, T., Tamamoto, H.: Adaptive keypose extraction from motion capture data. J. Inf. Process. **22**(1), 67–75 (2014)
4. Shibata, T., Tamamoto, H., Matsumoto, N., Miura, T., Yokoyama, Y.: Development of user-oriented interactive dance learning assistant system. IEICE Trans. Inf. Syst. (Japan. Edn.) D **J97-D**(5), 1014–1023 (2014). Japanese
5. Chan, J.C.P., Leung, H., Tang, J.K.T., Komura, T.: A virtual reality dance training system using motion capture technology. IEEE Trans. Learn. Technol. **4**(2), 187–195 (2011)
6. van der Linden, J., Schoonderwaldt, E., Bird, J., Johnson, R.: MusicJacket - combining motion capture and vibrotactile feedback to teach violin bowing. IEEE Trans. Instrum. Meas. **60**(1), 104–113 (2011)
7. Erwin, S., Altenmuller, E.: Mastering the violin: motor learning in complex bowing skills. In: Proceedings of the International Symposium on Performance Science, pp. 649–654 (2011)
8. Mitobe, K., Tomioka, M., Saito, M., Suzuki, M.: Development of a ubiquitous learning system for dexterous hand operation. In: USB Proceedings of International Symposium on Mixed and Augumented Reality, Atlanta, USA, pp. 299–300 (2012)
9. Mitobe, K., Kodama, J., Miura, T., Tamamoto, H., Suzuki, M., Yoshimura, N.: Developments of the learning assist system for dextrous finger movements. In: ACM SIGGRAPH ASIA, Seoul, Korea (2010). Article No. 30
10. Kosakaya, J., Kodama, N., Kawamorita, R.: Research of the automatic scoring system by taking sounds down in musical notations for traditional folk songs (Tsugaru Shamisen). IEICE Technical report, EA, vol. 108, no. 491, pp. 37–42 (2009). Japanese
11. Hamanaka, T., Sakamoto, D., Igarashi, T.: Aibiki: supporting shamisen performance with adaptive automatic score scrolling. IPSSJ SIG Technical report, vol. 2014-HCI-157, no. 69, pp. 1–6 (2014). Japanese
12. Shibata, T., Mitobe, K., Saito, M., Fujiwara, K., Tamamoto, H.: Development of a self-learning system for stroking shamisen plectrum using VR environment and evaluation of the user skill. In: Annual Conference on Proceedings of the Virtual Reality Society of Japan, vol. 19, pp. 216–219 (2014). Japanese
13. Kosaka, S., Shibata, T., Tamamoto, H., Katsura, H., Yokoyama, H.: A learning assistant system for a basic skill in samisen performance using features of a bachi motion. Forum Inf. Technol. **10**(4), 421–426 (2011). Japanese
14. Katsura, H.: The Formation of Schools of Shamisen Performers in Akita Prefecture, Memoirs of Faculty of Education and Human Studies, Akita University. The Humanities & the Social Sciences, vol. 70, pp. 1–7 (2013). Japanese

Action Learning to Single Robot Using MARL with Repeated Consultation: Realization of Repeated Consultation Interruption for the Adaptation to Environmental Change

Yoh Takada and Kentarou Kurashige$^{(\boxtimes)}$

Muroran Institute of Technology, Muroran, Hokkaido 050-8585, Japan
16043034@mmm.muroran-it.ac.jp, kentarou@csse.muroran-it.ac.jp

Abstract. We have proposed multi-agent reinforcement learning with repeated consultation (MARLRC) as a multi-agent reinforcement learning that agents can select the concerted action. In MARLRC, agents select a virtual action and share it with other agents several times in one robot decision-making. In this study, we focused on the problem that MARLRC does not take into account the environment that time constraints may occur in the decision-making. As an approach to solve this problem, we considered to introduce to determine the amount of time that can be used in decision-making and decision-making in predetermined time. We introduced the method to decision-making in time predetermined by MARLRC in this study.

Keywords: Multi-agent system · Reinforcement learning · Cooperative control

1 Introduction

In recent years, robots have become active in complex environments. In complex environments, the amount of information handled by a robot will increase explosively. Therefore, it seems that a conventional robot that has a mechanism to manage the whole cannot adapt to complex environments. An autonomous decentralized system (ADS) [1] has been paid attention to adapt these environments. An ADS does not have a mechanism to manage the whole. In an ADS, each component that make up the system acts autonomously. An ADS solves the overall system problem by the interaction of each element. An ADS distributes the information handled by a robot to each element. Therefore, a robot using an ADS has the potential to adapt to complex environments. This study deals with a multi-agent system (MAS) [2–6] which is one of ADSs. A MAS is a system consists of multiple agents. A MAS has more excellent problem-solving capacity, fault tolerance and flexibility than a single-agent system. Agents act individually and interact with other agents. A MAS can solve problems that are difficult or impossible for a single agent.

© Springer International Publishing Switzerland 2016
N. Kubota et al. (Eds.): ICIRA 2016, Part II, LNAI 9835, pp. 371–382, 2016.
DOI: 10.1007/978-3-319-43518-3_36

As a configuration using a MAS to a robot, there is a group robot configurations [7–9] and single robot configuration [10,11]. Group robot configuration sets agents to multiple robots. Single robot configuration sets multiple agents in each module of a robot. In this study, we aim to improve fault tolerance, flexibility, and processing speed by using a single robot configuration. In a single robot configuration, an action of a robot is composed from actions of each agent. Therefore, an action of each agent has a significant impact on other agents action. When individual agents make decisions without considering an action of other agent, a robot cannot solve the system-overall goal. For these characteristics, coordination of agents is of particular importance in a MAS of a single robot configuration.

In this study, we use reinforcement learning [12–17] to action learning of an agent to be set in each module. Thus, a robot can acquire actions by trial and error in an unknown environment. Studies using a multi-agent reinforcement learning [10,11] in a single robot have been done from the previous. In conventional methods, a reinforcement learning agent is applied to each robot module. In previous studies where an agent was specified for each actuator in a robot arm, each agent reduced the overall learning time by learning an action of a corresponding actuator in parallel. However, there is a problem that each agent does not explicitly cooperate in the conventional method. If each agent is in complex environments that have multiple optimal action for agents, each agent does not always select the same optimal action. Therefore, optimal actions of each agent are not necessarily an optimal action of a robot in these environments. To solve this problem, we have proposed multi-agent reinforcement learning with repeated consultation [18] (MARLRC) as a multi-agent reinforcement learning (MARL) to select a concerted action. In MARLRC, agents select a virtual action and share a virtual action with other agents several times in one robot decision-making. A virtual action is defined as action that does not output as an actual action of the robot in this study. An actual action is defined as an action that is output from the robot. Thus, agent can make decisions taking into account an action of selecting other agents. Therefore, MARLRC solves a problems of conventional researches.

In the proposed method, it does not take into account an environment in which the time constraints in a decision-making occur. If a time that a robot can be used in decision making is unlimited, a robot is able to select an optimum behavior over time. However, if a time constraints on a decision-making occur, a robot must select an action within a time that can be used in decision-making. In this study, we are considering to introduce to determine an amount of time that can be used in decision-making and decision-making in time determined as an approach to solve this problem. In this study, we aim at realization of a method to decision-making in time determined in MARLRC.

2   MARL of Repeated Consultation

MARLRC is a method that an agent selects a concerted action in multi-agent reinforcement learning. In this method, the state of an agent includes actions of other agents. Thus, agents can assess their behavior, including actions of other agents. In this method, agents select a virtual action and share a virtual action with other agents several times in one robot decision-making. A virtual action is defined as action that does not output as an actual action of the robot in this study. An actual action is defined as an action that is output from the robot. Agents can make decisions taking into account an action of selecting other agents by selecting a virtual action and sharing a virtual action with other agents several times.

An overview of this method is shown in Fig. 1. In this method, each agent selects a virtual action several times relative to decision-making at time t. An agent sends a virtual action to other agents after selecting a virtual action. This operation is defined as a "step". Steps are performed from the first step and step 1 to step N. First step is defined as the step that performed before step 1. Each agent does not output an actual action until the end of the step N. When all of agents has finished step n, agents proceeds to step $n + 1$. After step N, each agent outputs a selected virtual action as an actual output. This process requires a selection of a single action for a robot. An agent does not output an actual action until the end of step N.

The detail of this method is shown below. In this method, we used Q-learning as the learning method for each agent. An action evaluation value in Q-learning is represented as $Q_i(s, a)$. A s represents a state, and an a represents an action. The i-th agent represents agent i in this method. The state of an agent is denoted by s^i, and an action of an agent is represented by a^i. In order to output a concerted action with other agents, a state of each agent adds actions of other agents. Actions of other agents in agent i are represented by b^i. A state of an agent in this method is represented as $\{s^i, b^i\}$. Thus, an action evaluation value in this method is represented by $Q_i(\{s^i, b^i\}, a^i)$. In each step, an action selection is determined using the cooperative action evaluation value, which is defined in this method. The cooperative action evaluation value is represented as $Q_i^c(s^i, a^i)$. The cooperative action evaluation value is a evaluation of an action value including indicators for cooperating with other agents. By using the cooperative action evaluation, agents can assess their behavior, including actions of other agents. The cooperative action evaluation value is calculated by Eq. (1) using the evaluation of an action value and the action consultation degree. The action consultation degree is defined in this method, represented as $p_i(a^i, b^i)$. The action consultation degree is an indicator for cooperation with the other agent. The action consultation degree is calculated by Eq. (2), and it is updated using a Eq. (3) or (4).

$$Q_i^c(s^i, a^i) = \sum_{b^i \in B^i} (Q_i(\{s^i, b^i\}, a^i) \times p_i(a^i, b^i)) \tag{1}$$

$$p_i(a^i, b^i) = \begin{cases} \frac{1}{U}, & (Q_i(\{s^i, b^i\}, a^i) = \max(Q_i(\{s^i, b^i\}, a^i)) \\ 0, & (otherwise) \end{cases} \tag{2}$$

$$p_i(a^i, b^i) \leftarrow p_i(a^i, b^i) + \beta\{1 - p_i(a^i, b^i)\} \tag{3}$$

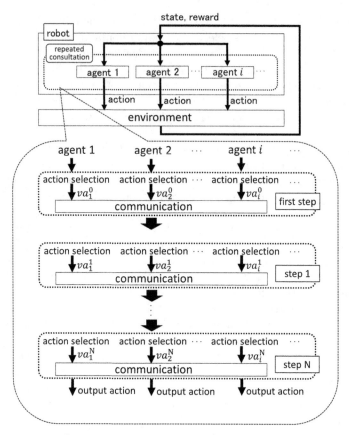

va_i^n :virtual action is taken by agent i in step n

Fig. 1. Overview of MARLRC

$$p_i(a^i, b^i) \leftarrow p_i(a^i, b^i) + \beta\{0 - p_i(a^i, b^i)\} \qquad (4)$$

The action consultation degree is the degree that other agents choose actions b^i for action a^i of agent i in a certain state. The action consultation degree is calculated by Eq. (2), based on the action evaluation value of each agent when first step. U is the number of maximum action evaluation values. The action consultation degree is calculated for all combinations of action of agent i and actions of other agents. When the value of the action consultation degree is almost 1, a combination of a^i and b^i is a cooperative combination. In contrast, when the value of the action consultation degree is almost 0, a combination of a^i and b^i is the combination that is not cooperative. At the beginning of each step except first step, the action consultation degree is updated using a Eq. (3) or (4). Eq. (3) is applied to the combination of b^i selected by other agents and a^i. Eq. (4) is applied to the combination of b^i non-selected by other agents and

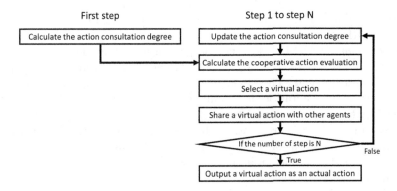

Fig. 2. Flow of the operation in each step

a^i. This increases the degree that the combination of b^i selected by other agents and a^i, and decreases the combination of b^i non-selected by other agents and a^i. β is a parameter that determines a degree to update the action consultation degree. When the value of β is almost 1, the agent can select a cooperated action with other agents in a small number of updates. However, the update of the action consultation degree will become sketchy. In contrast, when the value of β is almost 0, an agent must update the action consultation degree many times to select a cooperated action with other agents. However, it is possible to accurately perform the update of the action consultation degree. Thus, agents can learn an action has the high reward.

Finally, we describe an operation of each step. At the beginning of a "step," each agent updates their action consultation degrees using Eq. (3) or (4). It should be noted that in first step, each agent calculates their action consultation degrees using Eq. (2). Each agent then calculates the cooperative action evaluation value based on an action evaluation value and the action consultation degree using Eq. (1). An agent selects a virtual action and shares it with other agents. This is performed repeatedly from first step and step 1 to step N. When N steps have been completed, each agent outputs a virtual action as an actual action. The flow of the operation in each step is shown in Fig. 2.

3   Interruption of Repeated Consultation

MARLRC does not take into account the environment in which a time constraints in decision-making occur. In this study, a time constraint is a constraint by environment changing by progress at time. If a time that a robot can use to decision-making is unlimited, a robot is able to select a optimum behavior over time. However, if a time constraints on a decision-making occur, it seems that a robot can not choose the optimal action. A robot must select a better action within a time that can be used in decision-making. As an approach to solve this problem, we considered to introduce to determine the amount of time that can be used in decision-making and decision-making in determined time. In this study, we introduced the method to decision-making in the predetermined time in MARLRC.

Conventional MARLRC does not output an actual action until the end of the step N. It seems that an agent cannot make decision in determined time if step N does not end. Therefore, in this study we introduced the interruption of repeated consultation as the method to decision-making in determined time in MARLRC. In this method, each agent interrupts the repeated consultation if it meets the requirements of the interruption of repeated consultation. When repeated consultation was interrupted by each agent, each agent outputs a selected virtual action as an actual action. When agents got a reward and an environmental state by an action, agents make decision newly.

In this study, we configure a time of consultation interruption as the basis for the repeated consultation interruption in the determined time. The time of consultation interruption is a time of the criteria interrupt the repeated consultation. In MARLRC of introducing the interruption of repeated consultation and a time of consultation interruption, each agent measures a time required for a consultation from a time of consultation started. Agents measure a time required for repeated consultation every steps. When a time required to the repeated consultation has exceeded a time of consultation interruption, agents interrupt the repeated consultation. Therefore, agents can make a decision in a limited time if a time constraints on the decision-making occur by using the repeated consultation interruption.

4   Simulation Experiment of Leaching Task

4.1   Experimental Aim

This experiment aimed to verify whether agents could make a decision in the determined time by introducing the repeated consultation interruption. In addition, we compared the experimental results for the proposed method and normal MARLRC. We verified whether agents make a decision in the determined time and the learning results changes in the experiment by introducing an interruption of the repeated consultation.

4.2   Experimental Overview

In this study, we performed a simulation experiment of reaching task using a robot arm. We compared a time used in decision-making and the number of actions in each trial by using experimental results.

We used Gazebo [19, 20] as a simulation environment in this experiment. Gazebo is a three-dimensional robot simulator using a physics engine. This experiment is performed using the robot arm of the four-link. The robot arm is constituted by four links and four actuators. We define one action of the robot arm as one running on each actuator of the robot arm. The task is accomplished when the tip of the robot arm leaches the target point. The target point is a sphere with a constant radius. The coordinate and the radius of the target point is the same in all trials. Each agent gets a reward in that accomplishes the task. If the robot is not able to accomplish the task within 500 actions, each agent will get a negative reward as punishment. We define one trial as to accomplish the task or to reach the action number of the robot 500 action. After trial, the position of the robot arm returns to the initial state.

4.3   Experimental Setting

The parameters used in the experiments are shown in Table 1. The setting of the robot arm shown in Fig. 3(a). Each actuator is mounted to the joint connecting a link. The actuator is running to any of three angles. The robot arm performs a reaching action by operating respective actuators. The setting for the link is shown in Fig. 3(b). The actuator can take any of actions, $-\Delta\theta^\circ, 0^\circ, \Delta\theta^\circ$ in one action. The range of motion of the link is set to 0° from 90°.

Table 1. Parameter used in this experiment

Number of trials	1000
Reward	1.0
Negative reward	-100.0
α	0.1
γ	0.9
ϵ	0.01
N	50
β	0.01
θ_1	10°
θ_2	10°
θ_3	10°
θ_4	10°
$\Delta\theta$	10°
Link size	$1.0 \times 0.1 \times 0.1\,[\mathrm{m}^3]$
Link weight	$1.0\,[\mathrm{kg}]$
Center coordinates of the target point	$(2.0, 0.3, 0.0)$
Radius of the target point	$1.0\,[\mathrm{m}]$

4.4   Agent Configration

In this experiment, each agent is set to each link of the robot arm. Since the number of links of the robot arm is 4, the number of agents is four. Agents learn

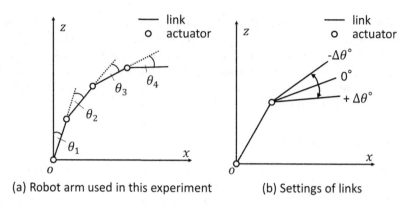

(a) Robot arm used in this experiment   (b) Settings of links

Fig. 3. Settings of the robot arm and links

a behavior of the corresponding link. In this experiment, we used Q-learning as the learning method for each agent. In MARLRC, agents are included in actions of other agents as a state. Thus, update equation the action evaluation value of agents i is shown by Eq. (5). S_t^i represents the environment state at time t. b_t^i represents the actions of the other agents in the agent i at time t. r_{t+1} represents a reward obtained at time $t+1$. α represents the learning rate. γ represents the discount rate. The learning rate and the discount rate are defined in Q-learning.

$$
\begin{aligned}
Q_i(\{s_t^i, b_t^i\}, a_t^i) \leftarrow &\, Q_i(\{s_t^i, b_t^i\}, a_t^i) \\
&+ \alpha \left\{ r_{t+1} + \gamma \max_{b^i, a^i} Q_i(\{s_{t+1}^i, b^i\}, a^i) - Q_i(\{s_t^i, b_t^i\}, a_t^i) \right\}
\end{aligned} \tag{5}
$$

We also used the ϵ-greedy method as the action selection method. ϵ-greedy method selects a random action with a probability of ϵ and an optimal action in an other probabilities. Thus, ϵ-greedy method can take an exploratory action and an optimal action well-balanced. In this experiment, each agent decides whether to select an optimal action or to select a random behavior in first step. If an agent has selected a random action in first step, an agent continues to select a virtual action that selected in first step.

4.5   Configration of Interruption of Repeated Consultation

In the proposed method, time constraints are generated for all of decision-making of a robot. Time constraints due to environmental change is not intended to be generated for all of decision-making. For clarity the results, time constraints for every decision are assumed to occur. In this experiment, we set the time of consultation interruption as 50 [ms]. To verify whether the decision-making within a determined time, we measure the consultation time in this experiment. We defined the consultation time as a time from starting to repeated consultation to the end of decision-making. We define the timing of the decision-making as when an agent finishes step N or the repeated consultation has been interrupted by agents. In this experiment, we measured the consultation time for each action and calculated the average at each trial. We defined the average consultation time as the consultation time of one trial.

4.6 Experimental Results

First, we compared the average consultation time in each trial in conventional method and proposed method. The average of consultation time in each trial is shown in Fig. 4(a) and (b). The horizontal axis represents the number of trials and the vertical axis represents the average of consultation time in each trial. In conventional method, it took about 150 ms to decision-making. On the other hand, proposed method was performed decision-making in 50 ms near the consultation interruption time.

Then, the number of action in each trial is shown in Fig. 4(c) and (d). The horizontal axis represents the number of trials and the vertical axis represents the number of action in each trial. In conventional method, the number of actions has converged in 300 trials from approximately 200 trials. On the other hand,

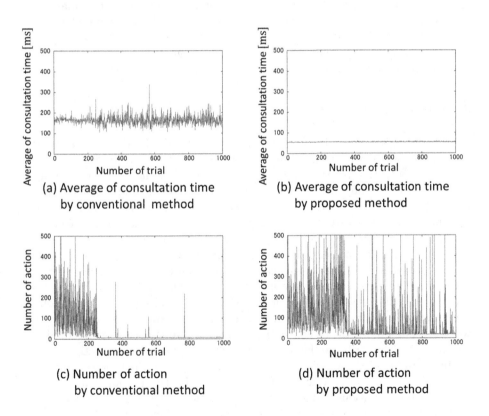

(a) Average of consultation time by conventional method

(b) Average of consultation time by proposed method

(c) Number of action by conventional method

(d) Number of action by proposed method

Fig. 4. Results in this experiment

the number of actions of proposed method is extremely larger than conventional methods one. The number of action is changing about 400 trials to border in proposed method. The number of actions after 400 trials is less than the number of action before 400 trials. However, the number of actions is not clearly converge.

4.7   Discussion

By these results, agents were able to make decision in 50 ms near a collegiate interruption time. Accordingly, we have realized the action selection in the determined time with the purpose of the present experiment. By contrast, the number of actions in each trial of proposed method has become very large value than conventional method. In addition, the number of actions in proposed method does not clearly converge compared with conventional method. It seems that the convergence of action was delayed by each agent could not cooperate satisfactorily by the interruption of the repeated consultation. In MARLRC, each agent selects a concerted action by performing a repeated consultation at the time of an action selection. A virtual action has been selected as an action that has not been enough consultation by the interruption of the repeated consultation. As a result, it seems that agents output an actual actions that is not coordinated. In addition, time constraints for every decision are assumed to occur for clarity the results. Thus, the repeated consultation was interrupted in all decision-making. For this reason, it seems that the learning result of agents was worse as compared with the conventional method result.

5   Conclusion

In this study, we focused on the problem that MARLRC does not take into account the environment that time constraints may occur in decision-making. As an approach to solve this problem, we considered to introduce to determine the amount of a time that can be used in decision-making and decision-making in the determined time. We introduced the method to decision-making in time determined by MARLRC in this study. Based on simulation experiments, we verified whether agents could make a decision in the determined time by introducing the repeated consultation interruption. From the experimental results, we have demonstrated that agents can perform decisions in accordance with the determined time by using the proposed method. Thus, we have confirmed that agents can make decision to the determined time by using the interruption of the repeated consultation. On the other hand, the learning result of agents was worse as compared with the conventional method result by each agent could not cooperate satisfactorily by the interruption of the repeated consultation.

In future research, we will improve learning results of the proposed method. There is room for improvement of the learning by the proposed method. In reviewing the conditions and method of interruption, we will close to the learning results of the proposed method on the results of the conventional method. In addition, we will consider introducing to determine the amount of time that can be used in decision-making. We will consider experiments using the proposed method in an environment where time constraints on decision-making occur. We will verify whether the agent can perform the decision-making in an environment where time constraints are generated. The other, we also consider experiments in other tasks. The current task is a very simple task. Therefore, it seems that the experimental results of the proposed method are not much different from the experimental results of single-agent system. Then, we are considering that experiments with increased number of links the robot arm or experiments of reaching task by an another robot. Moreover, we consider experiments using a real robot since a real robot experiments with MARLRC are not performed.

References

1. Liu, W., Gu, W., Sheng, W., Meng, X., Wu, Z., Chen, W.: Decentralized multi-agent system-based cooperative frequency control for autonomous microgrids with communication constraints. IEEE Trans. Sustain. Energy **5**, 446–456 (2014)
2. Ferber, J.: Multi-Agent System: An Introduction to Distributed Artificial Intelligence. Addison Wesley Longman, Boston (1999)
3. Busoniu, L., Babuska, R., De Schutter, B.: Multi-agent reinforcement learning: a survey. In: Proceedings of the 9th International Conference on Control, Automation, Robotics and Vision (ICARCV 2006), Singapore, pp. 527–532 (2006)
4. Panait, L., Luke, S.: Cooperative multi-agent learning: the state of the art. Auton. Agent. Multi-Agent Syst. **11**, 387–434 (2005)
5. Montano, B.R., Yoon, V., Drummey, K., Liebowitz, J.: Agent learning in the multi-agent contracting system [MACS]. Decis. Support Syst. **45**, 140–149 (2008)
6. Ren, H.: Reinforcement learning in cooperative multiagent systems. In: Eighteenth National Conference on Artificial Intelligence, pp. 326–331 (2002)
7. Innocenti, B., Lopez, B., Salvi, J.: A multi-agent architecture with cooperative fuzzy control for a mobile robot. Robot. Auton. Syst. **55**, 881–891 (2007)
8. Wang, Y., de Silva, C.W.: A machine-learning approach to multi-robot coordination. Eng. Appl. Artif. Intell. **21**, 470–484 (2008)
9. Ota, J.: Multi-agent robot systems as distributed autonomous systems. Adv. Eng. Inform. **20**, 59–70 (2006)
10. Varshavskaya, P., Kaelbling, L.P., Rus, D.: Automated design of adaptive controllers for modular robots using reinforcement learning. Int. J. Robot. Res. **27**(3–4), 1–26 (2008)
11. Iijima, D., Yu, W., Yokoi, H., Kakazu, Y.: Obstacle avoidance learning for a multi-agent linked robot in the real world. In: Proceedings of the 2001 IEEE International Conference on Robotics and Automation (ICRA), vol. 1, pp. 523–528 (2001)
12. Sutton, R.S., Barto, A.G.: Reinforcement Learning: An Introduction. MIT Press, Cambridge (1998)
13. Dayan, P.: Reinforcement learning. In: Encyclopedia of Machine Learning, pp. 849–851 (2010)
14. Kaelbling, L.P., Littman, M.L., Moore, A.P.: Reinforcement learning: a survey. J. Artif. Intell. Res. **4**, 237–285 (1996)
15. Watkins, C.J.C.H., Dayan, P.: Q-learning. Mach. Learn. **8**, 279–292 (1992)
16. Eric, M., Alexander, S., Arkin, R.C.: Robot behavioral selection using Q-learning. In: IEEE/RSJ International Conference on Intelligent Robots and Systems, vol. 1, pp. 970–977 (2002)

17. Milln, J.R., Posenato, D., Dedieu, E.: Continuous-action Q-learning. Mach. Learn. **49**, 247–265 (2002)
18. Chiba, S., Kurashige, K.: Action learning to single robot using MAS-a proposal of agents action desicion method based repeated consultation. In: The 10th Asian Control Conference 2015 (ASCC 2015), pp. 410–415 (2015)
19. Gazebo. http://gazebosim.org/
20. Koenig, N., Howard, A.: Design and use paradigms for Gazebo, an open-source multi-robot simulator. In: IEEE/RSJ International Conference on Intelligent Robots and Systems (IROS), Sendai, Japan (2004)

A Study on the Deciding an Action Based on the Future Probabilistic Distribution

Masashi Sugimoto[1]([✉]) and Kentarou Kurashige[2]

[1] National Institute of Technology, Kagawa College, Mitoyo, Japan
sugimoto-m@es.kagawa-nct.ac.jp
[2] Muroran Institute of Technology, Muroran, Japan
kentarou@csse.muroran-it.ac.jp

Abstract. In case of operating the robot in a real environment, its behavior will be probabilistic due to the slight transition of the robot's state or the error of the action that is taken at each time. We have previously reported that prediction of the state-action pair, is the prediction method to link the state and action of the robot for future the state and the action. From this standpoint, we have proposed the method that decides the action that tends to take in the future. In this paper, we will try to introduce the statistical approach to the prediction of the state-action pair. From this standpoint, we propose the method that decides the action that tends to take in the future, for the current action. In the proposed method, we will calculate the existence probability of the state and the action in the future, according to the normal distribution.

Keywords: Prediction of the state-action pair · Decision for the optimal action · Prediction distribution · Online SVR · Normal distribution

1 Introduction

For controlling robots in some dynamic environments, it may realize choosing the action that will be adopted the current results as state and action by predicting a future state using previous actions and states [1]. In case of operating the robot in a real environment, its behavior will be changed as a probabilistic due to the slight fluctuation of the robots' state or the error of the action that is taken at each time. In this case, stochastic technique will be necessary for handling the problems with unknown disturbances, will be particularly important [2–7]. However, these studies had been left the future states and actions of robots out of considered. In other words, it will be difficult to predict the future states by considering the current action. We have previously reported that prediction of the state-action pair, is the prediction method to link the state and action of the robot for future the state and the action. From this standpoint, the methods that had decided the action that tends to take in the future, had been proposed [8–10]. As mentioned above, the stochastic technique that contains state and action pair prediction method, will be effectively and necessary. In this paper,

© Springer International Publishing Switzerland 2016
N. Kubota et al. (Eds.): ICIRA 2016, Part II, LNAI 9835, pp. 383–394, 2016.
DOI: 10.1007/978-3-319-43518-3_37

the statistical approach will be introduced to the state-action pair prediction. From this standpoint, the method that decides the action that tends to take in the future is proposed, for the current action. In the proposed method, the existence probability of the state and the action in the future will be calculated, according to the normal distribution.

This paper is organized as follows: In Sect. 2, how to predict and decide an optimal action for robot applying state-action pair prediction technique that was introduced stochastic technique, will be motivated. Further, details about the stochastic state-action pair prediction will be stated. In Sect. 3, a verification experiment configuration will be described. In Sect. 4, the summary of this work is concluded.

2   An Approach for Introducing Normal Distribution Function to State-Action Pair Prediction

In this study, we will try to consider to obtain the optimal action to minimize the body pitch angle of the inverted pendulum (in Fig. 1) under the predictable unknown disturbance given continuously, using the state-action pair prediction that has been proposed by our former studies [8]. Therefore, in this paper, the system in Fig. 2 based on the former studies is considered. As shown in Fig. 2, $^{t-l}\hat{\mathbf{u}}(t+j)$ describes the prediction result of the control input $\mathbf{u}(t+j)$, when this input is predicted in time $(t-l)$. Therefore, this proposed method will be trying to compensate the current action that will be taken, using combination of the optimal control and the prediction result of state-action pair prediction. The structure of the prediction of a state-action pair is named as "N-ahead state-action pair predictor," is the internal structure has been described in Fig. 3. In this case, if N will be larger or more time distant than current time t, then the prediction error rate will be proportional to N, moreover, the cumulative prediction error cannot be ignored in the prediction depend on the time t. In case of a stochastic probability of an ordinary robot control, it has some peak of probabilistic curves. However, in this study, maximum peak of probability of the robot action will be focused on. From this viewpoint,

(a) The Structure

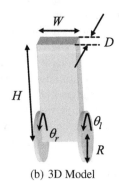

(b) 3D Model

Fig. 1. The structure of "NXTway-GS"

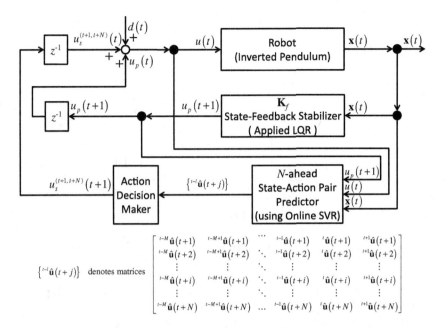

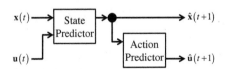

Fig. 2. A outline of the deciding the optimal action for the robot using the prediction of stochastic state-action pair

the normal distribution will be applied in this study. In this study, the existence probability of state and action in the future, according to the normal distribution will be calculated. Moreover, future action will be revised by state and action that have highest existence probability. In addition, it will be considered that increase the influence for the action as the latest time t, focused on the time that prediction is started [11].

On the basis of this idea, the distribution of existence probability of prediction series will be calculated, based on the time that prediction is started. In addition, its average and standard deviation will be focused on. The predictor will be predicting at each point in time, continuously. Therefore, from this mechanism, the values of prediction results will be obtained at each sampling time. Future values at each sampling time predicted and obtained by a stochastic state-action pair prediction. The image of these results are illustrated in Fig. 4(a). Note the "focus" area in Fig. 4(a).

Fig. 3. A structure of the N-ahead state-action pair predictor [1]

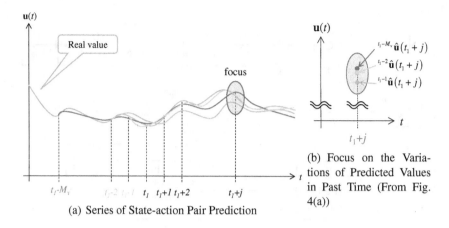

(a) Series of State-action Pair Prediction

(b) Focus on the Variations of Predicted Values in Past Time (From Fig. 4(a))

Fig. 4. Variations of predicted series using a state-action pair

Here, the prediction results at time (t_1+j) can be obtained from each previous time at $(t_1 - i)$ (the range of i will be given by $(0 \ll M_v \leq i \leq N)$) as illustrated in Fig. 4(b). In this zone at $(t_1 + j)$, the prediction results are distributed. To quantitatively measure these variations, the average and the standard deviation will be focused on. In brief, the average $\mu_j^{\mathbf{u}}$ and the standard deviation $\sigma_j^{\mathbf{u}}$ of the control input in time $(t_1 + j)$, using the prediction values that are predicted in each $(t_1 - i)$ (as past time), can be derived as follows:

$$\mu_j^{\mathbf{u}} = \mu \left(\frac{w_M}{w_\Sigma}{}^{t_1-M}\hat{\mathbf{u}}(t_1 + j), \frac{w_{M+1}}{w_\Sigma}{}^{t_1-M+1}\hat{\mathbf{u}}(t_1 + j), \ldots, \frac{w_1}{w_\Sigma}{}^{t_1+1}\hat{\mathbf{u}}(t_1 + j) \right)$$
(1)

$$\sigma_j^{\mathbf{u}} = \sigma \left(\frac{w_M}{w_\Sigma}{}^{t_1-M}\hat{\mathbf{u}}(t_1 + j), \frac{w_{M+1}}{w_\Sigma}{}^{t_1-M+1}\hat{\mathbf{u}}(t_1 + j), \ldots, \frac{w_1}{w_\Sigma}{}^{t_1+1}\hat{\mathbf{u}}(t_1 + j) \right)$$
(2)

Similarly, about the state $\mathbf{x}(t_1 + j)$, average $\mu_j^{x_k}$ and standard deviation $\sigma_j^{x_k}$ are derived. Here, $k \in \dim \mathbf{x}$.

$$\mu_j^{x_k} = \mu \left(\frac{w_M}{w_\Sigma}{}^{t_1-M}\hat{x}_k(t_1 + j), \frac{w_{M+1}}{w_\Sigma}{}^{t_1-M+1}\hat{x}_k(t_1 + j), \ldots, \frac{w_1}{w_\Sigma}{}^{t_1+1}\hat{x}_k(t_1 + j) \right)$$
(3)

$$\sigma_j^{x_k} = \sigma \left(\frac{w_M}{w_\Sigma}{}^{t_1-M}\hat{x}_k(t_1 + j), \frac{w_{M+1}}{w_\Sigma}{}^{t_1-M+1}\hat{x}_k(t_1 + j), \ldots, \frac{w_1}{w_\Sigma}{}^{t_1+1}\hat{x}_k(t_1 + j) \right)$$
(4)

Hereby, $\mu(\cdot)$ denotes the average of $(\cdot)$, and $\sigma(\cdot)$ denotes the standard deviation of $(\cdot)$. Moreover, $w_M, w_{M+1}, \ldots, w_1$ denote weights of weight moving average, and w_Σ denote sum of the weight of the weight moving average. From the above expression, the normal distribution function $N_j^{x_k}$ in time $t_1 + j$, and the function $N_j^{\mathbf{u}}$ for the control input $\mathbf{u}$ are derived, shown in below equations.

$$N_j^{x_k}(X_k) = \frac{1}{\sqrt{2\pi \left(\sigma_j^{x_k}\right)^2}} \exp\left\{ -\frac{\left(X_k - \mu_j^{x_k}\right)^2}{2\left(\sigma_j^{x_k}\right)^2} \right\} \tag{5}$$

$$N_j^{\mathbf{u}}(U) = \frac{1}{\sqrt{2\pi \left(\sigma_j^{\mathbf{u}}\right)^2}} \exp\left\{ -\frac{\left(U - \mu_j^{\mathbf{u}}\right)^2}{2\left(\sigma_j^{\mathbf{u}}\right)^2} \right\} \tag{6}$$

Here, X_k, U are random variables:

$$X_k = {}^{t-M}\hat{x}_k(t_1 + j), \ldots {}^{t-2}\hat{x}_k(t_1 + j), {}^{t-1}\hat{x}_k(t_1 + j) \tag{7}$$

$$U = {}^{t-M}\hat{\mathbf{u}}(t_1 + j), \ldots {}^{t-2}\hat{\mathbf{u}}(t_1 + j), {}^{t-1}\hat{\mathbf{u}}(t_1 + j) \tag{8}$$

From this normal distribution function, the higher value than a threshold ϑ can be shown below:

$$\Pr\left(N_j^{x_k} > \vartheta\right) = N_{j,l}^{x_k} \tag{9}$$

$$\Pr\left(N_j^{\mathbf{u}} > \vartheta\right) = N_{j,l}^{\mathbf{u}} \tag{10}$$

In case of these equations, l denotes numbers of states or actions higher than the threshold ϑ. Hereby, the compensation action that will be regarded in a future can be obtained if an action can correspond to the disturbance input before a future action will be created, based on probabilistic. In other words, the compensation action for normal distribution function $N_{j+1,l}^{\mathbf{u}}$ as follows:

$$N_{j+1,l}^{\mathbf{u}} = -\mathbf{k}_f \cdot N_{j,l}^{\mathbf{x}} \tag{11}$$

In above equation, $x_k \in \mathbf{x}, 1 \leq l \leq k$. From this equation, future action will be compensated the highest existence probability as the compensation action at each time, continuously.

3   Verification Experiment - Computational Simulation Using the Proposed Method

3.1   The Experimental Outline

In this verification experiment, the posture of a two-wheeled inverted pendulum "NXTway-GS", as an application that was stabilized using a computer simulation as a verification experiment. Here, we have obtained training sets while postural control. The training sets contain states and actions of NXTway-GS. In this experiment, the stabilizing the inverted pendulum will be considered between using a stochastic state-action prediction and ordinary LQR controller. The response of the proposed method using stochastic state-action prediction was compared with the control response of the conventional method using LQR. The experiment included 340 steps (actual stochastic predictable range was 3.00 [s] to 20.00 [s]).

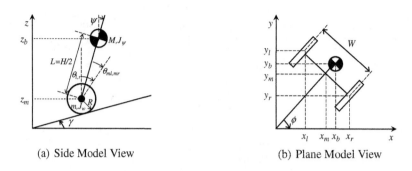

(a) Side Model View   (b) Plane Model View

Fig. 5. Model of side and plane view of NXTway-GS [12]

3.2 Configuration of Simulation 1 – for NXTway-GS Model

As shown in Fig. 5, "NXTway-GS" can be described as a two-wheeled inverted pendulum model. The coordinate system used in Sect. 3.3 is described in Fig. 5. Moreover, in Fig. 5, ψ denotes the body pitch angle and $\theta_{l,r}$ denotes the wheel angle (l and r indicate *left* and *right*, respectively. Further, $\theta = 1/2 \cdot (\theta_l + \theta_r)$.), and $\theta_{ml,mr}$ denotes the DC motor angle (l and r indicate *left* and *right*, respectively). The NXTway-GS's physical parameters are listed in Table 1.

3.3 Configuration of Simulation 2 – for NXTway-GS Modeling

NXTway-GS's motion equations can be derived, shown in Fig. 5. If the direction of the model is the x-axis positive direction at $t = 0$, the equations of motion for each coordinate can be given as follows [12]:

$$\left[(2m + M)R^2 + 2J_w + 2n^2 J_m\right] \ddot{\theta}$$
$$+ \left(MLR - 2n^2 J_m\right) \ddot{\psi} - Rg(M + 2m)\sin\gamma = F_\theta \tag{12}$$

$$\left(MLR - 2n^2 J_m\right) \ddot{\theta} + \left(ML^2 + J_\psi + 2n^2 J_m\right) \ddot{\psi} - MgL\psi = F_\psi \tag{13}$$

$$\left[\frac{1}{2}mW^2 + J_\phi + \frac{W^2}{2R^2}\left(J_w + n^2 J_m\right)\right] \ddot{\phi} = F_\phi \tag{14}$$

Here, $\mathbf{x}_1$ and $\mathbf{x}_2$ represent state variables. In addition, $\mathbf{u}$ denotes input:

$$\mathbf{x}_1 = \left[\theta \; \psi \; \dot\theta \; \dot\psi\right]^\top \tag{15}$$

$$\mathbf{x}_2 = \left[\phi \; \dot\phi\right]^\top \tag{16}$$

$$\mathbf{u} = \left[v_l \; v_r\right]^\top \tag{17}$$

From above equations, state equations of NXTway-GS can be derived using Eqs. (12), (13), and (14).

$$\frac{d}{dt}\mathbf{x}_1 = \mathbf{A}_1 \mathbf{x}_1 + \mathbf{B}_1 \mathbf{u} + \mathbf{S} \tag{18}$$

$$\frac{d}{dt}\mathbf{x}_2 = \mathbf{A}_2 \mathbf{x}_2 + \mathbf{B}_2 \mathbf{u} \tag{19}$$

Table 1. Physical parameters of NXTway-GS

Symbol	Value	Unit	Physical property
g	9.81	$[\text{m/s}^2]$	Gravity acceleration
m	0.03	$[\text{kg}]$	Wheel weight [12]
R	0.04	$[\text{m}]$	Wheel radius
J_w	$\frac{mR^2}{2}$	$[\text{kgm}^2]$	Wheel inertia moment
M	0.635	$[\text{kg}]$	Body weight [12]
W	0.14	$[\text{m}]$	Body width
D	0.04	$[\text{m}]$	Body depth
H	0.144	$[\text{m}]$	Body height
L	$\frac{H}{2}$	$[\text{m}]$	Distance of center of mass from wheel axle
J_ψ	$\frac{ML^2}{3}$	$[\text{kgm}^2]$	Body pitch inertia moment
J_ϕ	$\frac{M(W^2+D^2)}{12}$	$[\text{kgm}^2]$	Body yaw inertia moment
J_m	1×10^{-5}	$[\text{kgm}^2]$	DC motor inertia moment [12]
R_m	6.69	$[\Omega]$	DC motor resistance [12]
K_b	0.468	$[\text{V·s/rad.}]$	DC motor back EMF constant [12]
K_t	0.317	$[\text{N·m/A}]$	DC motor torque constant [12]
n	1	$[1]$	Gear ratio [12]
f_m	0.0022	$[1]$	Friction coefficient between body and DC motor [12]
f_W	0	$[1]$	Friction coefficient between wheel and floor [12]

In this verification experiment, only a state variable $\mathbf{x}_1$ will be used. Because $\mathbf{x}_1$ contains the body pitch angles as variables ψ and $\dot{\psi}$, which are important for self-balancing. That's why the plane motion ($\gamma_0 = 0, \mathbf{S} = \mathbf{0}$) will not be considered in this experiment:

$$\frac{d}{dt}\mathbf{x}_1 = \mathbf{A}_1\mathbf{x}_1 + \mathbf{B}_1\mathbf{u} \tag{20}$$

3.4 Configuration of Simulation 3 – Applying the Online SVR as a Predictor

In this study, an online SVR will be used as a predictor (see Fig. 3). In addition, in this experiment, RBF (as known as radial basis function) kernel is applied to this predictor. The RBF kernel of samples will be notated by $\mathbf{x}$, $\mathbf{x}'$, which represent the feature vectors in any input space, will be defined as follows:

$$k\left(\mathbf{x}, \mathbf{x}'\right) = \exp\left(-\beta \left\|\mathbf{x} - \mathbf{x}'\right\|^2\right) \tag{21}$$

Moreover, the predictor's parameters will be listed in Table 2. In the table, $i \in \{1, 2, 3, 4\}$.

3.5 Configuration of Simulation 4 – Applying the LQR as a Predictor

In this study, Linear-Quadratic Regulator (as known as LQR) is applied as shown in Fig. 3. The feedback gain $\mathbf{k}_f$ so as to minimize the quadratic cost function J will be calculated by this LQR given as Eq. (22).

$$J = \int_0^\infty \left[\mathbf{x}^\top(t)\mathbf{Q}\mathbf{x}(t) + \mathbf{u}^\top(t)\mathbf{R}\mathbf{u}(t) \right] dt \qquad (22)$$

In this verification experiment, matrices $\mathbf{Q}$ and $\mathbf{R}$ are as following:

$$\mathbf{Q} = \begin{bmatrix} 1 & 0 & 0 & 0 & 0 \\ 0 & 6 \times 10^5 & 0 & 0 & 0 \\ 0 & 0 & 1 & 0 & 0 \\ 0 & 0 & 0 & 1 & 0 \\ 0 & 0 & 0 & 0 & 4 \times 10^2 \end{bmatrix} \qquad (23)$$

$$\mathbf{R} = 1 \times 10^3 \cdot \begin{bmatrix} 1 & 0 \\ 0 & 1 \end{bmatrix} \qquad (24)$$

Mentioned above equation, $\mathbf{k}_f$, is a gain of optimal feedback, will be obtained by minimizing J. From these results, $\mathbf{k}_f$ will be applied as the action predictor [1]. Furthermore, the feedback gain $\mathbf{K}_f$ of the state-feedback stabilizer will be applied. However, in the verification experiment, the plane movement of the two-wheeled inverted pendulum will be not considered. Hence, $\phi = 0, \theta_{ml} = \theta_{mr}$, and $\mathbf{u} = u, \mathbf{d}(t) = d(t)$ were considered.

3.6 Simulation Conditions – Training Set Will Be Acquired

In this verification simulation, the action signal with an unknown periodic disturbance signal $\mathbf{d}(t)$ (Fig. 6) as a predictable, will be mixed. Further, $\mathbf{d}(t)$ will be given below equation.

$$d(t) = A_{d1} \sin\left(2\pi f_{d1} t\right) \qquad (25)$$

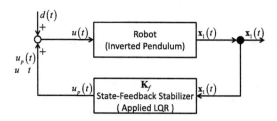

Fig. 6. Control input obtained by mixing the action and disturbance inputs

In this verification experiment, "NXTway-GS" model takes self-balancing. In addition, the properties of disturbance signal that we will provide as input and other conditions of a simulation are listed in Table 3.

Table 2. Learning parameters of online SVR

Symbol	Value	Property
C_i	300	Regularization parameter for predictor of x_i
ϵ_i	0.02	Error tolerance for predictor of x_i
β_i	30	Kernel parameter for predictor of x_i

Table 3. Parameters for the simulation of the proposed method

Symbol	Value	Unit	Physical property
ψ_0	0.0262	[rad.]	Initial value of body pitch angle
γ_0	0.0	[rad.]	Slope angle of movement direction
t_s	0.05	[s]	Sampling rate
N_s	60	—	Initial dataset length
$N_{\max}$	340	—	Maximum dataset length for the prediction
N	20	—	Step size of outputs for N-ahead state-action pair predictor's outputs
N_σ	10	—	Calculation range of the standard deviation of the predicted values
$t_{d,\text{start}}$	0.0	[s]	Start time of application of predictable disturbance
$t_{d,\text{finish}}$	25.0	[s]	Finish time of application of predictable disturbance
A_{d1}	1.0	[V]	Amplitude of predictable disturbance
f_{d1}	1.0	[Hz]	Frequency of predictable disturbance

3.7 Results of Simulation

In this experiment, the training sets provided to the NXTway-GS are based on predicted results generated by the proposed method. In addition, the NXTway-GS model performs stationary balancing on the basis of these sets. Figures 7 and 8 show the compensation results and ordinary results to compare of state $\mathbf{x}_1$, and Fig. 9 shows the compensation results of control input $\mathbf{u}$. In this section, the part of the graph that was obtained from the actual training sets is not considered. Thus, only those parts of the graph pertaining to the state prediction part shown in T (at $t = 3.00$ [s]) of Figs. 7, 8 and 9 will be analyzed.

3.8 Discussion on Simulated Results of the Proposed Method

Here, predicting the state prediction point is shown after $t = 3.00$ [s]. Therefore, we will only argue and focus on the part of the graph pertaining to state prediction part shown in T. According to Fig. 7, compensation result obtained using the proposed method (shown as the solid line) approach and oscillate to near zero, with time. From Fig. 8, $\theta(t)$ will be rotated backward continuously and can be confirmed. Therefore, it can be said that the wheel is moving while trying to decrease the body pitch angle $\psi(t)$. However, the wheel rotation angle is rotating to backward direction, continuously from 12 [s]. On the other hand, it will be confirmed that the slope of the wheel rotation angle is not sharp.

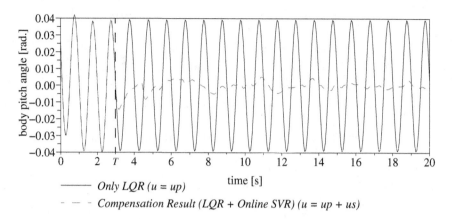

Fig. 7. Control response of body pitch angle ψ using LQR as the training set vs. control response of the proposed method

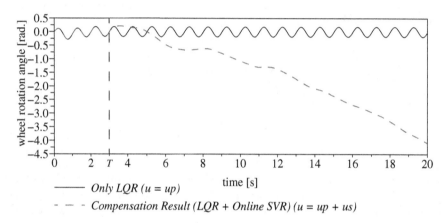

Fig. 8. Control response of wheel rotation angle θ using LQR as the training set vs. control response of the proposed method

And also, the control input $u(t)$ in Fig. 9, a compensation results obtained using the proposed method is generating a control input to drive the wheel backward, continuously. Also, this result is taken in advance states than the conventional method that using only LQR. As the reason, the compensate control input is combining current action and the action that takes future in advance. In this case, as the action that takes future, compensation control input will be using prediction result of the state-action pair prediction, directly. In other words, $\theta(t)$ and $u(t)$ are influenced by the sum of the compensation input and disturbance input $d(t)$. Therefore, the compensate control input generates the action that considered future disturbance. As the result, the effectiveness of the disturbance signal will be reduced. From this system characteristics, the control response of body pitch angle will be converging in the desirable state. In this experiment, we

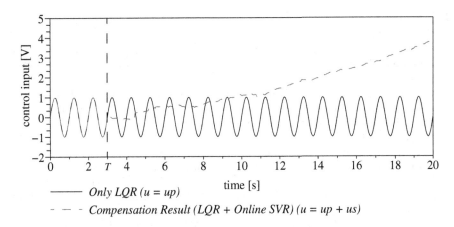

Fig. 9. Control response of the control input u using LQR as the training set vs. control response of the proposed method

only focused on the body pitch angle. From this viewpoint, it can be concluded that the simulation results will be retaining stable state, continuously. Moreover, this response of the wheel rotation angle will be re-improved if the distribution function will be reconsidered. Therefore, it will be concluded that the results of the verification experiment will be reasonable.

4 Conclusion

In this paper, the relationship between state and action of a robot of these stochastic property will be focused on. To achieve this problem, on the basis of former our works, we proposed the method that decides compensation action on each sampling times, based on prediction results that obtained by recent stochastic tendency. Moreover, in the proposed method, the LQR was applied to derive action predictor's gain, and the normal probabilistic function was applied to define the weight coefficients. Applying this proposed method, the compensated action for rapidly convergence was obtained. In other words, it was obtained that the body pitch angle of NXTway-GS was converged to zero, with time.

From the verification experimental results, the proposed method could be converged to a desirable state. To be more specific, the slope of the body pitch angle of NXTway-GS will be converged to zero, based on state and action prediction and decision. Accordingly, the proposed method that is included stochastic technique can be adapted to the situation of stochastic transition will be concluded. In addition, as a future work, we will confirm the response of the proposed system when the mixed a disturbance signal that will be described by an almost-periodic function.

References

1. Sugimoto, M., Kurashige, K.: A study of effective prediction methods of the state-action pair for robot control using online SVR. J. Robot. Mechatron. **27**(5), 469–479 (2015)
2. Cannon, M., Kouvaritakis, B., Raković, S.V., Cheng, Q.: Stochastic tubes in model predictive control with probabilistic constraints. In: Proceedings of 2010 American Control Conference, pp. 6274–6279 (2010)
3. Hashimoto, T.: Probabilistic constrained model predictive control for linear discrete-time systems with additive stochastic disturbances. In: Proceedings of IEEE 52nd Annual Conference on Decision and Control, pp. 6434–6439 (2013)
4. Blackmore, L., Ono, M., Bektassov, A., Williams, B.C.: A probabilistic particle-control approximation of chance-constrained stochastic predictive control. IEEE Trans. Robot. **26**(3), 502–517 (2010)
5. Gray, A., Gao, Y., Lin, T., Hedrick, J.K., Borrelli, F.: Stochastic predictive control for semi-autonomous vehicles with an uncertain driver model. In: Proceedings of IEEE 16th Annual Conference on Intelligent Transportation Systems (ITSC 2013), pp. 2329–2334 (2013)
6. Shah, S.K., Pahlajani, C.D., Lacock, N.A., Tanner, H.G.: Stochastic receding horizon control for robots with probabilistic state constraints. In: Proceedings of 2012 IEEE International Conference on Robotics and Automation (ICRA), pp. 2893–2898 (2012)
7. Weissel, F., Schreiter, T., Huber, M.F., Hanebeck, U.D.: Stochastic model predictive control of time-variant nonlinear systems with imperfect state information. In: Proceedings of 2008 IEEE International Conference on Multisensor Fusion and Integration for Intelligent Systems (MFI 2008), pp. 40–46 (2008)
8. Sugimoto, M., Kurashge, K.: The proposal for real-time sequential-decision for optimal action using flexible-weight coefficient based on the state-action pair. In: Proceedings of 2015 IEEE Congress on Evolutionary Computation, pp. 544–551 (2015)
9. Sugimoto, M., Kurashige, K.: The proposal for compensation to the action of motion control based on the prediction of state-action pair. In: Proceedings of the International Conference on Electronics and Software Science, pp. 35–48 (2015)
10. Sugimoto, M., Kurashige, K.: Future motion decisions using state-action pair predictions. Int. J. New Comput. Archit. Appl. **5**(2), 79–93 (2015)
11. Devcic, J.: Weighted Moving Averages: The Basics (2006). http://www.investopedia.com/articles/technical/060401.asp
12. Sugimoto, M., Yoshimura, H., Abe, T., Ohmura, I.: A study on model-based development of embedded system using Scilab/Scicos. In: Proceedings of the Japan Society for Precision Engineering 2010 Spring Meeting, Saitama, D82, pp. 343–344 (2010)

Rehabilitation Support System Using Multimodal Interface Device for Aphasia Patients

Takuya Mabuchi$^{(\boxtimes)}$, Joji Sato, Takahiro Takeda, and Naoyuki Kubota

Graduate School of System Design, Tokyo Metropolitan University, 6-6 Asahigaoka, Hino, Tokyo, Japan
mabuchi-takuya@ed.tmu.ac.jp

Abstract. In this research, develops a novel system using Multimodal Interface Device for supporting the rehabilitation of aphasia. The system supports diagnosis by conducting statistical analysis of the diagnostic result. So, it's necessary to share a check result by uploading large-volume data in a data server. It is possible to simply perform comparison between data sets and upload patient's data into a database based on previous statistics.

Keywords: Aphasia · Higher brain function disorder · Computational system rehabilitation · Rehabilitation support system

1 Introduction

Recently, the number of patients who recovered from serious brain damages caused by traffic or cerebrovascular accidents has increased. In general, the rehabilitation of the acquired brain damage is done by the following steps; (1) acute care and neurosurgery, (2) post-acute rehabilitation, (3) community-based rehabilitation, and (4) Longer-term community support [1]. The rehabilitation starts as soon as possible, in order to reduce the symptoms and to prevent further complications which can occur in the near future. The symptoms, caused by the brain damages can decrease quality of life (QOL) of the patient. We have a chance to recover these disorders by early and continuous rehabilitation. In the case of aphasia, about 33 % of the patients have recovered from aphasia in 12 to 18 month, but the rest of them still have a substantial aphasia [2]. In order to support those patients, we have developed a system to support rehabilitation at the early stage because starting rehabilitation is considered to be very important as early as possible. However, the plural dysfunctions which were caused by the brain damage, considered to be hard to treat. In order to support the rehabilitation specialists, we have to use an integrated system of information technology, network technology, robot and mechatronics technology, and intelligence technology [3]. If such system can be developed, an efficient rehabilitation would be achieved which is beneficial for the patients and to the caregivers as well.

There are several types of aphasia such as speech, hearing, writing, reading, understanding and recall problem. The combination of the above mentioned problems

© Springer International Publishing Switzerland 2016
N. Kubota et al. (Eds.): ICIRA 2016, Part II, LNAI 9835, pp. 395–404, 2016.
DOI: 10.1007/978-3-319-43518-3_38

also exist according to damaged part in their brain. It is considered to be a serious problem that aphasia disturbs the verbal communication skills of the patients. To gain more information about the types and stages of patients who struggle with aphasia, Japanese ST (Standard Therapist) generally used standard language test to measure the seriousness of aphasia (SLTA - Standard Language Test of Aphasia) [4]. The test evaluates the patient's "Hearing", "Speaking", "Reading", "Writing" and "Calculating" skills from 26 sub tests. The SLTA defines 6 stages in every sub tests. The result of the tests helps us decide what rehabilitation program should be performed by recording the answers. As this, The SLTA gives physical and psychological burden to the ST. Reentry, the burden to the ST become big problem, since the number aphasia have been increasing. Not only the evaluation, but a continuous rehabilitation is important to achieve the early recovery. On the other hand, the rehabilitation of aphasia mostly requires therapists to achieve progress, moreover, it is considered to be difficult to perform the rehabilitation therapy in their home without any experts.

We have proposed the concept of computational system rehabilitation (CSR) composed of (1) measurement, (2) data collection, (3) motion analysis, (4) user model building, (5) personal analysis, (6) program planning support, and (7) rehabilitation support [5–9]. Furthermore, we applied CSR to the rehabilitation on unilateral special neglect (USN or Hemispatial neglect) using Tablet PC [10].

This paper is organized as follows. In the first section, we explain the concept of CSR. In the second section, we explain the architecture of rehabilitation system. Next, we propose a rehabilitation support system by using multimodal interface device. In the fourth section, experimental results are shown and discussed. In the final section, we conclude this paper.

2 Rehabilitation for Aphasia

2.1 Computational System Rehabilitation

The total flow of the computational system rehabilitation (CSR); can be seen in Fig. 1. If the patients are doing the rehabilitation program at home, the choice of the exercises is very limited, and it is difficult for the patients to realize the progress of their rehabilitation. To make the rehabilitation program more efficient, instead of an actual human therapist a robot partner is used for monitoring and supporting a patient at home. A patient can perform the rehabilitation program through interaction with the robot. The robot can repeatedly show the rehabilitation programs to the patient, and can observe the behavior of the patients. The observed data can be useful and important information for the therapist to understand the state of rehabilitation, and to upgrade the rehabilitation program.

2.2 Rehabilitation to Patients with Aphasia

A patient who has been diagnosed with aphasia must have difficulties comprehending, repeating, or producing meaningful speeches. And this difficulty must not be caused by simple sensory or by lack of motivation or motor deficits. A damage to a region of

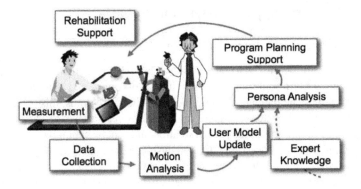

Fig. 1. Computational system rehabilitation

inferior left frontal lobe causes expressive aphasia (Broca's aphasia). This type of aphasia makes the patient hard to produce words on their own. On the other hand, the receptive aphasia (Welnicke's aphasia) is done by poor speech comprehension and production of meaningless speech.

The assessment of aphasia is more art than quantitative science [10, 11]. Lezak suggested a review of speech and language functions on the following aspects of verbal behaviors; (a) spontaneous speech, (b) repetition of words, phrase, and sentences, (c) speech comprehension, (d) naming, (e) reading, and (f) writing. The Halstead screening test requires the subjects to perform a series of tasks; (1) name common goods; (2) spell simple words; (3) identify individual letters and numbers; (4) read, write, enunciate, and understand spoken language, (5) identify body parts, (6) calculate simple arithmetic problems, (7) differentiate between right and left, and (8) copy simple geometric shapes. Based on the above mentioned methods, we propose a home rehabilitation system using tablet PC for a person who is diagnosed with aphasia.

In Japan, the above introduced (SLTA) language test is used to evaluate the types and stages of aphasia. The SLTA has five major evaluation indexes such as hearing, speaking, reading, writing and calculating abilities. And the major indexes consist of 26 miner indexes. For example, in the writing Japanese character, the therapist shows an illustrated card to patient, and asks to write the name of the illustrated item, then therapist scores the answer related to the solution time and correctness. The test takes about 60 or 90 min per evaluation. Recently, according to increasing number of aphasia patients, the test has become one of the biggest physical and psychological burden for therapist.

Rehabilitation of aphasia aims to improve the patients' linguistic ability. Generally, the rehabilitation is performed in a completely isolated and silent room in order to let the patient be more focused on the rehabilitation program. The speech therapist provides a task to the patient related to her/his linguistic ability. For example, therapist shows several illustrated cards, and tells the name of the illustration item, then asks the patient to choose the right card. Here, some tasks are similar to the SLTA test. The therapist needs to have linguistic ability in order to decide the suitable question for each person. We can realize that, it is supposed to be difficult to perform the right rehabilitation program by common family members who might have no experience in linguistics.

3 Rehabilitation Support System for Aphasia

3.1 Architecture of Rehabilitation System

Conventional rehabilitation has to face with two big problems, therapists feel physical and psychological burden during the rehabilitation program, and it is difficult to perform the program by the patients at home. To solve these problems, this paper develops a rehabilitation support system by using multimodal interface device. The system has four process; providing, measurement, evaluation and rehabilitation program planning, based on computational system rehabilitation design. Figure 2 shows the process of our proposed system. In the providing phase, the system offers task for patients using the actual device. Then in the measurement process, the system stores the solution time and the answers of the given task. It also calculates the scores based on SLTA. Based on the calculated scores and personal information, the system evaluates the linguistic ability of the patient. Then, in the rehabilitation program planning procedure, the system decides the next task from the existing task sets, which are stored in an inner or a cloud database for the patients based on evaluated ability. At the end, the system performs rehabilitation programs for aphasia patients.

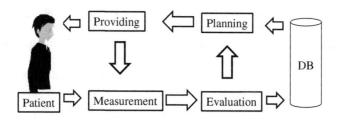

Fig. 2. Concept of our rehabilitation support system

In order to support the aphasia patients, the rehabilitation tools require multimodal interface. For example, a speaker which is necessary for listening tasks, microphone is necessary for speaking tasks. Moreover, to show the illustrated system, high resolution displays and pointing tools are needed.

In this study, we employ iPad as a suitable user interface. As shown in Figs. 3 and 4, iPad has high resolution display, and it is suitable to show illustration and other information. iPad is also equipped with speaker/microphone to record voice and hints, with sensors to gain information, with inner camera to detect faces, movements and expressions. There is also outer camera available to update illustrations, and Wi-Fi or Bluetooth modules to access database and perform tele-healthcare. In addition, the CPU in the iPad measures easily and perfectly the solution time of each tasks. Moreover, we can add more functions to the iPad, for example, handwriting recognition system can be installed [12]. By using the above introduced functions, we propose a rehabilitation support system.

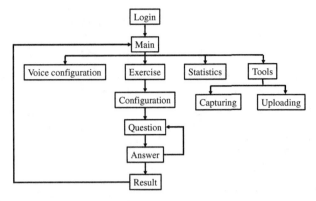

Fig. 3. Application construction of the proposed system.

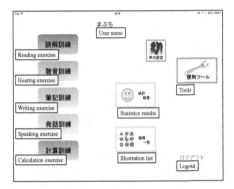

Fig. 4. Main screen.

3.2 Exercise Support Application

Figure 5(a) shows a screen shot of the writing exercise part. In this exercise, the system shows one illustrations on the left corner of the screen. And the system shows some buttons with Japanese characters. Then, patient touches the right button to match the name of the displayed picture. On each button Japanese characters are printed which are random words from the database except one which is the actual name of the picture. These buttons are shuffled randomly and placed in the right side of the screen. The system shows red circle as the correct answer and blue cross as the incorrect answer. Moreover, the system shows correct answer if the user made a mistake. The user can choose if he/she prefers to end the quiz or continue. Task level is set as the number of buttons, more difficult task comes with wider range of buttons and according to the difficulty level the displayed pictures can be generally or rarely seen. At the easiest level, patient only select completed terms as it is shown in Fig. 5(a). While in the hardest level patients need to select multiple buttons to complete the test shown in Fig. 5(b).

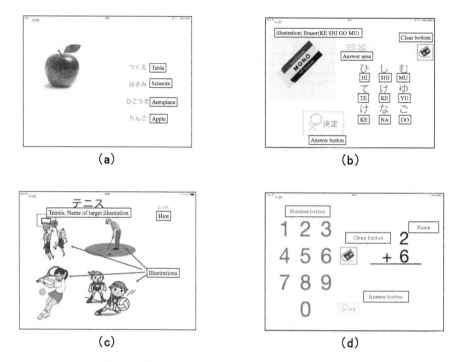

Fig. 5. The example of exercise (Color figure online)

Figure 5(c) shows a screen shot of the reading exercise. In this exercise, the system shows one text and some illustrations on the screen. Patient has to touch the picture which matches the text. As we have introduced it before, the system shows a red circle or a blue cross depending on the answer and shows the correct answer too. There can be a case when the patient cannot solve the given task, that time the program offers a "Hint" option to help out the user with the solution. If the "Hint" button is touched, half of the pictures are removed which makes finding the right answer easier.

Figure 5(d) shows the calculation exercise. In this exercise, the system displays one summing task and a number keyboard on the screen. The keyboards are placed on the left side of the screen because it's said, that an aphasia patient has problem using his/her right hand properly. Forms of the exercise are simple calculations such as additions, subtractions, multiplications and divisions.

4 Experimental Results

This experiment was cooperated with the Saitama Misato Sougou Rehabilitation Hospital. We explained the contents of the system at the hospital, and we also introduced, how to use the application in 2015. We installed the proposed application in iPad and made an aphasia patient test it, and we estimated the log. The patient used an example of writing exercise with the lowest level, he also tried out the writing exercise, the reading exercise and calculation exercise (Table 1).

Table 1. Experiment condition

Place	Saitama Misato Sougou Rehabilitation Hospital
Period	2015/10/16–2015/11/19
Condition	• It was explained about application
	• A use production is iPad
	• A patient exercised with ST

We collected the data during the experiment which can be seen in Table 2. Experiment 1, example of reading exercise 25 questions were asked and answered. Experiment 2, example of writing exercise 30 questions was asked and answered. Experimental 3, calculation exercise 10 questions $\times$ 3, 15 questions $\times$ 1 were asked and answered. We made logs about the length of the character strings and about the answering time.

Table 2. Experiment contents

Ex	Device	Date	Questions
1	iPad	2015/10/26	25
2		2015/10/16	30
3		2015/10/16–/11/16	$10 \times 3, 15 \times 1$

Figure 6 shows experimental result of Ex1, Fig. 6(a) shows a log file and (b) and (c) were made from the log file. The Fig. 6(a) shows answer time of each questions. It took about 20 s to answer the questions using the pictures about a flower, a picture about a sun-glass and a picture about a green Japanese paprika. The Fig. 6(b) shows average answer time using Kanji or Katakana. The average answer time of the question using Kanji was 8.3 s. The average answer time of the question using Katakana was 7.4 s. We could observe the difference in about 1 s. The patient mistook the picture of a figure skate into a sunglass. We could realize that recognizing words written with Katakana takes less time than recognizing the same word written with kanjis. The experiment was very useful because it made us realize the difficulties for aphasia patients to read complicated Japanese characters such as Kanjis.

Figure 7 shows experimental result of Ex2, Fig. 7(a) shows a log file and (b) and (c) were made from the log file. The Fig. 8(b) shows every answer times to each questions. We can create 4 classes according to this graph. In the first group, the patient is answering to every questions without any mistakes quickly. In the second group we can realize that, the patient is answering correctly but slowly. According to the 3rd group, the patient is giving incorrect answers but quickly. In the fourth group, the patient is answering incorrectly and slowly. The Fig. 7(c) shows average answer time correctly or incorrectly. The average time of the correct answers was 31.1 s. The average time of the incorrect answers was 43.4 s. We could realize that, the difference between the average correct and the average incorrect answers was about 12 s. We found out that incorrect answers were mainly from the calculation exercise and we generally got more correct answers from the reading or writing exercise.

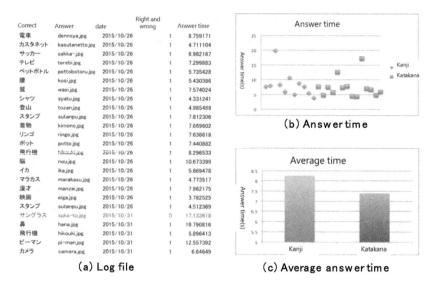

Fig. 6. Experimental result of Ex1

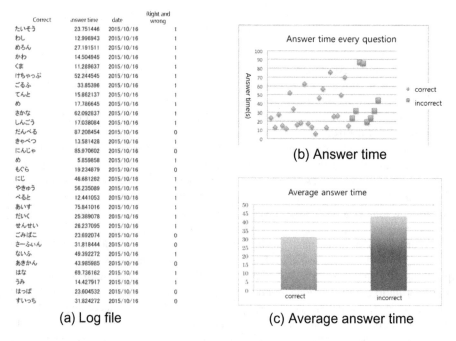

Fig. 7. Experimental result of Ex2

Figure 8 shows the Answer time of every patient. Patient #a, #b and #c could answer 10 questions in a short period of time. Patient #d answered 15 questions. Patient #a and #d could answer only the addition exercise. Patient #b and #c could answer only the subtraction exercise.

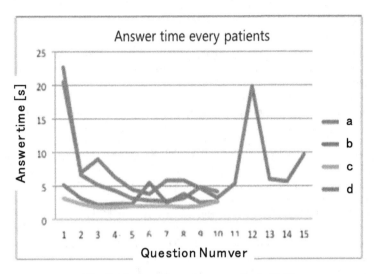

Fig. 8. Answer time of every patients (Ex3)

In the Case of patient, #a, answering the question was relatively slow at the first question but later the answering time became shorter. However, it took him a little bit longer to answer the 9th and the 10th questions. In the Case of patient b, it was also slow to answer the couple of first questions and patient #b had hard time with the 6th question because it was a calculation task with 2 digit numbers. But It became faster however the last 2 questions were calculations tasks but other than that, patient #b was quick, answering the questions. Patient #c answered every question quickly. Patient #d made mistakes in several questions such as question number 11, 12 and 15 and his answers were relatively slow especially at the final stage. We can say, that answering to questions took quite a long time for everyone. The answering took especially a long time to patient #a and #c, and at the final stage for patient #a and #d. We can consider, that accustoming to the iPad device takes time to the patients and this might have influenced the answering times.

5 Summary

In this research, we have developed a system based on the knowledge and experience of a doctor and a therapist. We proved the usefulness of our system by testing it in an actual rehabilitation environment. We could estimate several tendencies from an experimental result. It will be also necessary to get data continuously and estimate from now on. We developed a rehabilitation application with reading, hearing, writing and speaking exercise developed on iPad devices. From the experimental result, we considered that the system was useful to be used in rehabilitation programs.

As future work, we would like to propose a rehabilitation planning support system based on previous user models and personal analysis, and conduct experiments with more subjects and for longer period of time.

References

1. Rehabilitation following acquired brain injury, Royal College of Physicians, 23 July. https://www.rcplondon.ac.uk/sites/default/files/documents/rehabilitation-followingacquired-brain-injury.pdf
2. Johansson, M.B.: Aphasia and Communication in Everyday Life: Experiences of persons with aphasia, significant others, and speech-language pathologists. Doctoral thesis of Uppsala University (2012)
3. Obo, T., Kusaka, J., Kubota, N., Nitta, O., Matsuda, T.: Informationally structured space for computational system rehabilitation. In: Proceedings of 2014 IEEE International Conference on System, Man, and Cybernetics (2014)
4. Japan Society for Higher Brain Dysfunction: Standard Language Test of Aphasia Manual, 2nd edn. 6th impression. Shinko Medical Publisher (2013)
5. Kubota, N., Sotobayashi, H., Obo, T.: Human interaction and behavior understanding based on sensor network with iPhone for rehabilitation. In: Proceedings of the International Workshop on Advanced Computational Intelligence and Intelligent Informatics (2009)
6. Kubota, N., Yorita, A.: Topological environment reconstruction in informationally structured space for pocket robot partners. In: Proceedings of the 2009 IEEE International Symposium on Computational Intelligence in Robotics and Automation (CIRA 2009), Daejeon, Korea, pp. 165–170 (2009)
7. Kubota, N., Botzheim, J., Obo, T.: Human motion tracking and feature extraction for cognitive rehabilitation in informationally structured space. In: Proceedings of the 9th France-Japan and 7th Europe-Asia Congress on Mechatronics and the 13th International Workshop on Research and Education in Mechatronics (REM), Paris, France, pp. 464–471 (2012)
8. Botzheim, J., Obo, T., Kubota, N.: Human gesture recognition for robot partners by spiking neural network and classification learning. In: Proceedings of Joint 6th International Conference on Soft Computing and Intelligent Systems and 13th International Symposium on Advanced Intelligent Systems (SCIS & ISIS 2012), Kobe, Japan, pp. 1954–1958 (2012)
9. Hiwada, E., Aoki, O., Kubota, N.: Appearance and motion pattern generator of balls in a support system for diagnosis of unilateral spatial neglect. In: Proceedings of the 21st Fuzzy, Artificial Intelligence, Neural Networks and Computational Intelligence, pp. 47–50 (2011) (in Japanese)
10. Browndyke, J.N.: Aphasia Assessment. Neuropsychology Central, pp. 1–7 (2002)
11. Lezak, M.: Neuropsychological Assessment. Oxford University, New York (2012)
12. Mazec, MetaMoji Corporation. http://www.product.metamoji.com/en/mazec/

Post-operative Implanted Knee Kinematics Prediction in Total Knee Arthroscopy Using Clinical Big Data

Belayat Md. Hossain[1], Manabu Nii[1,2], Takatoshi Morooka[3], Makiko Okuno[3],
Shiichi Yoshiya[3], and Syoji Kobashi[1,2(✉)]

[1] Graduate School of Engineering, University of Hyogo, Kobe, Japan
kobashi@eng.u-hyogo.ac.jp
[2] WPI Immunology Research Center, Osaka University, Suita, Japan
[3] Department of Orthopaedic, Hyogo College of Medicine, Nishinomiya, Japan

Abstract. Total knee arthroscopy (TKA) is a very effective surgery for damaged knee joint treatment. Because, there are some TKA operation methods and TKA implant products, it is difficult to decide an appropriate one at the pre-operative planning. This study introduces a novel approach to assist surgeon for the pre-operative planning, and proposes a prediction method of post-operative knee joint kinematics. The method is based on principal component analysis (PCA) for characteristics extraction, and machine learning algorithms. The proposed method was validated by leave-one-out cross validation test in 46 osteoarthritis (OA) knee patients. The results show that the proposed method can predict the post-operative knee joint kinematics from the pre-operative one with a mean correlation coefficient of 0.69, and a root-mean-squared-error (RMSE) of 1.8 mm.

Keywords: Total knee arthroscopy · Kinematics · Surgery planning · Prediction · Clinical big data · Principal component analysis

1 Introduction

Total knee arthroscopy, TKA in short, is an effective surgical operation for the knee osteoarthritis (OA), the rheumatoid arthritis (RA), and others. TKA is now very common surgery, and it has been applied to over eighty thousand patients every year in JAPAN. Basically, it is an operation which replaces the damaged knee joint with an artificial knee implant. The knee implant mainly consists of the femoral component, the tibial component, and the tibial insert.

There are some methods of TKA surgical operation and various kinds of prosthesis of the knee implants. For example, cruciate retaining (CR), posterior stabilized (PS), and cruciate substituting (CS) are main prosthesis types of TKA implant. There are also some providers of TKA implants. Because, the anatomical structure and function of the knee joint are different from patient to patient, there should be an appropriate TKA implant for each individual.

It is known that the outcome of TKA operation strongly depends on prosthesis and operation methods. Some studies have investigated a relationship between the TKA

© Springer International Publishing Switzerland 2016
N. Kubota et al. (Eds.): ICIRA 2016, Part II, LNAI 9835, pp. 405–412, 2016.
DOI: 10.1007/978-3-319-43518-3_39

methods and the outcomes. However, a few researcher is studying a dependency of a combination of pre-operative clinical diagnostic test results and operation methods for the post-operative outcome.

And, there are few studies that predict quantitative outcomes of TKA operation at pre-operative planning. As a related study, KneeSIM (LifeModeler Inc., USA) is a commercial software to simulate knee kinematics based on a physical analysis [1]. However, it is mainly used to evaluate physical stress using finite element method (FEM) analysis. That is, surgeons are now selecting a TKA operation method and an implant product model without predicting the outcome numerically.

There are some criteria to decide the TKA operation method and implant product model in pre-operative planning, however, it is still unclear and surgeon has to decide subjectively in most cases. To overcome this problem, this study introduces a prediction method of post-operative implanted knee kinematics using clinical big data. The method analyses a relationship between the pre- and post-operative knee kinematics by means of principal component analysis (PCA) and generalized linear model (GLM) under a specified surgical method and implant product model. The proposed method has been validated using 46 TKA patient's data.

2   Clinical Data Collection and Problem Statement

This study recruited 46 OA knees that had been operated by using PS type knee implant (Vega, Aesculap, B/Braun, Germany). For each subject, an informed consent was obtained followed by a guideline of local Ethics committee in Hyogo College of Medicine (Hyogo, JAPAN).

Usually, knee joint kinematics has been measured by three rotation angles and three translations. They are defined using Grood coordinate system [2] as shown in Fig. 1. The angles are *flexion/extension* (*f/e*), *valgus/varus* (*v/v*) *rotation*, and *internal/external* (*i/e*) *rotation* angles. The translations are medial-lateral (M-L), anterior-posterior (A-P), and superior-inferior (S-I) translations along X, Y and Z axis, respectively.

To quantitate pre- and post-operative kinematics, we have measured the three angles and the translation of knee joint with passively flexing the knee joint between a flexion angle of 10° and 100° for every 10° under a supine position and non-load-bearing condition. The measurement was performed both at the first and last of TKA operation surgery. The first measurement shows the pre-operative kinematics, and the last one shows the post-operative kinematics. We have used a CT-free navigation system (Ortho-Pilot, B/Braun, Aesculap, Germany) for the measurement. This study focuses on A-P translation only as the pilot study, although we have simultaneously measured *v/v* and *i/e* angles, and M-L and S-I translations. The proposed method described below may also be applicable to other angles and translations as well.

Figure 2 shows a comparison of the measured A-P translations between the pre- and the post-operative knee joint kinematics. We can find various kinds of combination of the pre- and post-operative kinematics, and they are different from person to person. The objective of this study is to predict the post-operative knee kinematics from the pre-operative one. Predicting of the post-operative knee kinematics before TKA surgery will

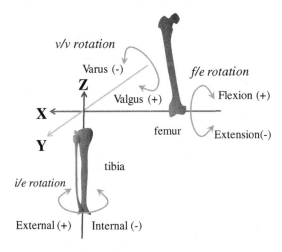

Fig. 1. Knee kinematics definition.

lead surgeons to select an appropriate TKA operation method and TKA implant product model, and hence it will also be very feasible to get informed consent of patients for planning TKA operation because patient will be aware of post-operative performance of the surgery through the prediction estimated by the proposed method.

3 Proposed Method

The proposed method constructs a prediction model using clinical big data that collect the pre- and post-operative knee joint kinematics of individuals. In the following, Subsect. 3.1 shows the training method of the prediction model, and Subsect. 3.2 shows the prediction method using the trained prediction model.

3.1 Learning from Clinical Big Data

To extract individual variability of knee joint kinematics, this method first applies PCA, which is a method to reduce the dimensionally of high-dimensional data using orthogonal transformation. Assume N_m be the number of measurement point (10 points in this experiment, *i.e.*, every 10 *f/e* degree from 10° to 100°), and N_s be the number of subjects. A matrix of the collected data is represented by Eq. (1).

$$
\mathbf{F}^{\mathbf{pre}} =
\begin{bmatrix}
f_1^{pre}(1) & f_2^{pre}(1) & \cdots & f_{N_s}^{pre}(1) \\
f_1^{pre}(2) & f_2^{pre}(2) & \cdots & f_{Ns}^{pre}(2) \\
\vdots & \vdots & \ddots & \vdots \\
f_1^{pre}(N_m) & f_2^{pre}(N_m) & \cdots & f_{N_s}^{pre}(N_m)
\end{bmatrix},
\tag{1}
$$

where $f_i^{pre}(t)$ is A-P translation of subject i at measurement point t. It is $N_m{\times}N_s$ matrix, and each column corresponds to each individual. First, it calculates the mean value of

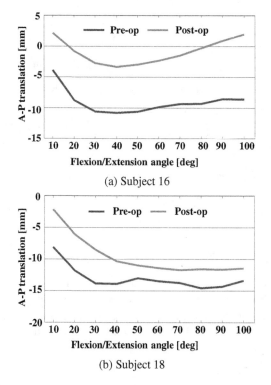

Fig. 2. Comparison of anterior and posterior translation between the pre- and the post-operative knee joint kinematics.

the subjects for each measurement point, $\boldsymbol{\mu}^{\mathbf{pre}} = \left[\mu^{pre}(1), \mu^{pre}(2), \cdots, \mu^{pre}(N_m) \right]^T$, and subtracts it from each row of $\mathbf{F}^{\mathbf{pre}}$.

Second, PCA is applied to the $\mathbf{F}^{\mathbf{pre}}$ by solving the following Eigenvalue problem.

$$\mathbf{F}^{\mathbf{pre}}\mathbf{V}^{\mathbf{pre}} = \lambda^{\mathbf{pre}}\mathbf{V}^{\mathbf{pre}}, \tag{2}$$

where $\mathbf{C}^{\mathbf{pre}}$ is a covariance matrix of $\mathbf{F}^{\mathbf{pre}}$. $\mathbf{V}^{\mathbf{pre}}$ and $\lambda^{\mathbf{pre}}$ are Eigenvector matrix and Eigenvalue matrix, respectively, that are represented by Eqs. (3) and (4).

$$\mathbf{V}^{\mathbf{pre}} = \begin{bmatrix} e_1^{pre}(1) & e_2^{pre}(1) & \cdots & e_{N_{Epre}}^{pre}(1) \\ e_1^{pre}(2) & e_2^{pre}(2) & \cdots & e_{N_{Epre}}^{pre}(2) \\ \vdots & \vdots & \ddots & \vdots \\ e_1^{pre}(N_m) & e_2^{pre}(N_m) & \cdots & e_{N_{Epre}}^{pre}(N_m) \end{bmatrix}, \tag{3}$$

where each column, $e_j^{pre} = \left[e_j^{pre}(1), e_j^{pre}(2), \cdots, e_j^{pre}(N_m) \right]^T$, is j-th Eigenvector, and the columns are sorted by the contribution rate. N_{Epre} is the maximum dimension number of reduced matrix, and is determined by the cumulative contribution rate.

$$\lambda^{\mathbf{pre}} = \begin{bmatrix} \lambda_1^{pre}(1) & \lambda_1^{pre}(2) & \cdots & \lambda_1^{pre}(N_s) \\ \lambda_2^{pre}(1) & \lambda_2^{pre}(2) & \cdots & \lambda_2^{pre}(N_s) \\ \vdots & \vdots & \ddots & \vdots \\ \lambda_{N_{Epre}}^{pre}(1) & \lambda_{N_{Epre}}^{pre}(2) & \cdots & \lambda_{N_{Epre}}^{pre}(N_s) \end{bmatrix}, \tag{4}$$

where each column corresponds to each individual. That is, the number of feature vector dimension is reduced from N_m to N_{Epre}.

So far, pre-operative knee joint kinematics of the individual is represented by N_{Epre} Eigenvalues. Similarly, post-operative knee joint kinematics is represented by N_{Epost} Eigenvalues, mean vector $\mathbf{\mu}^{\mathbf{post}}$, and Eigenvector matrix $\mathbf{V}^{\mathbf{post}}$. In summary, pre-operative and post-operative knee kinematics of subject j are represented by $\lambda^{\mathbf{pre}}(j) = \left[\lambda_1^{pre}(j), \lambda_2^{pre}(j), \cdots, \lambda_{N_{Epre}}^{pre}(j) \right]^T$ and $\lambda^{\mathbf{post}}(j) = \left[\lambda_1^{post}(j), \lambda_2^{post}(j), \cdots, \lambda_{N_{Epost}}^{post}(j) \right]^T$, respectively.

To predict the post-operative knee joint kinematics, the method estimates the post-operative Eigenvalues, $\lambda^{\mathbf{post}}(j)$, from the pre-operative Eigenvalues, $\lambda^{\mathbf{pre}}(j)$. It is a kind of mapping problem, is formulated by Eq. (5).

$$f : \lambda^{\mathbf{pre}} \rightarrow \lambda^{\mathbf{post}}. \tag{5}$$

The mapping function, f, is obtained using training dataset and a machine learning algorithm. There are various kinds of machine learning algorithm to obtain the mapping function, and this study employs GLM and artificial neural networks (ANN).

3.2 Prediction of Post-operative Knee Kinematics

Consider a pre-operative knee joint kinematics of subject x be $\mathbf{f}^{\mathbf{pre}}(x)$. Eigenvalues of the pre-operative knee joint kinematics is calculated by Eq. (6).

$$\lambda^{\mathbf{pre}}(x) = (\mathbf{f}^{\mathbf{pre}}(x) - \mathbf{\mu}^{\mathbf{pre}})\mathbf{V}^{\mathbf{pre}^T} \tag{6}$$

Using the mapping function defined by Eq. (5), $\lambda^{\mathbf{post}}(x)$ is estimated. Finally, the post-operative knee kinematics is predicted by Eq. (7).

$$\mathbf{f}^{\mathbf{post}}(x) = \lambda^{\mathbf{post}}(x)\mathbf{V}^{\mathbf{post}} + \mathbf{\mu}^{\mathbf{post}} \tag{7}$$

4   Experimental Results

The proposed method was applied to 46 OA patients mentioned in Sect. 2. We have utilized a statistical software package (R, Ver. 3.2.2) [3] to implement the proposed method. Leave-one-out cross validation (LOOCV) test was also carried out to evaluate performance of the proposed technique. The maximum dimension number of reduced matrix was 5 dimensions at pre-operative knee joint kinematic modelling, and 3 dimensions at post-operative knee joint kinematic modelling.

Figure 3 shows the experimental result of Subject 1 which was chosen randomly. The prediction model was trained by GLM using the remaining 45 subjects. Basically, a comparison among the pre-operative, the post-operative, and the predicted post-operative knee joint kinematics is shown in the Fig. 3. The correlation coefficient between the measured and the predicted post-operative knee joint kinematics is 0.77, and the root-mean-squared-error (RMSE) is 0.7 mm.

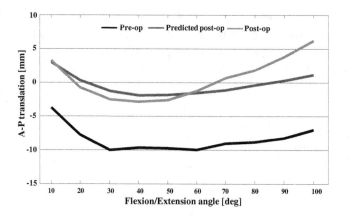

Fig. 3. Prediction result of subject 1

For each subject, the proposed method was applied in which the subject was used as evaluation data, and rest of the subjects were used as training data according to a procedure of LOOCV test. Figures 4 and 5 show the histogram of the correlation coefficient and RMSE. The mean correlation coefficient and mean RMSE are 0.69 and 1.8 mm, respectively. The same experiments were performed using ANN, and calculated the mean correlation coefficient is 0.55, and the mean RMSE is 2.4 mm.

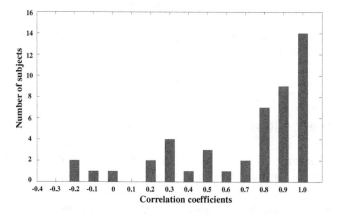

Fig. 4. Histogram of correlation coefficients with GLM.

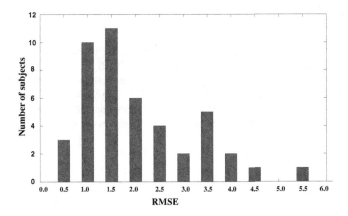

Fig. 5. Histogram of root-mean-squared error with GLM.

According to the experimental results, it is obvious that the proposed method can predict the post-operative knee joint kinematics with a mean correlation coefficient of 0.69, and a mean RMSE of 1.8 mm by utilizing GLM. And, this study also reveals that GLM outperforms over ANN in predicting post-operative knee joint kinematics. Although, there are a few uncorrelated subjects in the experimental results, the differences of pre-operative clinical test between the uncorrelated subjects and the others will be investigated to improve the prediction performance in future.

5   Conclusion

This study discusses a personalized-prediction method of post-operative knee joint kinematics in TKA surgical operation. The proposed method appreciates clinical big data, and constructs the prediction model using PCA and machine learning algorithms. The experimental results obtained by utilizing 46 OA TKA patients shows that the proposed method can predict the post-operative knee joint kinematics from the pre-operative one with a mean correlation coefficient of 0.69, and a mean RMSE of 1.8 mm using GLM.

The prediction model was constructed under the specified surgical operation method and the specified TKA implant product model. By applying this method to the other surgical operating methods and the other product models, we can construct the prediction model for each of them. After that, surgeons can compare the expected post-operative knee kinematics among operation methods and product models to decide for individual patients.

The remaining works are to investigate clinical diagnosis factors of uncorrelated subjects. Using them, the prediction performance may be improved. In addition to this, in future, this method will be applied to other kinematics indexes as well, such as i/e and v/v rotation angles, and M-L and S-I translations.

References

1. Mihalko, W.M., Williams, J.L.: Computer modeling to predict effects of implant malpositioning during TKA. Orthopedics **33**(10), 71–75 (2010)
2. Grood, E.S., Suntay, W.J.: A joint coordinate system for the clinical description of three-dimensional motions, application to the knee. J. Biomech. Eng. **105**(2), 136–144 (1983)
3. R Core Team: A language and environment for statistical computing. R Foundation for Statistical Computing, Vienna, Austria. ISBN 3-900051-07-0 (2013). http://www.R-project.org/

Camera-Projector Calibration - Methods, Influencing Factors and Evaluation Using a Robot and Structured-Light 3D Reconstruction

Jonas Lang$^{(\boxtimes)}$ and Thomas Schlegl

Regensburg Robotics Research Unit, Faculty of Mechanical Engineering,
Ostbayerische Technische Hochschule Regensburg, Regensburg, Germany
{jonas2.lang,thomas.schlegl}@oth-regensburg.de

Abstract. As camera and projector hardware gets more and more affordable and software algorithms more sophisticated, the area of application for camera-projector configurations widens its scope. Unlike for sole camera calibration, only few comparative surveys for projector calibration methods exist. Therefore, in this paper, two readily available algorithms for the calibration of those arrays are studied and methods for the evaluation of the results are proposed. Additionally, statistical evaluations under consideration of different influencing factors like the hardware arrangement, the number of input images or the calibration target characteristics on the accuracy of the calibration results are performed. Ground truth comparison data is realized through a robotic system and structured light 3D scanning.

Keywords: Projector-camera calibration · Evaluation methods · Reprojection error · Reconstruction error · Extrinsic error · Robotic measurement · Structured light · 3D scanning

1 Introduction

Camera calibration has a long-term research history. Development of new algorithms and surveys on the current state of the art are a popular subject in the community. For projector calibration, the situation is slightly different. On the one hand, the development of algorithms is growing due to inexpensive hardware and new use cases like smart rooms, 3D-scanning and augmented reality (like our project SmartWorkbench (SWoB), where projectors are used to support a worker with advices and markers in a semi-automated, collaborative, human-robot workplace [1]). On the other hand, it is often difficult to compare the results of different methods. The hardware and the spatial arrangement of the components varies heavily. Furthermore, the prior influence of the intrinsic parameters on the extrinsics is not always sufficiently taken into consideration. Finally, the evaluation criterion is differing, although the reprojection error is

© Springer International Publishing Switzerland 2016
N. Kubota et al. (Eds.): ICIRA 2016, Part II, LNAI 9835, pp. 413–427, 2016.
DOI: 10.1007/978-3-319-43518-3_40

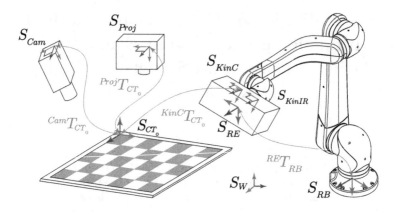

Fig. 1. Utilized hardware and coordinate system conventions

frequently used, which, in the eyes of the authors of this paper and others [2], is often not a good basis for comparisons.

Another aspect is user-friendliness. With the transition from experts and developers to less experienced end users, a conflict between the accuracy of and the effort for the calibration procedure arises. Thus, apart from mathematical evaluations, the complexity for the user of different toolboxes is considered, too. Although many methods are proposed by the scientific community, it is often difficult to easily use and compare them, because there is no readily available implementation of the algorithm or no sufficient code, pseudo-code or formulas to implement the algorithm oneself. To be a guidance for future users, the paper only considers toolboxes that are freely available and ready for instant use without heavy user modifications. Finally, an illustration of the general topic area is shown in Fig. 1.

1.1 State of the Art

Projector calibration is generally comparable to camera calibration, for example using Zhang's algorithm [3], where the projector is modeled as a inverse pin-hole camera. However, problems arise when solving the correspondence between world coordinates and projector pixel coordinates. Therefore, an additional camera and different methods are used to resolve this issue. Many current methods first calibrate the camera and then, in an additional step, the projector with checkerboard, ARTag [4,5] or point [6,7] patterns or known 3D objects [8]. The acquisition of the markers for camera and projector calibration can be performed sequentially or simultaneously, whereas for the latter one, a prewarp is often performed, so that printed and projected markers do not interfere with each other. To integrate the calibration of both components into one unified step, structured light techniques [9,10], vanishing points [11] or color channel information [12] are used. Contrary to the aforementioned methods, [13] use an uncalibrated camera to calibrate the projector. Additionally, there are some readily available

MATLAB or OpenCV implementations without corresponding papers [14–16] and commercial solutions like DAVID4 [17].

The most common evaluation criterion for cameras and projectors alike is the reprojection error. While it is the most meaningful item for single camera calibration results, better possibilities arise when calibrating a stereo camera or projector setup. Therefore, a rectification error is introduced by [2], which helps to estimate the quality of the intrinsic and extrinsic calibration utilizing the characteristics of the stereo setup. Comparable to this, epipolar lines [18] or visual hull reconstruction [19] can be exploited to assess the results. Using the calibrations results of both, camera and projector, and structured light patterns, 3D objects or planes can be reconstructed and compared to the known ground truth [9,20]. The usage of an additional measurement machine or a robot leads to the possibility to further evaluate the extrinsic parameters through known locations and orientations of the components.

2   Evaluation Methodology

The following chapter describes the methodology which is used to assess the results of the camera-projector calibration. An outline of the system hardware and coordinate system conventions is depicted in Fig. 1. The main algorithms for the testing originate from Audet's ProCamCalib toolbox [4], whereas Moreno's toolbox [9] serves as an additional comparison method. The flowchart for the former and a short overview over the used hardware is depicted in Fig. 2.

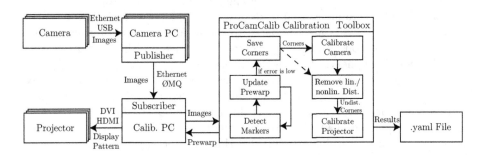

Fig. 2. Flowchart for the hardware setup and the ProCamCalib algorithm [4]

First, the stability or variation of the parameters respectively are evaluated. Therefore, the calibration process is repeated several times without changing the spatial arrangement of the components. However, for each calibration run the calibration target poses and, to a minor degree, the ambient lighting may change. This was chosen deliberately as it simulates the real use case for such calibrations instead of a synthetic environment with constant target poses. Additionally, the number of input images, the size of the calibration target and the number of feature points on the calibration target are varied. Afterwards, a statistical evaluation (mean value, standard deviation (SD) σ, normalized SD σ_n, variance) of the results of all repeated runs is performed. Generally, σ_n is used as the most important evaluation criterion as it is a suitable factor to assess the variability and therefore reliability of a certain calibration parameter.

As an additional study, the common viewing area (VA) between camera and projector ($\delta_{VA} = \frac{VACam}{VAProj}$) is varied. Therefore, the projector stays at its position and the camera is relocated with a robot arm (see Fig. 5 for the arrangement of the components).

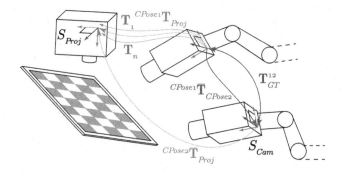

Fig. 3. Robot-based evaluation of the extrinsic calibration results

For further direct testing of the calibration results, two different possibilities for the generation of ground truth comparison data are proposed. First, a structured-light technique is used to reconstruct the known spatial arrangement of planes and the results are compared to the widely used calibration method by Moreno and Taubin [9]. Second, the known relative position and orientation of calibration components through the kinematics of a robot can be utilized to evaluate the extrinsic accuracy of the projector-camera calibration results (see Fig. 3). To lower the noise and to account for varying results, the calibration process is repeated several times. The different components of the translation vector are simply averaged. For averaging the rotation matrices, they are transformed into their quaternion form [21,22]. However, thereby, two problems arise. The averaged quaternion rotation $\bar{q}$ is not a unit quaternion and q and $-q$ represent the same rotation but account to different results in the standard averaging equation. The first is solved by dividing $\bar{q}$ by its norm (1), the second can be neglected as all rotations only differ by a small amount and thus, no opposite q values will occur. The equation to compare the robot-based ground truth data with the results of the calibration process is shown in (2), where $\mathbf{T}$ are homogenous transformation matrices.

$$\bar{q} = \frac{1}{n} \sum_{i=1}^{n} q_i \qquad \bar{q}_n = \frac{\bar{q}}{\|\bar{q}\|} \tag{1}$$

$$^{CPose_1}\mathbf{T}_{Proj} \left(^{CPose_2}\mathbf{T}_{Proj}\right)^{-1} = {}^{CPose_1}\mathbf{T}_{CPose_2} \stackrel{?}{=} \mathbf{T}_{GT}^{12} \tag{2}$$

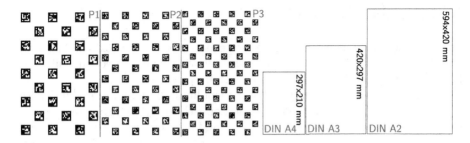

Fig. 4. Used calibration patterns with black, printed ARTag markers and gaps for the white, projected ones (left) and different calibration target sizes (right)

Figure 4 shows the different number of feature points on and the different sizes of the calibration target. The targets are printed on a laser printer and applied to leveled acrylic glass boards using spray adhesive. During the experiments, a light to medium gray background produced better results than the white one, as the contrast between the projected, white markers and the background is increased.

For first testing, a simple Logitech C270 USB webcam in combination with a Asus P1 projector is used. Further evaluation utilizes hardware of an existing experimental system of the work group, the SmartWorkbench (SWoB) (see Fig. 5). A top mounted monochrome industrial camera and the color sensor of a Kinect v2 are tested along with a Panasonic PT-EZ570EL (see Sect. 3 for further hardware details). A Gomtec (now: ABB) Roberta 6-axis lightweight robot can be used for accurate positioning of the vision components.

Fig. 5. Hardware arrangement for the calibration setups in Subsects. 3.2 and 3.3 with projector (1), industrial camera (2), Kinect v2 (3) and calibration target area (4) on the left and a image of the calibration process on the right

3   Experimental Results

In this section, first, some general remarks concerning the toolboxes of Audet [4] and Moreno [9] will be given. Then, in the following three subsections, different hardware setups and evaluation approaches are presented.

Concerning speed, flexibility and user-friendliness, both toolboxes differ quite a lot. In Moreno's toolbox, the acquisition of each image (series) takes about 5 to 15 s, depending on the number of structured light patterns and the wait and exposure time. Additionally, dedicated clicking to start the acquisition is necessary. During each series, the target has to be steady and the whole calibration pattern has to be in view. Furthermore, only one camera and one projector can be calibrated at a time and only USB webcams are natively supported.

In contrast, images for Audet's toolbox are taken in a free running mode without dedicated triggering. Thus, the pattern may be moved and held by the operator and the algorithm only takes new images when the scene is steady enough and the target location has changed in comparison to previous input images. In addition, parts of the target may be occluded or outside the camera's field of view. Multiple cameras and multiple projectors can be calibrated at the same time and the toolbox comes with support for different frame grabbing modules. To fulfill our requirements, we have expanded it by an universal frame grabbing module based on the ØMQ publisher-subscriber interface, which provides the possibility to distribute the vision components over different computers (see Fig. 2). This can be a help for connection issues with big hardware arrays.

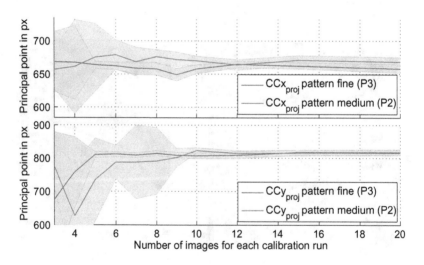

Fig. 6. Variation (mean/SD σ) of the principal point parameters with different calibration patterns and numbers of input images

3.1 Webcam Setup

As a first evaluation setup, a Logitech C270 USB webcam (1280 × 960 px) and
a Asus P1 projector (1280 × 800 px) are placed with collinear optical axes and
the calibration is performed. Figure 6 depicts results for Audet's toolbox where
the number of data points on the target as well as the number of input images
for each calibration run are varied. Each run is repeated 15 times and the mean
and SD values are examined. The fine pattern P3 with 150 data points allows
for more accurate results with a smaller number of input images in comparison
to the coarser pattern P2 with only 96 data points (see Fig. 4).

The reprojection error is a measure for the fitting of 2D image points onto
3D world points an vice versa using the calibration model. With only little input
information (e.g. 2-3 pattern images) it is easy to create a model which perfectly
fits these data points, so the reprojection error is small. The more images and
pattern data are used, the more difficult it is to fit a general model which suffices
all pattern locations.

As can be seen in Fig. 7, a small number of input images leads to small
camera reprojection errors but varied and thereby also wrong camera calibration
parameters which results in high projector and stereo calibration errors. After
reaching a minimum between 10 and 15 images, all reprojection errors are slowly
increasing with 15 to 40 input images, although the accuracy and quality of the
calibration is improved. Therefore, the reprojection error should not be used as
a single evaluation criterion for camera-projector setups and additional methods
can be utilized to correctly assess the calibration results.

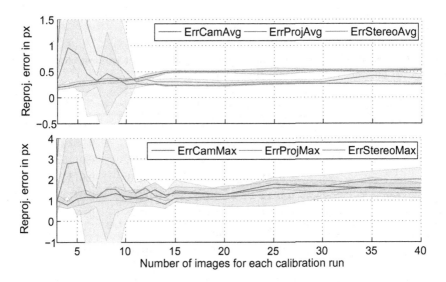

Fig. 7. Variation (mean/SD σ) of the reprojection error with calibration pattern P1
and different numbers of input images

3.2   Kinect v2 Setup

In this setup, the color sensor of a Kinect v2 (1920×1080 px) is calibrated together with a Panasonic PT-EZ570EL (1920×1200 px) projector. Both components have a comparable field of view and are mounted with skewed optical axes (see Fig. 5). A test series comparable to the one in Subsect. 3.1 with a differing number of input images, a varied number of feature points (see Fig. 4) and multiple repetitions of the calibration process is performed. Subsequently, the results are evaluated statistical. An exemplary behavior for the focal length FCx_{Cam} is depicted in Fig. 8. Other intrinsic and extrinsic calibration parameters show similar results.

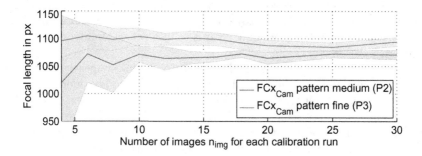

Fig. 8. Variation (mean/SD σ) of the focal length parameters with different calibration patterns and numbers of input images

Table 1. Normalized SD σ_n (15 calibration runs) of different calibration parameters with varied difference of the common viewing area δ_{VA}

Parameter	n_{Img}	$d_{VA} \approx 0.25$	$d_{VA} \approx 0.33$	$d_{VA} \approx 0.50$	$d_{VA} \approx 1.00$
FCx_{Cam}	12	0.0038	0.0042	0.0041	0.0097
	24	0.0026	0.0034	0.0046	0.0075
FCx_{Proj}	12	0.0232	0.0156	0.0117	0.0115
	24	0.0224	0.0153	0.0120	0.0134
$TransX$	12	0.1942	0.1074	0.0734	0.0374
	24	0.1738	0.0883	0.0592	0.0272
$RotX$	12	0.0456	0.0341	0.0196	0.0181
	24	0.0374	0.0323	0.0180	0.0166

Furthermore, the common viewing area δ_{VA} between camera and projector is varied. Therefore, the projector stays at the same position as in the previous experiment, while the Kinect v2 camera is moved closer to the surface of the table using the robot (see Fig. 5). The calibration process is again repeated for several times with a varied number of images n_{img}. The smaller the common viewing area δ_{VA}, the worse the results for the intrinsic projector parameters and the extrinsics. Detailed numbers are shown in Table 1 and their illustrations can be observed in Fig. 9. As the camera gets closer to the target location area (see Fig. 5), markers occupy more pixels in the image and therefore errors in the detection process are reduced which has a positive influence on the accuracy of the intrinsic camera parameters, as can be seen in Fig. 9c, where lower values of the common viewing area δ_{VA} lead to smaller variation.

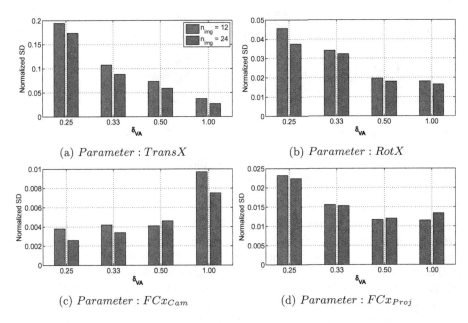

(a) *Parameter* : $TransX$ (b) *Parameter* : $RotX$

(c) *Parameter* : FCx_{Cam} (d) *Parameter* : FCx_{Proj}

Fig. 9. Normalized SD σ_n with 15 runs of the calibration process and n images each. The common viewing δ_{VA} is varied.

In addition, experiments with differently sized calibration targets are performed. The relative variation for the parameters can be observed in Table 2. The principal point parameter CCy_{Proj} and the distortion coefficients KCx strike the eye as their variation is very large compared to other parameters, which is explainable as they alternate around zero. Generally, the errors are smaller with more input images and a larger target, which is also illustrated in Fig. 10. The projector parameters in comparison to the camera parameters altogether have a higher variation (i.e. are prone to errors to a higher degree). This lies in the nature of this kind of calibration algorithm, where the results of the camera calibration are used to calibrate the projector. Thus, errors of the former propagate to the latter. The normalized SD σ_n also varies for different translation and rotation components with significantly lower errors in the $TransX$ and $RotY$ components.

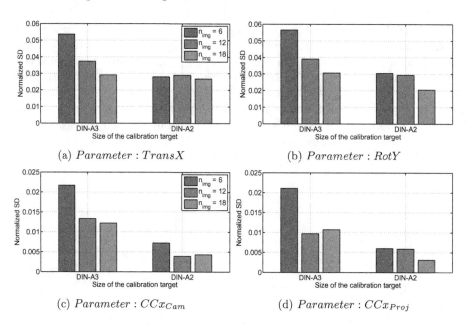

(a) *Parameter* : *TransX*

(b) *Parameter* : *RotY*

(c) *Parameter* : *CCx_{Cam}*

(d) *Parameter* : *CCx_{Proj}*

Fig. 10. Normalized SD σ_n with 15 runs of the calibration process and n images each. The size of the calibration target is varied.

Finally, the extrinsic results of the calibration process are evaluated. Therefore, the projector stays stationary and the Kinect is moved by the robot. Through accurate robot movements, a ground truth comparison data set is generated. As outlined in Sect. 2, to suppress noise induced by the nature of the calibration process, it is repeated 15 times with 20 input images each (see also Fig. 3). Equations (3) and (4) show the results which are obtained through the calibration process on the left side in comparison to the robot generated ground truth on the right side. The rotation matches pretty well whereas concerning the translation, errors up to 9 mm can be noticed.

$$
{}^{CPose_1}\mathbf{T}_{CPose_2} = \begin{bmatrix} 0.9998 & -0.0017 & 0.0191 & -5.128 \\ 0.0018 & 1.0000 & -0.0051 & 6.974 \\ -0.0192 & 0.0052 & 0.9998 & 98.784 \\ 0 & 0 & 0 & 1 \end{bmatrix} \quad \mathbf{T}_{GT}^{12} = \begin{bmatrix} 1 & 0 & 0 & 0 \\ 0 & 1 & 0 & 0 \\ 0 & 0 & 1 & 100 \\ 0 & 0 & 0 & 1 \end{bmatrix}
$$

$$(3)$$

$$
{}^{CPose_1}\mathbf{T}_{CPose_3} = \begin{bmatrix} 0.9999 & -0.0093 & -0.0132 & -7.129 \\ 0.0093 & 1.0000 & -0.0035 & 103.773 \\ 0.0133 & 0.0034 & 0.9999 & 9.790 \\ 0 & 0 & 0 & 1 \end{bmatrix} \quad \mathbf{T}_{GT}^{13} = \begin{bmatrix} 1 & 0 & 0 & 0 \\ 0 & 1 & 0 & 100 \\ 0 & 0 & 1 & 0 \\ 0 & 0 & 0 & 1 \end{bmatrix}
$$

$$(4)$$

Table 2. σ_n (for 15 calibration runs) of the calibration parameters with varying calibration target sizes, pattern P2 and a varying number of images n_{img}

Targetsize	DIN A3			DIN A2		
n_{img}	6	12	16	6	12	16
FCx_{Cam} (px)	0.0211	0.0097	0.0107	0.0060	0.0059	0.0031
FCy_{Cam} (px)	0.0202	0.0136	0.0105	0.0063	0.0055	0.0033
CCx_{Cam} (px)	0.0217	0.0133	0.0122	0.0072	0.0039	0.0043
CCy_{Cam} (px)	0.0260	0.0280	0.0153	0.0104	0.0104	0.0100
$KC1_{Cam}$	0.2323	0.2240	0.1776	0.1257	0.1146	0.0839
$KC2_{Cam}$	0.6050	0.6806	0.3329	0.4244	0.4704	0.3803
FCx_{Proj} (px)	0.0665	0.0315	0.0247	0.0173	0.0164	0.0155
FCy_{Proj} (px)	0.0564	0.0344	0.0219	0.0199	0.0155	0.0143
CCx_{Proj} (px)	0.0538	0.0345	0.0285	0.0300	0.0297	0.0263
CCy_{Proj} (px)	0.7343	0.8446	0.4497	2.8156	0.6924	1.1478
$KC1_{Proj}$	4.9806	1.3849	2.0705	3.7787	2.0871	1.3359
$KC2_{Proj}$	2.9487	2.8592	1.8868	3.7521	1.2903	2.1857
$TransX$ (mm)	0.0536	0.0374	0.0292	0.0279	0.0290	0.0268
$TransY$ (mm)	0.2146	0.2256	0.1683	0.1870	0.2989	0.2349
$TransZ$ (mm)	0.2146	0.2256	0.2831	0.0758	0.0430	0.0635
$RotX$	0.1760	0.1568	0.1587	0.0662	0.1039	0.0865
$RotY$	0.0565	0.0391	0.0306	0.0304	0.0294	0.0205
$RotZ$	0.2144	0.1927	0.1347	0.0900	0.0531	0.0718

3.3   Industrial Camera Setup

In the following experimental setup, a high resolution, monochrome 2D Camera (Teledyne DALSA Genie TS-M2500, 2560×2048 px) with a 16 mm lens is used alongside the same projector as in Subsect. 3.2. The hardware setup is depicted in Fig. 5. The evaluation of the corresponding calibration results is realized via structured light 3D reconstruction of planes. Ground truth comparison data is provided via the commonly used Moreno toolbox and an acrylic glass board ($1000 \times 800 \times 20$ mm) is used as a scanning target. The scans are performed using 3D scanning software [23] with 10 horizontal and 10 vertical projection patterns. Subsequently, identical images but different camera-projector calibration data sets are used to reconstruct the point cloud data which is analyzed in CloudCompare. There, a plane is fitted into the point cloud and the distances between the cloud (i.e. the scan results) and the mesh (i.e. the fitted plane) are computed. The results are visualized in an histogram and the root mean square (RMS) is calculated as an evaluation criterion. This procedure is repeated for five plane poses in space and the RMS of all corresponding scanned planes are averaged to the final RMS_{mean} result. The calibrations using both toolboxes were performed with comparable targets (see Table 4) and 25 images each. The resulting intrinsic and extrinsic calibration parameters are shown in Table 3.

Table 3. Calibration results for setup 3 with toolboxes from Moreno [9] and Audet [4] using 25 images each

Camera	FCx (px)	FCy (px)	CCx (px)	CCy (px)	$KC1$	$KC2$	$KC3$	$KC4$
Moreno	2704.15	2705.91	1280.27	1024.84	−0.0784	0.1131	0.0039	−0.0004
Audet	2680.05	2682.69	1294.69	1028.57	−0.0764	0.0844	0.0019	0.0007
Projector	FCx (px)	FCy (px)	CCx (px)	CCy (px)	$KC1$	$KC2$	$KC3$	$KC4$
Moreno	2507.09	2505.42	1051.98	−3.0723	−0.0156	−0.0024	0.0048	0.0031
Audet	2460.64	2470.09	1041.25	26.2375	−0.0211	−0.0032	0.0088	0.0029
Extrinsics	Translation (mm)			Rotation (Euler vector)				
Moreno	$\begin{pmatrix} 101.829 & -170.076 & 572.612 \end{pmatrix}$			$\begin{pmatrix} -0.0113 & -0.7063 & -1.5644 \end{pmatrix}$				
Audet	$\begin{pmatrix} 106.902 & -205.596 & 542.228 \end{pmatrix}$			$\begin{pmatrix} -0.0142 & -0.7211 & -1.5397 \end{pmatrix}$				

The results for $Pose_{Plane1}$ are visualized in Fig. 11. Using the calibration results obtained through Moreno's and Audet's toolboxes, 1184973 and 1034984 3D points were reconstructed. The final results can be taken from Table 4. Overall, the calibration results of Moreno's toolbox lead to better results than the parameters of Audet's toolbox. For the latter one, increasing the number of feature points on the target improves the 3D reconstruction results.

Table 4. Plane reconstruction evaluation results using the calibration parameters of the toolboxes of Moreno [9] and Audet [4]

Method	Moreno	Audet	Audet
n_{img}	25	25	25
Target FP	8×7	9×6	12×8
RMS_{Plane_1}	0.4119	1.4285	1.0267
RMS_{Plane_2}	0.3957	1.4054	0.9864
RMS_{Plane_3}	0.4879	1.8905	1.2311
RMS_{Plane_4}	0.4683	1.6404	1.1857
RMS_{Plane_5}	0.3988	1.8073	1.2019
RMS_{mean}	0.4325	1.6344	1.1264

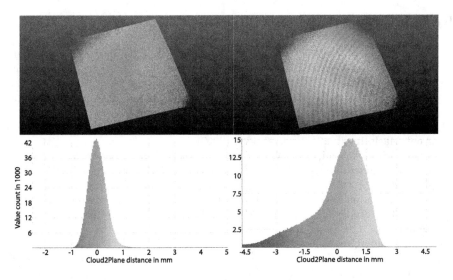

Fig. 11. Distance between reconstructed point clouds and fitted planes and their corresponding histograms with 256 classes (left: Moreno, right: Audet)

4   Conclusion and Outlook

The paper outlined the general topic of camera-projector calibration. A statistical evaluation procedure and two error measures aside from the classical reprojection error were introduced and used to assess different algorithms, hardware setups and influencing factors. Those results can be a guideline when designing and calibrating camera-projector systems.

For further research, additional methods can be tested in detail, as well as additional spatial arrangements of the hardware like varying skew between the optical axes. Further evaluation criterions can be examined, for example, straight line testing to evaluate the distortion, a measurement of the pattern locations as assessment as well as an additional constraint for the algorithm and also projecting on certain, known geometries using an optical, qualitative, instead of the previous quantitative validation. The statistical evaluation can be expanded to Moreno's and other toolboxes and the introduced structured light reconstruction to more sophisticated 3D objects than planes. Concerning the testing of the extrinsics, instead of averaging quaternions, there is also the possibility to use Lie algebra or a least squares optimization algorithm.

Finally, the results of the calibration will be used for the project SmartWorkbench to support the worker at an augmented workplace with human-robot-collaboration (see Fig. 5).

References

1. Niedersteiner, S., Pohlt, C., Schlegl, T.: Smart workbench: a multimodal and bidirectional assistance system for industrial application. In: IECON 2015–41st Annual Conference of the IEEE Industrial Electronics Society, pp. 2938–2943 (2015)
2. Bradley, D., Heidrich, W.: Binocular camera calibration using rectification error. In: Computer and Robot Vision (CRV), pp. 183–190 (2010)
3. Zhang, Z.: A flexible new technique for camera calibration. IEEE Trans. Pattern Anal. Mach. Intell. 22(11), 1330–1334 (2000)
4. Audet, S., Okutomi, M.: A user-friendly method to geometrically calibrate projector-camera systems. In: IEEE Computer Society Conference CVPR Workshops 2009, pp. 47–54 (2009)
5. Fiala, M.: Automatic projector calibration using self-identifying patterns. In: IEEE Computer Society Conference on Computer Vision and Pattern Recognition, p. 113 (2005)
6. Ben-Hamadou, A., Soussen, C., Daul, C., Blondel, W., Wolf, D.: Flexible calibration of structured-light systems projecting point patterns. Comput. Vis. Image Underst. 117(10), 1468–1481 (2013)
7. Ouellet, J., Rochette, F., Hebert, P.: Geometric calibration of a structured light system using circular control points. In: Fourth International Symposium on 3D Data Processing, Visualization and Transmission (3DPVT 2008), 2008, pp. 1–8 (2008)
8. Resch, C., Naik, H., Keitler, P., Benkhardt, S., Klinker, G.: On-site semi-automatic calibration and registration of a projector-camera system using arbitrary objects with known geometry. IEEE Trans. Vis. Comput. Graph. 21(11), 1211–1220 (2015)
9. Moreno, D., Taubin, G.: Simple, accurate, and robust projector-camera calibration. In: Proceedings of the 2012 International Conference on 3D Imaging, Modeling and Processing, pp. 464–471. IEEE Computer Society (2012)
10. Yamazaki, S., Mochimaru, M., Kanade, T.: Simultaneous self-calibration of a projector and a camera using structured light. In: 2011 IEEE Computer Society Conference on Computer Vision and Pattern Recognition Workshops (CVPRW), pp. 60–67 (2011)
11. Orghidan, A.R., Gordan, C.M., Vlaicu, D.A., Salvi, B.J.: Projector-camera calibration for 3D reconstruction using vanishing points. In: 2012 International Conference on 3D Imaging (IC3D), pp. 1–6 (2012)
12. Li, T., Zhang, H., Geng, J.: Geometric calibration of a camera-projector 3D imaging system. In: Image and Vision Computing New Zealand (IVCNZ), 2010, pp. 1–8 (2010)
13. Garcia-Dorado, I., Cooperstock, J.: Fully automatic multi-projector calibration with an uncalibrated camera. In: 2011 IEEE Computer Society Conference on Computer Vision and Pattern Recognition Workshops (CVPRW), pp. 29–36 (2011)
14. Alvaro, C.: ofxCv2 CamProjCalib Addon. https://github.com/alvarohub/ofxCv2. Accessed 11 Apr 2016
15. Lanman, D., Taubin, G.: 3D Scanning - Calibration. http://mesh.brown.edu/byo3d/source.html. Accessed 10 Apr 2016
16. Falcao, G., Hurtos, N., Massich, J.: Projector-Camera Calibration Toolbox. https://code.google.com/archive/p/procamcalib/. Accessed 11 Apr 2016
17. DAVID Group: DAVID 3D Scanner Calibration. http://www.david-3d.com/en/support/david4/calibration. Accessed 11 Apr 2016

18. Salvi, J., Armangue, X., Batlle, J.: A comparative review of camera calibrating methods with accuracy evaluation. Pattern Recogn. **35**(7), 1617–1635 (2002)
19. Datta, A., Kim, J.S., Kanade, T.: Accurate camera calibration using iterative refinement of control points. In: 2009 IEEE 12th International Conference on Computer Vision Workshops (ICCV Workshops), pp. 1201–1208 (2009)
20. Zhang, X., Zhang, Z., Cheng, W.: Iterative projector calibration using multi-frequency phase-shifting method. In: 2015 IEEE 7th International Conference on Cybernetics and Intelligent Systems (CIS), pp. 1–6 (2015)
21. Markley, F.L., Cheng, Y., Crassidis, J.L., Oshman, Y.: Averaging quaternions. J. Guidance Control Dyn. **30**(4), 1193–1197 (2007)
22. Hartley, R.I., Trumpf, J., Dai, Y., Li, H.: Rotation averaging. Int. J. Comput. Vis. **103**(3), 267–305 (2013)
23. Moreno, D., Taubin, G.: 3D Scanning Software. http://mesh.brown.edu/calibration/software.html. Accessed 12 Apr 2016

Robot Hand

Automatic Grasp Planning Algorithm for Synergistic Underactuated Hands

Lei Hua, Xinjun Sheng[✉], Wei Lv, and Xiangyang Zhu

State Key Laboratory of Mechanical System and Vibration,
Shanghai Jiao Tong University, Shanghai 200240, China
xjsheng@sjtu.edu.cn

Abstract. Over the last decades, effective grasp planning algorithms, which can handle grasps of common objects for robotic hands, have been extensively investigated. Among these, synergistic actuation, a promising approach, attracts lots of attentions. However, synergies will reduce the dexterity and manipulability of robotic hands. To obtain high grasp quality, the synergy parameters need to be optimized for different objects. This paper focuses on the automatic grasp planning algorithm for synergistic underactuated hands. Our major efforts are the quantitative description of the grasp quality and the optimization of synergy parameters for different objects. Based on the proposed virtual object model, the grasp quality function, equilibrium equations and contact constraints are defined. Then an automatic grasp planning algorithm is established for synergistic underactuated hands. Finally, numerical examples are presented on a robotic hand model, and simulation results show that the proposed automatic grasp planning algorithm can be used to grasp different objects.

Keywords: Grasp · Underactuated hand · Syneygy · Virtual object model

1 Introduction

Over the last decades, some multi-fingered robotic hands have been developed to imitate the human hand. Typical prototypes include the UTAH/MIT hand [1], the DLR hand [2], the Robonaut Hand [3] and several others. As the number of the joints and actuators of robotic hands begins to approach that of the human hand, effective control strategies [4,5], which can handle grasps with multi-fingered hands, have been extensively investigated. And dexterous and robust control algorithm is still a challenging problem in grasp and manipulation tasks [6].

To simplify the control algorithm, the movement of the hand is considered as a unit rather than separated single joints. In consideration of this matter, Napier has proposed grasp taxonomy [7], later developed by Cutkosky [8]. Their work had a profound impact for the field of grasping. Using the grasp taxonomy concept, Miller et al. defined a set of grasp starting positions and pregrasp shapes

© Springer International Publishing Switzerland 2016
N. Kubota et al. (Eds.): ICIRA 2016, Part II, LNAI 9835, pp. 431–442, 2016.
DOI: 10.1007/978-3-319-43518-3_41

to search for the best grasps of given objects within the grasping simulator–
GraspIt! [9]. Aleotti and Caselli proposed a pregrasp planning algorithm to mimic
the human hand motion and to reconstruct human hand trajectories based on
NURBS curves and a data smoothing algorithm [10].

Recently, studies in neuroscience show that only a few parameters, called
synergies or eigengrasps, suffice to describe hand postures [11] in grasping tasks.
These studies provide new methods on the control strategies for robotic hands.
Two [12] and three [13] postural synergies were used to control the Anatomically
Correct Testbed Robotic Hand for complex robotic manipulation tasks – writing
and musical piano performance. Based on the kinematic structure of the robotic
hand and on the taxonomy of the grasps of common objects, postural synergies
configuration subspace of the UB Hand IV is derived [14]. In [15], the characteristics of synergistic hands were analyzed and the low-dimensional posture
subspaces have been applied to derive effective pre-grasp shapes for a number of
complex robotic hands.

Our work also focuses on the analysis and application of synergy actuation for
robotic hands. The major work are the quantitative description of the grasp quality and the optimization of synergy parameters for robotic hands, especially for
synergistic underactuated hands. Through the analysis of the proposed virtual
object model, a grasp energy function and grasp constraints are defined. Then
an automatic grasp planning algorithm is established for synergistic underactuated hands. The rest of this paper is organized as follows. In Sect. 2, the concept
and application of synergies are briefly introduced. The virtual object model
is described in Sect. 3, and the grasp quality function and grasp algorithm are
established based on the proposed model. The numerical simulations are presented in Sect. 4. The conclusion is presented in Sect. 5.

2 Synergies and Application

As we have discussed in our literature review, a low-dimensional input, can
approximately imitate most common grasp postures. For a robotic hand, with n_q
joints, controlled through a limited dimension set of input synergies n_z ($n_z \leq n_q$),
the relationship between hand postures and input synergies is

$$p = Sz \tag{1}$$

Where $p = \begin{bmatrix} q_1 & q_2 & \cdots & q_{n_q} \end{bmatrix}^{\mathrm{T}} \in \Re^{n_q}$ is the hand posture configuration and $z = \begin{bmatrix} z_1 & z_2 & \cdots & z_{n_z} \end{bmatrix}^{\mathrm{T}} \in \Re^{n_z}$ is the synergy weighting vector. The matrix $S \in \Re^{n_q \times n_z}$
is the synergy matrix.

Aleotti and Caselli [10] collected a large set of data containing 57 grasping
postures from five subjects. Linearized Principal Component Analysis of this
data reveals that the first two principal components account for more than 80 %
of the variance, which means that the posture synergies can be linearly combined
to reconstruct most common grasp postures. For simplicity, we let the hand
postures be linearly spanned using posture synergies. In this paper, we will

refer to the principal components of grasp postures as posture synergies. Each posture synergy s_i is a n_q-dimensional vector and also be thought of as direction of motion in joint space:

$$s_i = [s_{i1}\, s_{i2} \cdots s_{in_q}]^{\mathrm{T}} \tag{2}$$

Choose a posture basis $S = [s_1 \cdots s_d]$ comprising $d(d \leq n_z)$ posture synergies. And thus hand postures belong to the subspace defined by this basis $p \in \Re(S)$. Therefore, the Eq. (1) can be rewritten as

$$p = Sz = \begin{bmatrix} s_{11} & \cdots & s_{d1} \\ & \vdots & \\ s_{1n_q} & \cdots & s_{dn_q} \end{bmatrix} \begin{bmatrix} z_1 \\ \vdots \\ z_d \end{bmatrix} \tag{3}$$

Where z_i is the weighting coefficient of posture synergy s_i. Since the given posture synergies s_i are constant, hand configuration based on synergistic control only depends on the weighting coefficients.

Synergy-based control is inspired by human biological systems, and usually posture synergies are obtained from human hand grasping data. For general robotic hand models, we have derived posture synergies attempting to define a posture basis similar to the one obtained from human hand grasping data. The mapping rules from human hand to other robotic hands can often be made directly based on the similar structures. For example, the MCP and IP joints can be mapped to the proximal and distal joints of robotic fingers. In this paper, we use the Barret hand as the simulation model. This robotic hand has a spread joint which is different from human hand. In this case, we map the spread angle to the abduction of human fingers. Later we will verify that this mapping relation can produce good grasp results.

3   Grasp Planning Based on the Virtual Object Model

The synergy-based control allows us to operate robotic hands using lower dimension input. But limited control input will reduce the dexterity and manipulability of robotic hands. And inter-finger coordination has an obvious influence on the motion and force controllability of the objects. That is to say, the weighting coefficients will impact on the grasp quality and object characteristics. The grasp planning tasks can be thought as an optimization problem of hand configuration. Therefore, the main issue to our approach is the optimization of the weighting coefficients. Next we will construct the optimization function relating to the grasp quality and the weighting coefficients.

All fingers of synergistic underactuated hands will not be in touch with objects synchronously, which is the main reason that influences the grasp quality. To quantitatively describe the grasp quality, we propose the virtual object model as an approximation of the object grasped by synergy-based hands with n contact points. For the sake of simplicity, this paper just considers precision grasps. Each finger has a rigid contact point with the object, and the rolling is out of consideration.

3.1   The Virtual Object Model

The virtual object model with n contact points is shown in Fig. 1. This virtual object model consists of spatial spring models connecting every pair of contact points, and each end of the spring models is joined at the contact points by frictionless ball joints. The proposed virtual object model has two features: (1) the virtual object model consists of spring models and is deformable; (2) the virtual object model is statically determinate once a stable grasp is achieved.

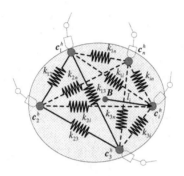

Fig. 1. The virtual object model.

The first feature means that the lengths of springs between contact positions in the virtual object model can be changed. This feature is especially useful to characterize the grasp quality of synergistic underactuated hands. The contact positions on the object are fixed relative to the actual object center once fingers have a contact with the object. We use the virtual object model to replace the actual object during the grasping tasks. In the virtual object model, fingers, contacting the object, will continue to move until all fingers have a contact with the object (each spring has a compression). Contact positions on the virtual object model are fixed to the fingertips. Thus contact position errors between the actual object and the virtual object exist. These position errors or spring forces in the virtual model can be used to quantitatively describe the grasp quality. The second feature indicates that all spring forces, namely internal force, can be determined by statics. The dimension of the internal force for n contact points is $(3n - 6)((2n - 3)$ for planar grasp) while the dimension of the spring models is $n(n - 1)/2$. The virtual model is said to be statically determinate only if these two dimensions are equal. We find that this constraint is satisfied for two-fingered, three-fingered and four-fingered grasp. For five-fingered grasp, we must take away an extra spring model from the virtual object model.

As mentioned above, the contact position errors or spring forces can be used to quantitatively describe the grasp quality. Next we will establish the relation between the spring forces and contact forces. With reference to Fig. 1, a reference frame (body frame) $\{O\} = \{B; x_b, y_b, z_b\}$ is fixed on the object center. On the ith contact point, frames $\{C_i^h\} = \{c_i^h; x_{hi}, y_{hi}, z_{hi}\}$ and $\{C_i^o\} = \{c_i^o; x_{oi}, y_{oi}, z_{oi}\}$ are attached to the fingertips and the object, respectively. Let $l_i \in \Re^3$ be the vector from point B to contact point c_i^h and k_{ij} be the coefficient of the spring between point c_i^h to point c_j^h. The symbol $\lambda_i \in \Re^c$ is the contact force of the ith contact point where $c \le 6$ only considers the constrained directions. In this paper, the hard finger (HF) contact model is considered and thus $c = 3$. Contact forces can be decomposed into two parts: λ_{in} (internal force) which causes tensions and λ_e (resultant force) which causes desired object motions:

$$\lambda = \lambda_{in} + \lambda_e \tag{4}$$

Where $\lambda = \left[\lambda_1^{\mathrm{T}} \cdots \lambda_n^{\mathrm{T}}\right]^{\mathrm{T}} \in \Re^{nc}$ collects the contact forces.

Define $t = \left[t_{12} \cdots t_{1n}\, t_{23} \cdots t_{(n-1)n}\right]^{\mathrm{T}} \in \Re^{\frac{n(n-1)}{2}}$ which collects all spring forces t_{ij} in the virtual model. For five or more fingered grasp, we let the spring forces of the extra spring models be 0 to make the virtual model statically determinate. According to the virtual object model in Fig. 1, we can obtain

$$\lambda_{in} = \begin{bmatrix} \sum\limits_{j=2}^{n} e_{1j}^{\mathrm{T}} t_{1j} \\ \vdots \\ \sum\limits_{j=1}^{i-1} e_{ij}^{\mathrm{T}} t_{ji} + \sum\limits_{j=i+1}^{n} e_{ij}^{\mathrm{T}} t_{ij} \\ \vdots \\ \sum\limits_{j=1}^{n-1} e_{nj}^{\mathrm{T}} t_{jn} \end{bmatrix} = -Et \tag{5}$$

Where $e_{ij} \in \Re^3$ is the unit vector from point c_i^h to point c_j^h, $e_{ij} = -e_{ji}$ and $e_{ij} = \frac{l_j - l_i}{\|l_j - l_i\|}$. The matrix $E \in \Re^{nc \times \frac{n(n-1)}{2}}$ is the linear map between the contact forces and spring forces in the virtual model:

$$E = \left[E_1^{\mathrm{T}} \cdots E_i^{\mathrm{T}} \cdots E_n^{\mathrm{T}}\right]^{\mathrm{T}} \tag{6}$$

Where

$$E_1 = \left[e_{12} \cdots e_{in}\, O\right], \quad E_i = \left[O\, e_{i1}\, O \cdots e_{i(i-1)}\, O\, e_{i(i+1)} \cdots e_{in}\right]$$

The spring forces are obtained using Eq. (5) as

$$t = -E^+ \lambda_{in} = -E^+(\lambda - \lambda_e) \tag{7}$$

Where E^+ is a left inverse, $E^+ = (E^{\mathrm{T}} E)^{-1} E^{\mathrm{T}}$.

Furthermore, the resultant force λ_e produces no tensions, namely $E^+ \lambda_e = 0$. Thus, the spring forces are given by

$$t = -E^+ \lambda \tag{8}$$

Similarly, the relation between the resultant force λ_e and the external force $\omega_e \in \Re^6$ can be obtained as

$$\omega_e = -G\lambda_e \tag{9}$$

Where $G \in \Re^{6 \times nc}$ is the grasp matrix.

3.2   Energy Function

Since the spring forces in the virtual model can be used to quantitatively describe the grasp quality, we can consider the potential energy in the object as the energy function. Therefore, we need to calculate the contact positions on the object and on the virtual model, respectively. Since all contacts on the object are rigid, object motions lead to a variation of the contact positions. The contact positions on the object can be approximately expressed as a function of object motions

$$c^o = G^{\mathrm{T}} u \tag{10}$$

Where $u \in \Re^6$ denotes the vector that describes the motion of frame O relative to the inertial frame. The $c^o = \begin{bmatrix} c_1^o & \cdots & c_n^o \end{bmatrix}^{\mathrm{T}}$ is a vector collecting motion components of contact points on the object constrained by the contact model.

For the virtual object model, the contact positions are equal to the fingertip positions which are only determined by the weighting coefficients for given hand models and posture synergies, namely

$$c^h = f_{c^h}(a) \tag{11}$$

Where $c^h = \begin{bmatrix} c_1^h & \cdots & c_n^h \end{bmatrix}^{\mathrm{T}}$ collects motion components of fingertips constrained by the contact model. The $f_{c^h}(a)$ is the forward kinematic equation of the given underactuated hand.

As described in the preceding section, the contact position errors between the actual object and virtual object lead to the compressions of the spring models and to spring forces. According to the principle of virtual work, the compressed length vector of the spring models is described as:

$$x = E^{\mathrm{T}}(c^h - c^o) \tag{12}$$

The spring force in the virtual object model is

$$t = K_B x = K_B E^{\mathrm{T}}(G^{\mathrm{T}} u - f_{c^h}(a)) \tag{13}$$

in which $K_B \in \Re^{\frac{n(n-1)}{2} \times \frac{n(n-1)}{2}}$ is a symmetric and positive definite matrix defining the stiffness matrix of the virtual object model. Thus the energy function can be obtained as

$$Q = (t - t_0)^{\mathrm{T}} W (t - t_0) \tag{14}$$

Where $W \in \Re^{\frac{n(n-1)}{2} \times \frac{n(n-1)}{2}}$ is a symmetric and positive weight matrix and t_0 is the desired spring force and usually set null.

In most cases, the hand configuration that minimizes the energy function creates an optimal grasp of the object. The value Q can be seen as a quantitative description of the grasp quality.

3.3 Grasp Constraints

We can find some desired hand configurations minimizing the energy function Eq. (14). But these computed results may not satisfy the actual situation. The object may be squeezed out from the hand or the fingers slide on the object. Therefore, we must add some constraints to find an optimal result. Here equilibrium constraints and contact constraints are considered.

The second feature of the virtual object model indicates that all spring models in the virtual model are statically determinate, namely the contact forces required at the grasp points can produce a desired resultant force and spring forces. According to Eqs. (4)–(9), we have

$$\lambda = -[G^+ \; E] \begin{bmatrix} \omega_e \\ t \end{bmatrix} \tag{15}$$

Where $G^+ = G^T(GG^T)^{-1}$ denotes the pseudoinverse of the grasp matrix G.

To ensure the nonslippage of the fingertips and to obtain a set of logical solution, the contact force λ_i at the ith contact point is constrained to lie within the friction cone $FC_{c_i^h}$, namely $\lambda_i \in FC_{c_i^h}$.

As mentioned above, the HF model is used for the contact model. For simplicity, we use the linearized formulation [16] which we will briefly review here. The friction cone can be approximated as an m-sided polyhedral cone $FC_{c_i^h}'$. Let $d_1(c_i^h), \cdots, d_m(c_i^h)$ be the edge vectors of $FC_{c_i^h}'$, which are normalized so that $d_j(c_i^h) \cdot n_i = 1 (i = 1, \cdots, n, j = 1, \cdots, m)$, where n_i is the unit normal vector of the object surface at the ith contact point. Thus, any valid contact force $\lambda_i \in FC_{c_i^h}'$ can be represented as

$$\lambda_i = \sum_{j=1}^m \alpha_{ij} d_j(c_i^h) = D_i \alpha_i, \alpha_{ij} \geq 0 \tag{16}$$

Where $\alpha_{ij}(i = 1, \cdots, n, j = 1, \cdots, m)$ are a set of nonnegative coefficients and $\alpha_i = \begin{bmatrix} \alpha_{i1} \cdots \alpha_{im} \end{bmatrix}^T \in \Re^m$ is the vector collecting all $\alpha_{ij}(j = 1, \cdots, m)$. Let μ_i be the friction coefficient. The matrix D_i collects all edge vectors of the m-sided polyhedral cone as

$$D_i = [d_1(c_i^h) \cdots d_m(c_i^h)], \quad d_j(c_i^h) = [1 \; \mu_i \cos(2j\pi/m) \; \mu_i \sin(2j\pi/m)]^T$$

Constraint (16) is for a single contact force λ_i. Now we collect constrains for the complete system in matrix form:

$$\lambda = D\alpha, \alpha \geq 0 \tag{17}$$

Where the matrix D collects the D_i in block diagonal form and the vector α collects the α_i in block column form.

Algorithm 1. Optimization algorithm for grasp planning

Require:
  Object parameters; Initial configuration; Synergies z
  while Iterations < MaxIterations **do**
    repeat
      Obtain a new vector z by Sim. Annealing Algorithm
      repeat
        for All hand joints **do**
          New_joints=Generate_joint(Current_joints)
        end for
        if collisions detected **then**
          Corresponding finger stops to move
        end if
      until All fingers have a contact with the object
      Compute c^h, c^o and contact force λ
      if constraints (15) and (17) are satisfied **then**
        legalState = true
      else
        legalState = false
      end if
    until legalState == true
    Compute the energy function Q_New using (14)
    if Q_New<Q_Saved **then**
      Q_Saved=Q_New, a_Saved=a
    end if
    Iterations = Iterations + 1
  end while
  return Q_Saved, a_Saved

3.4 Optimization Algorithm

After choosing the formulation of the quality function Q and grasp constraints (15) and (17), the optimization is performed. The complete optimization procedure is presented in Algorithm 1.

For this algorithm, we set the following conventions. The vector z is the target of the optimization, and each new vector is generated using the simulated annealing algorithm. The function Generate_joint, for generating new joint values, is performed. We have implemented this algorithm using our virtual integrated environment (VIE). The VIE is based on delta3d (v2.8), which is an open-source 3D simulation and game engine developed by Naval Postgraduate School [16]. The 3D hand models can be imported into the VIE, and then be assembled. And contact and collision detection are also performed using the VIE.

4 Numerical Example

The grasp planning algorithm has been applied to a robotic hand model which has the same structure as the Barrett hand (Barrett Technology, Inc.), shown in Fig. 2.

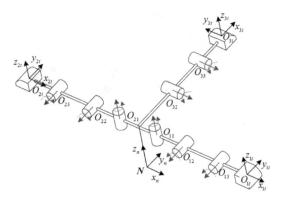

Fig. 2. Robotic hand model for simulation.

The hand model has eight joints and three identical fingers. Each finger has two joints. One of the fingers, referred to as 3, is fixed on the palm, while the other two fingers can rotate synchronously around the z axes. For finger i, let us define q_{i3} $(i = 1, 2, 3)$ the rotation value of the distal link with respect to the proximal link, q_{i2} $(i = 1, 2, 3)$ the rotation value of the proximal link with respect to the palm, and q_{i1} $(i = 1, 2)$ the spread value around the z axes. In the numerical simulations, let all structure parameters be equal to those of the real Barrett hand.

For the Barrett hand, there are four actuators. The proximate link is actuated and the distal link is coupled at a fixed rate with the proximal link for each finger. An additional actuator controls the rotation of the two fingers around the z axes. In this paper, we just use two actuators to control the hand: all proximal links and distal links are controlled by one actuator (z_1) and the synchronous rotation of the two fingers is controlled by another one (z_2). The relation between the joints and the synergies are expressed by the following relationships:

$$q_{i2} = q_{i3} = z_1, \quad i = 1, 2, 3$$

$$q_{11} = -q_{21} = z_2 = az_1$$

For three-fingered precision grasp, there are three spring models in the virtual object model. The stiffness matrix of the virtual object model is $\boldsymbol{K}_B = k_b \boldsymbol{I}_3$ where $k_b = 1000 \, \text{N/m}$, and $\boldsymbol{I}_3$ represents the 3×3 identity matrix. The desired spring force t_0 is set null. The weight matrix is $\boldsymbol{W} = \boldsymbol{I}_3$. For external force, we just consider the gravity of the objects. The contact model is HF type. The friction coefficient at the ith contact position is $\mu_i = 0.4$. Let the contact force be the nonnegative span of the evenly distributed six edge-vectors of the friction cone.

In order to test the effectiveness of our control algorithm for the task of dexterous robotic grasp planning, we have applied the proposed algorithm to the hand model described above to grasp different objects. Table 1 shows the grasped objects and their geometrical parameters. For each grasp simulation, the hand and object are fixed in the environment and the relative position between the

Table 1. Grasped object models and parameters.

Object	Sphere	Cylinder	Cube	Bottle
Model	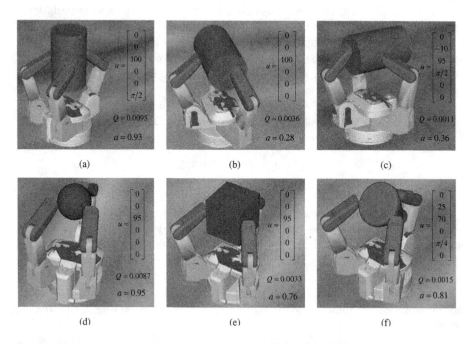			
Sizes	r=25mm	h=100mm r=32mm	d=60mm	h=80mm R=32mm r=12mm
Mass	60g	100g	80g	60g

hand and the object is given. The number of iterations is 2000. The grasp results, grasp quality and optimized amplitude are presented in Fig. 3. Figure 3(a)–(c) show the grasp results of a cylinder object with different relative positions. u is the relative position between the hand and object. Figure 3(d)–(f) show the grasp results of different objects (sphere, cube and bottle). Figure 4 shows the changing processes of the grasp quality and the amplitude for the grasp simulation in Fig. 3(a). As the simulation progresses, we can observe that the grasp state is sampled more often in the vicinity of the good quality.

Fig. 3. Simulation results based on the proposed algorithm. (a)–(c) Simulation results of a cylinder object with different relative positions. (d)–(f) Simulation results of different objects (sphere, cube and bottle) with given relative positions.

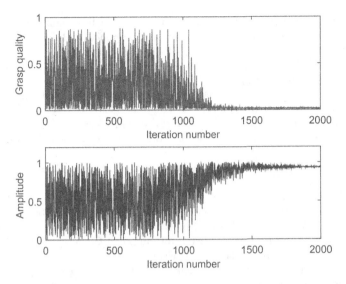

Fig. 4. The changing processes of the grasp quality and the amplitude.

These examples demonstrate that the proposed automatic grasp algorithm is valid for robotic hands, especially for synergistic underactuated hands. The obtained grasp configuration is stable and high-quality for different objects.

5   Conclusion

This paper presents an automatic grasp planning algorithm for synergistic underactuated hands. The main efforts are to establish a quantitative description of the grasp quality and to optimize the input parameters with synergy actuation. Based on the proposed virtual object model and contact models, the grasp quality function and grasp constraints, which ensure the obtained grasp results are stable and reasonable, are defined. Numerical simulation results show that the proposed optimization method can be used to obtain stable and high-performance grasp for synergistic underactuated hands.

Acknowledgements. This work is supported in part by the National Basic Research Program (973 Program) of China (Grant No. 2011CB013305) and the National Natural Science Foundation of China (Grant No. 51375296).

References

1. Jacobsen, S.C., Wood, J.E., Knutti, D.F., Biggers, K.B.: The Utah/MIT dextrous hand: work in progress. Int. J. Robot. Res. **3**(4), 21–50 (1984)
2. Grebenstein, M., Chalon, M., Friedl, W., et al.: The hand of the DLR Hand Arm System: designed for interaction. Int. J. Robot. Res. **31**(13), 1531–1555 (2012)

3. Bridgwater, L.B., Ihrke, C.A., Diftler, M.A., et al.: The Robonaut 2 hand - designed to do work with tools. In IEEE International Conference on Robotics and Automation, Minnesota, USA, pp. 3425–3430. IEEE Press (2012)
4. Bohg, J., Morales, A., Asfour, T., et al.: Data-driven grasp synthesis - a survey. IEEE Trans. Robot. **30**(2), 289–309 (2014)
5. Roa, M.A., Surez, R.: Grasp quality measures: review and performance. Auton. Robots **38**(1), 65–88 (2015)
6. Ala, R.K., Dong, H.K., Shin, S.Y., et al.: A 3D-grasp synthesis algorithm to grasp unknown objects based on graspable boundary and convex segments. Inf. Sci. **295**, 91–106 (2015)
7. Napier, J.R.: The prehensile movements of the human hand. J. Bone Jt. Surg. **38**(4), 902–913 (1956)
8. Cutkosky, M.R.: On grasp choice, grasp models, and the design of hands for manufacturing tasks. IEEE Trans. Robot. Autom. **5**(3), 269–279 (1989)
9. Miller, A.T., Knoop, S., Christensen, H.I., Allen, P.K.: Automatic grasp planning using shape primitives. In: IEEE International Conference on Robotics and Automation, Taipei, pp. 1824–1829. IEEE Press (2003)
10. Aleotti, J., Caselli, S.: Grasp recognition in virtual reality for robot pregrasp planning by demonstration. In: IEEE International Conference on Robotics and Automation, Orlando, pp. 2801–2806. IEEE Press (2006)
11. Santello, M., Flanders, M., Soechting, J.F.: Postural hand synergies for tool use. J. Neurosci. **18**(23), 10:105–10:115 (1998)
12. Rombokas, E., Malhotra, M., Matsuoka, Y.: Task-specific demonstration and practiced synergies for writing with the ACT hand. In IEEE International Conference on Robotics and Automation, Shanghai, pp. 5363–5368. IEEE Press (2011)
13. Zhang, A., Malhotra, M., Matsuoka, Y.: Musical piano performance by the ACT Hand. In: IEEE International Conference on Robotics and Automation, Shanghai, pp. 3536–3541. IEEE Press (2011)
14. Ficuciello, F., Palli, G., Melchiorri, C., Siciliano, B.: Postural synergies of the UB Hand IV for human-like grasping. Robot. Autonom. Syst. **62**(4), 515–527 (2014)
15. Allen, P.K., Ciocarlie, M., Goldfeder, C.: Grasp planning using low dimensional subspaces. In: Balasubramanian, R., Santos, V.J. (eds.) The Human Hand as an Inspiration for Robot Hand Development. Springer Tracts in Advanced Robotics, vol. 95, pp. 531–563. Springer, Cham (2014)
16. Zhu, X., Wang, J.: Synthesis of force-closure grasps on 3-D objects based on the Q distance. IEEE Trans. Robot. Autom. **19**(4), 669–679 (2003)
17. McDowell, P., Darken, R., Sullivan, J., Johnson, E.: Delta3D: a complete open source game and simulation engine for building military training systems. J. Defense Model. Simul. Appl. Method. Technol. **3**(3), 143–154 (2006)

Feasibility of a Sensorimotor Rhythm Based Mobile Brain-Computer Interface

Shiyong Su, Xiaokang Shu, Xinjun Sheng[(✉)], Dingguo Zhang,
and Xiangyang Zhu

State Key Laboratory of Mechanical System and Vibration, Shanghai Jiao Tong
University, 800 Dongchuan Road, Minhang District, Shanghai, China
xjsheng@sjtu.edu.cn
http://bbl.sjtu.edu.cn/

Abstract. Recently, mobile BCI has gained increasing attention. In the present study we investigated whether electroencephalogram (EEG) signals can be reliably measured when the subject is in walking condition. 5 subjects were recruited to perform motor tasks (making their fists with left or right hand) in two sessions, which included the control session (sitting in the armchair) and the experimental session (walking on the treadmill). Two sessions were conducted in the same environment. The mean classification accuracies are 70 % (seated) and 66 % (walking), so a bit lower classification accuracies when walking compared to that when sitting in the condition of performing motor execution for most subjects. Moreover, the motor and sensory cortices were activated and obvious ERD of contralateral area can be seen when performing motor execution no matter the subject was sitting or walking, and no obvious artifacts according to the presented time-frequency analysis. This leads us to conclude that it is possible to establish a mobile BCI system based on portable EEG devices.

Keywords: Mobile EEG · Classification accuracy · Application · Feasibility

1 Introduction

Brain-computer interface (BCI) is a kind of communication and control channel between human and computer or other electronic devices, by which people can express ideas or manipulate the device without the need of language or action, bypassing the brain's normal output pathways of peripheral nerves and muscles [1]. People who suffer from amyotrophic lateral sclerosis (ALS), brainstem stroke, brain or spinal cord injury, and many other diseases are enabled to have the ability to communicate with the external environment, and effectively improve their quality of life. As the development of BCIs, the research into BCI have been widened. More and more researchers are interested in mobile BCI, which is more representative of the complex and unpredictable situations that form the daily life.

© Springer International Publishing Switzerland 2016
N. Kubota et al. (Eds.): ICIRA 2016, Part II, LNAI 9835, pp. 443–452, 2016.
DOI: 10.1007/978-3-319-43518-3_42

Mobile EEG systems have been developed for a few years now, which not only means the systems are more portable and less costly, but also the widely application in people's daily life. One of the first devices able to record brain activity in mobile mode was the MindWare [2,3]. The device consists of a single EEG sensor comprising a dry electrode, which makes the MinSet [2] easy and fast to apply. Power for the device is supplied by an AAA battery, which allows a nonstop 10-hour recording. The drawback of the system is that the single sensor can only be placed on the forehead, making the system very inflexible and limited in its application [2]. Another low-cost pocket-sized device has been developed by Avatar EEG Solutions Inc. The system comes without electrodes but features eight channels, which can be configured in several different bipolar or monopolar montages. The system is powered by two rechargeable AA lithium batteries, which should suffice for 24 h of continuous recordings. The applications of this system are not only primarily in sleep research, but also in ambulatory neuroscience studies [2]. The group of Debener [12] presented a low-cost, small, and wireless 14-channel EEG system suitable for field recordings. They investigated whether a single-trial P300 response can be reliably measured with this system, while subjects were freely walking outdoors. Twenty healthy participants performed a three-class auditory oddball task, which included rare target and non-target distractor stimuli presented with equal probabilities of 16 %. Data were recorded in a seated (control condition) and in a walking condition, both of which were realized outdoors. P300 single-trial analysis was performed with regularized stepwise linear discriminant analysis and revealed above chance-level classification accuracies for most participants (19 out of 20 for the seated, 16 out of 20 for the walking condition), with mean classification accuracies of 71 % (seated) and 64 % (walking). At the same time, there are many groups investigating removing motion-related artifacts from EEG signals during walking in real-world environments, such as a method for removing motion artifacts occurred by body movement using inertial sensors like accelerometers and gyroscopes.

Studies investigating the mobile EEG are typically concern about the P300 [4] or steady-state visual evoked potential (SSVEP) [4], and few of them focus on the mobile BCI based on the sensory motor rhythm (SMR). In this paper we will examine whether it is possible to establish a mobile BCI system based on motor related activations during free movement, and discuss the application of this kind of mobile EEG.

2   Methods

2.1   Subjects

5 healthy subjects were recruited and participated in this experiment, 4 male, 1 female, with mean age of 23 years old. They all reported no past or present neurological or psychiatric conditions, and were informed about the whole experiment process. The study was approved by the Ethics Committee of Shanghai Jiao Tong University. All participants signed the informed consent forms before participating in the experiment.

2.2   EEG Recording

32 channel EEG signals were recorded using a BrainAmp system (Brain Products, German), and the electrodes were placed according to the 10/20 system. The ground electrode was located on the forehead. Meanwhile, the reference electrode was located on the mastoid behind the right ear. The sampling rate is 200 Hz, a bandwidth filter with 0 Hz to 50 Hz was applied to the original signals.

2.3   Experimental Paradigm

Subjects were required to do two sessions, one of which was the control session, after that was the experimental session. Each session contained 2 runs, and each run included 40 trials. A total of 160 trails were performed by each subject in the two sessions, lasting for about an hour.

There were 30 min for rest between two sessions, and both sessions were conducted in the same experimental room. In the control session, subjects were asked to sit in a comfortable armchair, with both forearms and hands rested in the armrest, as shown in Fig. 1(a). The subjects' task was to perform motor execution. In the motor execution task, the subjects were informed to clench their fists with right or left hands over and over again indicated from the cue until the crossing disappeared. In the experimental session, subjects were asked to walk on the treadmill at the speed of 0.1 m/s, as illustrated in Fig. 1(b). In the process of constant speed walking, subjects were informed to do the same motor execution as the control session.

The procedure of a single trail was showed as follows. At the beginning of each trail, there was blank in the screen. After 0.5 s, a fixation cross appeared to inform subjects to be prepared for the subsequent motor execution. Then at the 3rd second, a red cue bar pointing either left or right was presented, which superimposed on the fixation cross and lasted for 1.5 s. The subjects began to clench their hands with right or left hands after the appearance of the red cue bar. At the 8th second, the fixation cross disappeared and subjects stopped the motor execution at the same time for relax. The relax time lasted for 1.5 s. During the relax period, subjects could have a rest or blinked their eyes. At last, a random time period of about 0 to 2 s was inserted after the relaxation period to further avoid subjects' adaptation, after that the next trail started [5]. The whole trail structures were presented in Fig. 1(c).

2.4   Algorithms

Decoding algorithms for both actual and imagined hand clenching was mainly based on Common Spatial pattern (CSP) [6,7]. The raw EEG signal is represented as X_k with dimensions $ch \times len$, where ch is the number of recording electrodes, and len is the number of sample points. The normalized spatial covariance of the EEG can be obtained from

$$C_k = \frac{X_k X_k^T}{trace(X_k X_k^T)} \tag{1}$$

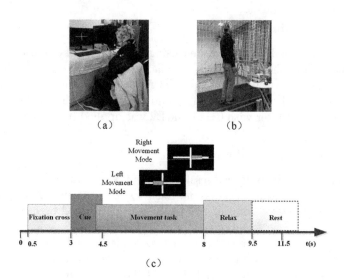

(a)   (b)

(c)

Fig. 1. (a) and (b) show experimental paradigm of two sessions. In the control session, subjects were asked to sit in a comfortable armchair, with both forearms and hands rested in the armrest. In the experimental session, subjects were asked to walk on the treadmill at the speed of 0.1 m/s. (c) shows the whole trail structures. (Color figure online)

where X_k^T denotes the transpose of the matrix X_k, and $trace(X_k X_k^T)$ is the sum of the diagonal elements of the matrix $X_k X_k^T$. Let

$$C_l = \sum_{k \in S_l} C_k \qquad C_r = \sum_{k \in S_r} C_k \qquad (2)$$

where S_l and S_r are the two index sets of the separate classes.

The projection matrix W could be gained from augmented generalized decomposition problem, $(C_l + C_r)W = \lambda C_r W$. The rows of W are called spatial filters, and the columns of W^{-1} are called spatial patterns. To the k-th trial, the filtered signal $Z_k = W X_k$ is uncorrelated. In this work, the log variance of the first three rows and last three rows of Z_k corresponding to largest three eigenvalues and smallest three eigenvalues are chosen as feature vectors, and linear discriminative analysis (LDA) was used as the classifier.

3   Results

The time period for off-line analysis (motor execution when sitting and walking) was chosen from the 4th second to the 7th second when the trail began, and the frequency band was chosen to cover the alpha and beta band of 8 Hz to 30 Hz using a 4th-order butterworth filter. A 10×10 fold cross validation was adopted to evaluate the classification accuracy between left and right.

Figure 2 showed the classification accuracies between left and right motor performance of 5 subjects respecting to the control session and the experimental session. The mean classification accuracy was 70 % in the control session and 66 % in the experimental session. There was no more than 10 % reduction of classification accuracies in the experimental session compared to the control session for S2, S3, s4, S5. As for S1, nearly 10 % higher classification accuracy in the experimental session than that in the control session. S3 achieved a reliable control above 75 % when walking on the treadmill and approximately 70 % when sitting in the armchair.

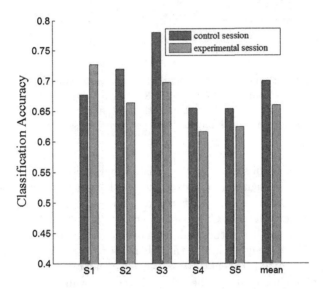

Fig. 2. Classification accuracies between left and right motor performance in two sessions. S1, S2, S3, S4, S5 represent the 5 subjects in the experiments individually. The red bars indicate the classification accuracies of the control session, and the green bars indicate the classification accuracies of the experimental session. (Color figure online)

In order to better understand the effect of walking on BCI performance, time/frequency decomposition of each session was carried out to construct the spatial-spectral-temporal structure corresponding to each task [5]. Event related spectrum perturbation (ERSP) at the critical channels C3 and C4 [8,9], and topograph of CSP patterns of subject 3 when performing motor execution with right hand during the control session and the experimental session, were compared as showed in Fig. 3. The configuration of nonsignificant parts which were wiped out under bootstrap significance was set as p = 0.01 with the tool of eeglab [10] and fieldtrip [11]. From the significance plot in (a), there was an obvious ERD distribution in the alpha and beta frequency domain in C3 during the control session, which indicated an activation in the left motor cortex when performing the motor execution with right hand. And the CSP pattern of right hand

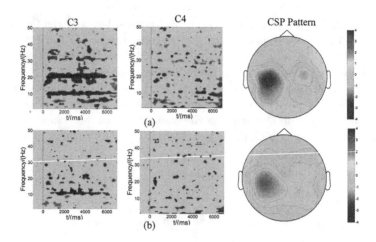

Fig. 3. Event Related Spectrum Perturbation (with bootstrap significance p = 0.01), and topograph of CSP pattern from subject 3 when making his fist with right hand during two sessions. (a) Motor execution with right hand in the controlled session. (b) Motor execution with right hand in the experimental session.

movement in the control session were consistent with time/frequency plot in C3 and C4. Plot (b) represented motor execution with right hand in the experimental session. As can be seen from it, there was a narrower ERD frequency domain in C3 and lower activation in CSP pattern compared to that in the control session, which may result from the walking movement that related to the quality of the signal. What is significant is that there were no obvious artifacts in the time/frequency plot and CSP pattern in the experimental session, indicating the feasibility of task classification when the subject was walking.

For the purpose of evaluating the signal recording of mobile BCI in detail, the classification accuracies between idle state and motor state of 5 subjects during the control and experimental sessions were showed in Fig. 4. The mean classification accuracy was 80 % in the control session and 70 % in the experimental session. There was more than 20 % reduction of classification accuracy in the experimental session compared to the controlled session for S1, and three subjects showed about 10 % discount of classification accuracies in the experimental session. As for S5, the classification accuracy in the experimental session is nearly 6 % higher than the control session.

To better explain the reduction of classification accuracies in the experimental session compared to that in the control session for most subjects. Time/frequency decompositions of idle state in two sessions were carried out to construct the spatial-spectral-temporal structure. Event related spectrum perturbation (ERSP) at the critical channels C3 and C4 [8,9], and topograph of CSP patterns of subject 3 when the subject was in idle state during the control session and experimental session, were shown in Fig. 5. Plot (a) and (b) respectively represented the time/frequency features and spatial-spectral-temporal structures of

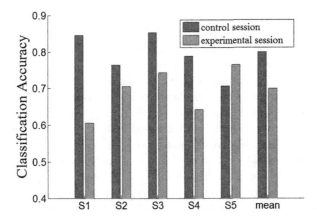

Fig. 4. Classification accuracies between idle state and motor state in the control and the experimental sessions. S1, S2, S3, S4, S5 represent the 5 subjects in the experiments individually. The red bars indicate the classification accuracies of the control session, and the green bars indicate the classification accuracies of the experimental session. (Color figure online)

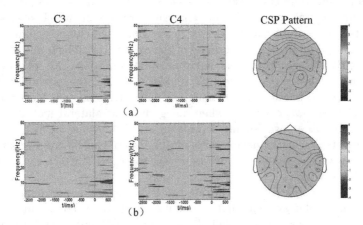

Fig. 5. Event Related Spectrum Perturbation (with bootstrap significance $p = 0.01$), and topograph of CSP pattern from subject 3 when in the idle state during two sessions.(a). Idle state in the controlled session. (b). Idle state in the experimental session.

idle state in the control state and the experimental state. What we can see from the plot is that there were tiny activations of idle state no matter in the control session or in the experimental session. Compared to the control session, the activation was more obvious in the experimental session, which might result from walking. When subjects were walking, their brain would be activated to some extent and led to the reduction of classification accuracies in the experimental session.

4   Discussion

The classification accuracy between left and right motor execution in the control session and the experimental session showed that there did have effects on the signal quality when the subject was walking to some extent. There raised the question whether the reduction in classification accuracies in the experimental session compared to that in the control session resulted from the poorer performance during walking caused by an increased amount of artifact compared to a seated condition. An alternative explanation of the lower classification accuracies in the experimental session is that the brain was probably busy with several more tasks than just the experimental one, like an increase of neck muscle activity, keeping posture, walking without to stumble, and keeping track of the walking path, which resulting in a weaker activation in the brain and smaller power of ERD in the time/frequency plot. Meanwhile, when subjects were walking in the experimental session, movement-related motion artifacts which might reside in slow frequency range, might influence the SMR BCI performance as the signals were between 8 and 30 Hz. Some studies have implied that in realistic daily life scenarios, brain activity can be used as input for a computer-based interaction with the environment. Also, the classification accuracies obtained from the across-condition training and testing analysis were similar, which suggests that consistent brain patterns were observed across both, seated and walking recording conditions. Apparently a retraining of the classifier system is not necessarily needed for BCI usage in different environmental conditions [12]. Although the classification accuracy of mobile BCI is relatively low, the possibility of taking use of the affected signal to classify the motor task existed. Meanwhile, the hemisphere brain was activated when doing the contralateral motor execution no matter the subject was sitting or walking and no obvious artifacts according to the presented time/frequency plots and CSP patterns. So from this point of view, there is great potential to investigate the mobile BCI which can be applied in people's daily life.

The goal of mobile BCI is to bring the BCI out of the restricted lab environment and develop systems that can be used in daily life environments. Many research have engaged to investigated the mobile EEG and have designed some different modes of low-cost, wireless EEG systems suitable for field recordings, among which some focused on variants of the auditory oddball paradigm and aimed for a binary switch BCI. Others developed multi-class paradigms aiming for an auditory alternative of the visual P300 matrix speller [12]. What is said above investigated the mobile BCI about P300 [12] or SSVEP [4], and analyzed the signal in time domain. The difference between this paper and other group is that here discussed the effect of walking movement EEG signal in motor cortex with motor execution and analyzed the signal in time and frequency domain. The results indicated that there is potential to achieve signal processing and task classification when the subject is walking. Therefore, the feasibility of mobile BCI is proved.

This kind of mobile BCI can become a life changing resource for persons with limited physical mobility or compromised brain function. For example, it is possible to envisage mobile EEG as a diagnostic tool which can link certain brain states or abnormalities in brain activity to neurological or psychiatric symptoms. Mobile EEG could further be used to monitor brain activity in outpatients to ensure treatment success, continued symptom control, or to indicate that a change in treatment might be required [2]. What's more, mobile EEG can also be applied to game playing for normal people.

5   Conclusion

There were a bit lower classification accuracies when walking compared to that when sitting in the condition of performing motor execution for most subjects. But it is demonstrated that it's effective to classify the motor task and analysis the signals when the subject is walking. The feasibility of mobile EEG have been demonstrated and this is a promising starting point to develop the mobile EEG later.

Acknowledgments. This work is supported by the National Basic Research Program (973 Program) of China (Grant No. 2011CB013305), and the Science and Technology Commission of Shanghai Municipality, and the National Natural Science Foundation of China (Grant No. 51375296).

References

1. Wolpaw, J.R., Birbaumer, N., McFarland, D.J., Pfurtscheller, G., Vaughan, T.M.: Brain-computer interfaces for communication and control. Clin. Neurophysiol. **113**(6), 767–791 (2002)
2. Kranczioch, C., Zich, C., Schierholz, I., Sterr, A.: Mobile EEG and its potential to promote the theory and application of imagery-based motor rehabilitation. Int. J. Psychophysiol. **91**(1), 10–15 (2014)
3. Chi, Y.M., Wang, Y.-T., Wang, Y., Maier, C., Jung, T.-P., Cauwenberghs, G.: Dry and noncontact EEG sensors for mobile brain-computer interfaces. IEEE Trans. Neural Syst. Rehabil. Eng. **20**(2), 228–235 (2012)
4. Wang, Y.-T., Wang, Y., Cheng, C.-K., Jung, T.-P.: Developing stimulus presentation on mobile devices for a truly portable ssvep-based BCI. In: 35th Annual International Conference of the IEEE Engineering in Medicine and Biology Society (EMBC), pp. 5271–5274. IEEE (2013)
5. Yao, L., Meng, J., Zhang, D., Sheng, X., Zhu, X.: Selective sensation based brain-computer interface via mechanical vibrotactile stimulation. PloS one **8**(6), e64784 (2013)
6. Fukunaga, K.: Introduction to Statistical Pattern Recognition. Academic press, San Diego (2013)
7. Ramoser, H., Muller-Gerking, J., Pfurtscheller, G.: Optimal spatial filtering of single trial EEG during imagined hand movement. IEEE Trans. Rehabil. Eng. **8**(4), 441–446 (2000)
8. Colon, E., Legrain, V., Mouraux, A.: Steady-state evoked potentials to study the processing of tactile and nociceptive somatosensory input in the human brain. Neurophysiol. Clin./Clin. Neurophysiol. **42**(5), 315–323 (2012)

9. Severens, M., Farquhar, J., Desain, P., Duysens, J., Gielen, C.C.A.M.: Transient and steady-state responses to mechanical stimulation of different fingers reveal interactions based on lateral inhibition. Clin. Neurophysiol. **121**(12), 2090–2096 (2010)
10. Delorme, A., Mullen, T., Kothe, C., Acar, Z.A., Bigdely-Shamlo, N., Vankov, A., Makeig, S.: EEGLAB, SIFT, NFT, BCILAB, and ERICA: new tools for advanced EEG processing. Comput. Intell. Neurosci. **2011**, 10 (2011)
11. Oostenveld, R., Fries, P., Maris, E., Schoffelen, J.-M.: Fieldtrip: open source software for advanced analysis of MEG, EEG, and invasive electrophysiological data. Comput. Intell. Neurosci. **2011**, 9 (2010)
12. De Vos, M., Gandras, K., Debener, S.: Towards a truly mobile auditory brain-computer interface: exploring the p300 to take away. Int. J. Psychophysiol. **91**(1), 46–53 (2014)

Brain Gamma Oscillations of Healthy People During Simulated Driving

Min Lei[1]($\boxtimes$), Guang Meng[1], Wenming Zhang[1], Joshua Wade[2],
and Nilanjan Sarkar[2]($\boxtimes$)

[1] State Key Laboratory of Mechanical System and Vibration,
Fundamental Science on Vibration, Shock and Noise Laboratory,
School of Mechanical Engineering, Institute of Vibration, Shock and Noise,
Shanghai Jiao Tong University, Shanghai, China
{leimin,gmeng,wenmingz}@sjtu.edu.cn
[2] Robotics and Autonomous Systems Laboratory,
Department of Mechanical Engineering,
Vanderbilt University, Nashville, TN 37212, USA
{joshua.w.wade,nilanjan.sarkar}@vanderbilt.edu

Abstract. Driving was a complex human behavior not only including limb movements but also involving a lot of neuropsychological processes, such as perception, attention, learning, memory, decision making, and action control. Gamma rhythm had been linked with several cognitive functions such attention, memory and perception. In order to explore the understanding of the brain under the real environments, a driving simulator environment might be a best tool that stimulates the brain dynamic activity. The purpose of this study was to investigate cerebral gamma oscillatory differences in the frontal, temporal, parietal, and occipital regions associated with psychological processes during simulated car driving. Neurophysiological signals of 5 healthy volunteers were recorded by using electroencephalography (EEG) during resting and simulated driving, respectively. Oscillatory differences in the gamma band were calculated by comparison between "resting" and "driving". "Resting" was the baseline, and oscillatory differences in the different regions during "driving" showed an increase in comparison with a baseline. The results indicated that brain oscillatory dynamics could play a role in cognitive processing, and might mediate the interaction between excitation and inhibition.

Keywords: Electroencephalogram (EEG) · Gamma oscillation · Brain oscillatory dynamics · Driving simulator · Neuropsychological process

1 Introduction

Driving behavior involved a lot of psychological processes, such as attention, inhibition, concentration, motor control, visuomotor integration, learning, memory, etc. Oscillatory brain activities in different frequency bands, such as delta (1–4 Hz), theta (4–8 Hz), alpha (8–13 Hz), beta (13–30 Hz), gamma (30–70 Hz) bands, had attracted considerable interest in regard to their role in psychological aspects of human life. In electroencephalography (EEG) study using driving simulators, most studies of EEGs

© Springer International Publishing Switzerland 2016
N. Kubota et al. (Eds.): ICIRA 2016, Part II, LNAI 9835, pp. 453–458, 2016.
DOI: 10.1007/978-3-319-43518-3_43

focused on fluctuations in the power of EEG signals in the theta, alpha, beta and delta over different scalp areas, such as the prefrontal, frontocentral, parietal, even occipital regions [1]. Laukka et al. (1995) used driving simulator to study concentration by EEG theta band increasing in the medial prefrontal cortex [2]. Schier (2000) found that alpha band decrease in the posterior area during attention [3]. Kim et al. (2004) used the oscillatory differences of alpha, beta and theta band to analyze optimal driving conditions [4]. Jäncke et al. (2008) studied the alpha oscillation in the lateral prefrontal cortex to explore the cognitive control in the fast driving state [5]. Lin et al. (2011) showed that during dual-task driving states, theta and beta bands in the frontal region increased and in the motor area, alpha and beta bands decreased [6]. Zhao et al. (2012) studied the relative power of theta, alpha, beta and delta bands during driving simulation [7]. Sonnleitner et al. (2014) showed the correlation between auditory distraction during driving and the alpha spindle rate to assess the distraction state of drivers [8]. However, few studies had reported the oscillatory differences of gamma band over different brain areas during driving conditions. The gamma-band oscillation had been reported to be related to several cognitive functions, such as perceptual binding, selective attention, short- and long-term memory [9], and could also be linked with the excitation and inhibition in the brain [10]. It could still a challenging issue whether the gamma-band oscillations play a role in cognitive processing during natural environments. In addition, previous brain oscillatory studies with driving simulators had almost been based on event-related driving tasks, such as fixed speed fast driving, right turn, left turn, audio, etc. [11–15]. Little was still known about the role of the brain oscillations during a full normal driving processing, like driving to work. In the present study, we used a realistic driving environment with a steering wheel, an accelerator, and a brake to stimulate brain activities. We hypothesized that oscillatory differences over different brain areas associated with concentration and attention required for driving could be detected. The purpose of this paper was to detect the role of the oscillatory activities associated with these psychological processes when a health person was manipulating a simulated driving car.

2 Materials and Methods

2.1 Participants

Five healthy young volunteers (aged between 16 and 27 years, mean age 21 ± 4.1833 years) participated in the current study. They are all right-handed males with normal or corrected vision. None of the subjects had a history of neurological or psychiatric disease. One had some experience with the driving game used in the test but had no driver's license. Others had the valid driver's licenses but first manipulated the driving simulator. And one was native Chinese speaker, others native English speakers. The experiment protocol had been approved by Vanderbilt University's Institutional Review Board.

2.2   Experiments

The driving simulation system was set up in an ordinary room [16, 17]. The language in the driving game was English. The subjects were seated on a comfortable chair to perform simulated driving operations like a real driving (see Fig. 1a). A computer screen was placed in front of the subjects, and a Driving Experiment Interface System gave driving epochs like a real driving including start up, accelerating, straight, turn-left, turn-right, stop, decelerating, etc. The simulated driving experiment consisted of a driving session for 5 min and the resting for 3 min. In the driving session, the subjects manipulated the wheel, accelerator and brake to drive the car in the street.

To explore oscillatory differences associated with psychological processes during driving, EEG activity was recorded in 128 Hz sampling frequency according to the international 10–20 system. There were 14 electrodes located at the scalp positions of the frontal (AF3, AF4, F7, F8, F3, F4, FC5, FC6), temporal (T7, T8), parietal (P7, P8) and occipital (O1, O2) regions (see Fig. 1b). We specifically analyzed brain areas that played important roles in concentration, attention, motor control, and visuomotor integration (see Table 1). Table 1 showed which Brodmann area (BA) was closest for each electrode position and its BA function.

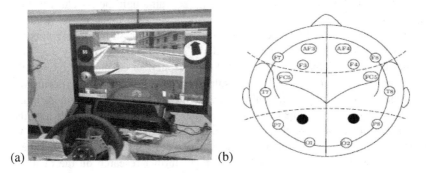

(a)   (b)

Fig. 1. The novel virtual reality driving environment and the sites of EEG electrodes [16, 17].

Table 1. Electrode and its nearest BA center

Electrode	AF3	AF4	F7	F8	F3	F4	FC5	FC6	T7	T8	P7	P8	O1	O2
Site	9L	9R	47L	47R	8L	8R	44L	44R	42L	42R	39L	39R	17L	17R
Function	Motor tertiary	Motor tertiary	Motor tertiary	Motor tertiary	Motor	Motor	Motor	Motor	Audition	Audition	Sensation tertiary	Sensation tertiary	Vision	Vision

2.3   Methods

Considering of the 128 Hz sampling frequency, the spectral powers of EEG signals in the gamma (30–64 Hz) band were calculated by fast Fourier transform using Matlab software. The length of the analysis time window was 2 min for driving and resting, respectively. The analysis data were the 2 min EEG signals after the 30 s beginning. This could give an overall detection of the brain oscillatory activities within 2 min.

3   Results and Discussions

The gamma oscillatory differences were demonstrated in the frontal, temporal, parietal and occipital regions between resting and driving (see Fig. 2). Table 2 gave the t-test analysis of Gamma power values, including resting vs driving, and the hemispherical dominance analysis during resting and driving, respectively. The increase of gamma oscillations was showed during driving (see Fig. 2a), especially in the FC6 channel of the frontal region ($p < 0.05$, see Table 2). Driving not only included human behaviors, such as limb actions, visual cognition, but also involved some cognitive function, such as attention, memory, and inhibition control. According to the Brodmann areas (see Table 1), the power values of the gamma oscillations might relate to behavioral performance of somatosensory-driven inhibition response. This result was consistent with that in the literature [10]. Besides, in the case of resting, the whole cortex had a certain right hemispherical dominance (see Fig. 2b), especially in the occipital region (p value was 0.0223 less than 0.05, see Table 2). During driving, the whole cortex had still some right hemispherical dominance (see Fig. 2c), although the p values in the posterior region of cerebral cortex were obviously larger than those in the front region. Meanwhile, the F7, F8, FC5 and FC6 in the frontal region showed a significant right hemispherical dominance (p values were 0.0494 and 0.0283, respectively, see Table 2). And the O1 and O2 had no obvious right hemispherical dominance (p was 0.3265 larger than 0.05, see Table 2). These results were different from those during resting. With regards to the inhibition control, the driving behavior required more inhibitory processes. The p values of the differences between resting and driving indicated that the driving behavior could directly stimulate the brain oscillatory activity. In other words, the driving simulator was suitable to investigate some psychological cognitive functions in the brain functions. These results might further research on the relationships

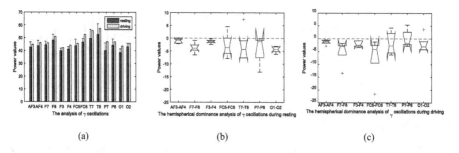

| (a) | (b) | (c) |

Fig. 2. The power analysis of the gamma oscillations in EEG signals. (a) for all 14 channels, the power values of the gamma oscillations during resting and driving; (b) the hemispherical dominance analysis during resting; (c) the hemispherical dominance analysis during driving.

Table 2. The t-test analysis of Gamma power values of EEG for health subjects

Channel	The hemispherical dominance analysis (p value)		p value
	Resting	Driving	Resting vs. Driving
AF3	0.3462	0.3096	0.1937
AF4			0.1953
F7	0.0823	**0.0494**	0.2227
F8			0.1859
F3	0.1674	0.1776	0.0743
F4			0.0637
FC5	0.1640	**0.0283**	0.2444
FC6			**0.0491**
T7	0.2711	0.3964	0.1389
T8			0.2322
P7	0.1295	0.5094	0.0904
P8			0.2287
O2			0.1787

among gamma oscillations, psychological cognition, neurophysiological indices, and human behavioral performance.

4 Conclusions

The gamma oscillatory differences were detected in the frontal, temporal, parietal, and occipital regions during simulated driving. According to the functions of these regions, the differences could be interpreted as correlates of an activated cortical area associated with cognitive functions. The results in these brain regions were consistent with those reported in prior EEG, MEG and fMRI studies of simulated driving. The oscillatory differences of EEG in gamma band in cognitive functions were demonstrated by using a driving simulator. It could be an evidence of a coupling between action performance and sensory sensitivity. The results indicated that brain oscillatory dynamics could play a role in cognitive processing, and might mediate the interaction between excitation and inhibition.

Acknowledgments. This work was funded by the Science Fund for Creative Research Groups of the National Natural Science Foundation of China (Grant No. 51421092), the National Natural Science Foundation of China (Grant No. 10872125), the Natural Science Foundation of Shanghai, China (Grant No. 06ZR14042), the Research Fund of State Key Laboratory of Mechanical System and Vibration, China (Grant No. MSV-MS-2010-08), the Research Fund from Shanghai Jiao Tong University for Medical and Engineering Science, China (Grant No. YG2013MS74), the NSF Project of USA (Grant Nos. 0967170, 1264462), and the NIH project of USA (Grant Nos. 1R01MH091102-01A1, 1R21MH103518-01).

References

1. Borghini, G., Astolfi, L., Vecchiato, G., Mattia, D., Bahiloni, F.: Measuring neurophysiological signals in aircraft pilots and car drivers for the assessment of mental workload, fatigue and drowsiness. Neurosci. Biobehav. Rev. **44**, 58–75 (2014)
2. Laukka, S.J., Jarvilchto, T., Alexandrov, Y.I., Lindqvist, J.: Frontal midline-theta related to learning in a simulated driving task. Biol. Psychol. **40**, 313–320 (1995)
3. Schier, M.A.: Changes in EEG alpha power during simulated driving: a demonstration. Int. J. Psychophysiol. **37**, 155–162 (2000)
4. Kim, J.Y., Park, J.S., Lee, H.Y., Yoo, S.Y.: Analysis of psychologically optimal driving state through measurement of physiological signals. J. Korean Soc. Emot. Sensibility **7**, 27–35 (2004)
5. Jäncke, L., Brunner, B., Esslen, M.: Brain activation during fast driving in a driving simulator: the role of the lateral prefrontal cortex. Neuroreport **19**(11), 1127–1130 (2008)
6. Lin, C.-T., Chen, S.-A., Chiu, T.-T., Lin, H.-Z., Ko, L.-W.: Spatial and temporal EEG dynamics of dual-task driving performance. J. Neuroeng. Rehabil. **8**, 11–23 (2011)
7. Zhao, C.L., Zhao, M., Liu, J.P., Zheng, C.X.: Electroencephalogram and electrocardiograph assessment of mental fatigue in a driving simulator. Accid. Anal. Prev. **45**, 83–90 (2012)
8. Sonnleitner, A., Simon, M., Kincses, W.E., Buchner, A., Schrauf, M.: Alpha spindles as neurophysiological correlates indicating attentional shift in a simulated driving task. Int. J. Psychophysiol. **83**, 110–118 (2012)
9. Fries, P.: Neuronal gamma-band synchronization as a fundamental process in cortical computation. Annu. Rev. Neurosci. **32**, 209–224 (2009)
10. Ray, S., Maunsell, J.: Do Gamma oscillations play a role in cerebral cortex? Trends Cogn. Sci. **19**(2), 78–85 (2015)
11. Kim, H., Hwang, Y., Yoon, D., Choi, W., Park, C.: Driver workload characteristics analysis using EEG data from an urban road. IEEE Trans. Intell. Transp. Syst. **15**(4), 1844–1849 (2014)
12. Calhoun, V.D., Pearlson, G.D.: A selective review of simulated driving studies: combining naturalistic and hybrid paradigms, analysis approaches, and future directions. Neuroimage **59**, 25–35 (2012)
13. Wang, Y., Chen, S., Lin, C.: An EEG-based brain-computer interface for dual task driving detection. Neurocomputing **129**, 85–93 (2014)
15. Zhang, H., Chavarriaga, R., Khaliliardali, Z., Gheorghe, L., Iturrate, I., Millán, J.: EEG-based decoding of error-related brain activity in a real-world driving task. J. Neural Eng. **12**, 066028 (2015)
16. Wade, J., Bian, D., Zhang, L., Swanson, A., Sarkar, M., Warren, Z., Sarkar, N.: Design of a virtual reality driving environment to assess performance of teenagers with ASD. In: Stephanidis, C., Antona, M. (eds.) UAHCI 2014, Part II. LNCS, vol. 8514, pp. 466–474. Springer, Heidelberg (2014)
17. Lei, M., Meng, G., Zhang, W.M., Sarkar, N.: Sample entropy of electroencephalogram for children with autism based on virtual driving game. Acta Phys. Sin. **65**(10), 108701 (2016)

Virtual Environments for Hand Rehabilitation with Force Feedback

Wei Lv, Xinjun Sheng$^{(\boxtimes)}$, Lei Hua, and Xiangyang Zhu

State Key Laboratory of Mechanical System and Vibration,
Shanghai Jiao Tong University, Shanghai 200240, China
xjsheng@sjtu.edu.cn

Abstract. Within recent years, an increasing number of researches that leverage virtual reality (VR) applications to improve motor rehabilitation have been conducted. These computer-based tools to various degrees help with motor rehabilitation for hand/arm impairment. Compared to the conventional therapy, people are more engaged and interested when they are being trained with the VR, leading to equal or higher improvements in terms of functional use of the hand/arm. In this paper, a VR-based platform for hand/arm rehabilitation is developed using delta3d, an open-source game and simulation engine of high-level interfaces to graphics and physics libraries. The platform provides not only visual information but also force feedback to users. Contrasted to previous similar researches, the contact force is directly extracted from the physics engine through virtual sensors. The effectiveness of virtual sensors is experimentally validated, and the relationship between joint angles and contact forces over time is presented. Finally, a training paradigm for hand impairment patients is proposed for future validation.

Keywords: Virtual environment · Hand rehabilitation · Force feedback · delta3d

1 Introduction

Arm/hand impairment or upper extremity amputation caused by stroke, accidents or war lead to diminished quality of patient's life. Meanwhile, robotic hands are becoming increasingly capable of reproducing the function of human hands, and have the potential to eventually become substitutes for them as prostheses. Within recent years, dramatic improvements have been made in the mechanical design and control of myoelectric prostheses [1].

However, several problems have been found in the research of motor rehabilitation. Firstly, it takes long time and costs much to produce prototype hardware. If there were any design defects, it would take longer time and cost more to modify the design. Secondly, evaluating control strategies need to be conducted after a prosthesis has already been ready to use. Thirdly, it is inefficient and ineffective to train subjects to learn how to use advanced prostheses properly only with oral instructions and static pictures. Recently, virtual reality has been suggested

© Springer International Publishing Switzerland 2016
N. Kubota et al. (Eds.): ICIRA 2016, Part II, LNAI 9835, pp. 459–470, 2016.
DOI: 10.1007/978-3-319-43518-3_44

as a method to quickly develop and evaluate control strategies and prototype devices [2,3]. The computer-based application has also been used for training and rehabilitating hand or arm function in stroke patients. According to a comprehensive review [4], people with disabilities are capable of motor learning within virtual environments, and movements learned in VR transfer to real-world equivalent motor tasks in most cases.

Acosta et al. [5] compared effectiveness between playing video games and training with avatar. The results of the study showed greater improvement, in this case on the reaching ability, during avatar feedback compared with video game feedback, suggesting the need for custom-designed applications that target the specific impairments. Lange et al. [6] focused on the use of the commercial video games and devices as neurological injury rehabilitation tools. They developed applications that use the Microsoft Kinect to track subjects motion more accurately. The results showed that subjects are more motivated and interested over the course of training. Their most recent Kinect-based rehabilitation game JewelMine [7] consists of a set of training exercises which encourage the players to reach out of their base of support. Popescu et al. [8] and Jack et al. [9] developed a virtual reality system for rehabilitation following stroke. Their system uses two inputs, a CyberGlove and a force feedback glove, allowing user interaction with a virtual environment. Their system consisted of four rehabilitation routines designed to exercise one specific parameter of hand movement: range, speed, fractionation or strength. Both objective measurements and subjective evaluation showed better effects compared to non-computer tasks.

While visual or audio feedback is common in a virtual reality application, haptic feedback is a more important factor that could increase the degree of realism and the users sense of immersion. There is increasing evidence that haptic information is an effective addition towards the accomplishment of certain treatment objectives such as increasing joint range of motion and force [10]. For example, in [8,9], they used a Rutgers Master II-ND (RMII) force feedback glove; and in [5], an Arm Coordination Training 3D system was employed to provide haptic interface. The force information was calculated using Hookes deformation law based on graphical features within the virtual environment, or solved from kinematic equations.

This paper describes the software and hardware components of a simulator based on delta3d, a well-known open source library for gaming and simulation. The major work is about providing subjects with force information extracted from the physics engine and visual feedback as well. In the training phase, the visual feedback is given to subjects through a display and the force information obtained from the physics engine in delta3d is sent to subjects through an electrotactile stimulator.

2   Platform Overview

The platform described in the paper is based on delta3d engine. delta3d is a widely-used, community-supported, open-source game and simulation engine appropriate for a wide variety of simulation and entertainment applications [11]. The decision to use delta3d was taken after evaluating several alternatives. The platform features: (1) a humanoid virtual hand under full control of physics engine; (2) supporting both manipulation and grasping; (3) several built-in virtual hand prototypes for algorithm evaluation; (4) supporting extensions of virtual hands; (5) several control methods, including EMG signals and data glove signals; (6) open APIs and allowing to extend new control methods; (7) real-time interfaces for users to interact with the virtual environment.

The platform consists of four layers (see Fig. 1). The lowest layer is composed of delta3d and other utility libraries for plotting or RS232 communication. In order to achieve high extendibility and scalability, an abstraction of delta3d APIs and some general process logic was designed. In this layer, basic functionality modules are provided, such as an extendable virtual hand helper. The third layer is application layer, where there are concrete virtual hands derived from the helper, and specific control strategies based on a keyboard, a data glove or an EMG-signal arm band. These modules cooperate with each other in higher-level business logics, providing complex functions like grasping objects under control of the data glove. And the force information is fed back to patients within this layer. The top layer, a graphics user interface (GUI) implemented with Qt, accepts user's input and provides direct visual display of the virtual environment.

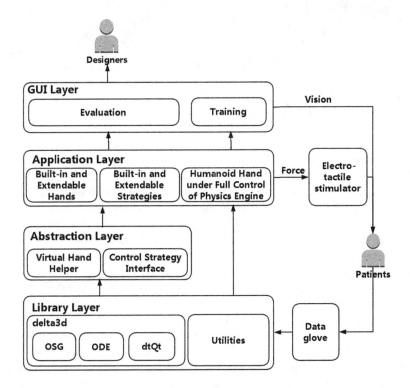

Fig. 1. The four-level architecture.

3   Virtual Hand Modeling

delta3d is not a model editor, and thus one cannot create complex 3D models in delta3d. Instead, delta3d uses OpenSceneGraph as a base of graphics rendering, which supports more than 50 different formats. Most common 3D data formats, such as COLLADA, LightWave (.lwo), Wavefront (.obj), 3D Studio Max (.3ds) and many others are able to be imported into the scene [12]. OpenSceneGraph provides its own native ASCII .osg format and binary .ive format as well. The best practice would be creating all hand models in 3D creation tools like Pro/E, UniGraph NX, and then converting them into .osg or .ive format. Figure 2 shows different kinds of built-in virtual hand prototypes in the current platform.

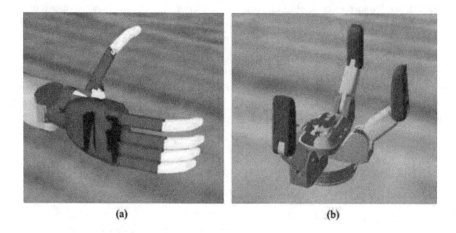

(a)   (b)

Fig. 2. Different built-in virtual hands.

In order to make imported models behave realistically, a physics engine is needed to handle all physics simulation calculation. delta3d uses ODE as its physics engine and provides glue code for coordination between rendering and physics. In the physics world, the model should be wrapped by a bounding geometry whose properties are really controlled by the physics engine. There are several types of bounding geometry, such as a box, a sphere, a cylinder or a triangle mesh. Using a simple geometry primitive like a box can have a lower computational complexity at the expense of accuracy, while using a triangle mesh can fit the shape of 3D models more accurately at a prohibitive cost of computational time.

These bounding geometries can be connected with each other by different types of joints, such as the hinge joint for revolution and the prismatic joint for translation. Besides the user-defined joint types, contact joints will be automatically created and attached to bodies once they interpenetrate into each other, or in another words, collide. In order to keep the joint at a desired angle, a PD controller is designed to control the joints.

4   Force Calculation in ODE

In general, a rigid body has six degrees of freedom, which can be represented as position r and rotation q. According to Newton's law, without any constraints or joints, the ith free rigid body's equations of motion is expressed as

$$\begin{bmatrix} M_i & 0 \\ 0 & I_i \end{bmatrix} \begin{bmatrix} \ddot{r}_i \\ \ddot{q}_i \end{bmatrix} = \begin{bmatrix} f_i \\ \tau_i - \dot{q}_i \times I_i \dot{q}_i \end{bmatrix} \tag{1}$$

The above equations can be simplified as

$$m_i \dot{v}_i = e_i \tag{2}$$

where $m_i = diag(M_i, I_i)$; $v_i = [\dot{r}_i{}^T, \dot{q}_i{}^T]^T$; and $e_i = [f_i{}^T, (\tau_i - \dot{q}_i \times I_i \dot{q}_i)^T]^T$

For a system consisting of N rigid bodies, system variables are defined as the velocity states $v = [v_1{}^T, v_2{}^T, \cdots, v_N{}^T]^T$; the block diagonal system mass matrix $M = diag(m_1, m_2, \cdots, m_N)$; and the system effort vector that represents external forces, $E = [e_1{}^T, e_2{}^T, \cdots, e_N{}^T]^T$. The unconstrained system dynamics are then given as:

$$M\dot{v} = E \tag{3}$$

The Eq. (3) describe a system with N free rigid bodies. As to constraints between bodies, ODE uses velocity constraint so that a body's velocity is allowed to have some values but not others. This can avoid non-linear constraints and is of the form

$$Jv = \begin{bmatrix} J_{11} & J_{12} & J_{13} & J_{14} & J_{15} & J_{16} \\ \vdots & \vdots & \vdots & \vdots & \vdots & \vdots \\ J_{m1} & J_{m2} & J_{m3} & J_{m4} & J_{m5} & J_{m6} \end{bmatrix} v = \begin{bmatrix} c_1 \\ \vdots \\ c_m \end{bmatrix} (1 \le m \le 6) \tag{4}$$

where J is the constraint Jacobian matrix and c is the constraint vector with $c_i = 0$ or $c_i \ge 0$ representing equality constraints or contact inequality constraints respectively. Using a vector of Lagrange multipliers λ, the velocity constraints are added to the equations of motion as

$$M\dot{v} = E + J^T \lambda \tag{5}$$

To solve Eq. (5), ODE takes a first-order Euler method as $\dot{v}\Delta t = v^{(n+1)} - v^{(n)}$, where Δt is the time step and $v^{(n)}$ is v value at the nth step. The difference equation for Eq. (5) is given as

$$(1/\Delta t)Mv^{(n+1)} - J^T \lambda = (1/\Delta t)Mv^{(n)} + E \tag{6}$$

Considering that the constraints are still satisfied at the $(n + 1)^{th}$ step, so that we have $Jv^{(n+1)} = c$. And the Eq. (6) can be rearranged by left multiplying by JM^{-1} as

$$JM^{-1}J^T \lambda = (1/\Delta t)c - J[(1/\Delta t)v^n + M^{-1}E] \tag{7}$$

ODE solves Eq. (7) using an iterative Projected Gauss Seidel algorithm and the constraint forces λ are obtained. Given λ and Eq. (6), the rigid body velocities $v^{(n+1)}$ at next time step are computed.

More details about Jacobian constraint matrix can be found in [14]. And in [15,16], the researchers presented a deep discussion about math in ODE, and proposed several methods to improve the ODE's performance.

5   Force Feedback

ODE solves constraint equations every simulation step, generating proper constraint forces or torques to ensure every physics body to stay at the right places or at a right velocity. A handy API called *dJointSetFeedback* can be used to obtain such information, and it takes a *JointID* and a *dJointFeedback* structure as arguments. The *dJointFeedback* structure includes four members, *f1* and *f2*, the forces exerted on body1 and body2; *t1* and *t2*, the torques exerted on body1 and body2. These forces and torques can be accessed any time after a *dJointFeedback* structure is attached to a certain joint through *dJointSetFeedback*.

But contact joint (or contact constraint) is different from any other types. Generally speaking, contact joints would be created at the very beginning of a simulation step and added to the joint group; and then they are calculated together with other joints to obtain positions and velocities of related physics bodies; and finally, at the very end of the same simulation step, the whole contact joint group would be removed, so that it is not possible to access the exactly same contact joints any more.

Considering how ODE works on contact joints, we proposed a special method, which could ensure association with those contact joints after they are created, and hold the information for future requests even after they are deleted. The first thing is to modify the *nearCallback* function, hooking *dJointFeedback* structures up to contact joints. The *nearCallback* function is where ODE executes contact detections between each pair of physics bodies. Once a contact is detected between two bodies, ODE will exert a proper force or torque on them to avoid further interpenetration. However, in the default *nearCallback* function, there is no *dJointFeedback* structure attached to contact joints because of memory management issues [13]. In order to manage the memory properly, we designed a virtual contact sensor, in which there is an STL vector container to hold all contact joint feedback structures generated in one simulation step. And a function called *PrePhysicsStepUpdate* will be executed periodically before the next simulation step is taken, in which the vector container is always being properly cleared.

In this paper, we access the contact sensors in the frame updating loop. However, the very nature of the discrete time-step simulation algorithm of the ODE and other physics engines imply that a complex-shape physics body in close contact to another one will not result in a smooth value of the simulated forces. Instead, they will interpenetrate a bit during one simulation step, triggering a constraint-violation and a resulting spike of high force, which turns out to start a fast motion in the opposing direction. The pair of contacted bodies then loose contact, with very low or zero contact forces during the next few simulation steps, until they collide again and the process repeats. A typical plot demonstrating this situation is shown in Fig. 3 (thin line). The obvious solution to this problem is to reduce the simulator time-step as far as necessary to reduce object inter-penetration depth and to avoid the spiking, but the performance penalty would be prohibitive. Instead, we choose to reduce the length of a time-step to 2 ms, and at the same time, to apply a simple moving average smoothing to reduce the spiking and approximate a time-averaged value of the normal forces (thick line), resulting in a smoother and more stable approximation to the contact forces.

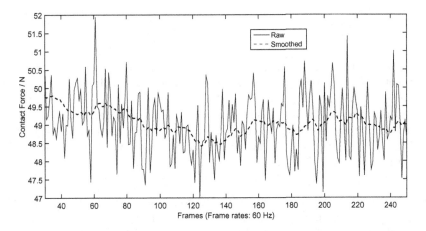

Fig. 3. Force measured when virtual hand is grasping an object. The thin line is raw data, while the thick line is smoothed data. The raw data is sampled at a frequency of 60 Hz, but the physics updating rate is 500 Hz.

6   Interfaces to the Real Hardware Devices

Another key feature of this platform is the integration of bidirectional interfaces to the virtual environment and the patients. This includes real-time control of the virtual hand via the 5DT Data Glove 14 Ultra [17] and the keyboard, and real-time force feedback to the user via an electrotactile simulation. In this way, a user-in-the-loop training or simulation can be done.

6.1   Tele-operation via Data Glove

The data glove we use has 14 sensors, which are positioned as in Fig. 4(a). With data glove APIs, we take joint data from the sensors on the glove and then perform a simple mapping from the glove sensors to the corresponding virtual

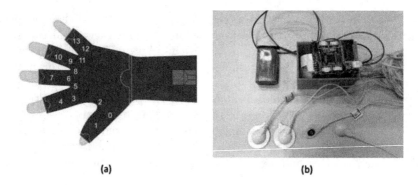

Fig. 4. (a) Sensor mappings for the 5DT Data Glove 14 Ultra. (b) The Electrotactile stimulator.

hand joint. Additional mapping is performed for those joints that have no direct equivalent sensors on the glove, such as the farthest joint of the index finger. The joint data have been calibrated and scaled before they are accessed, and thus in order to use the data as set points, they need to be re-scaled to corresponding joint movement range. Since no sensor measures the farthest joint angle of the index, middle, ring or little finger, an estimation based on middle joint is taken as follows:

$$\theta_{DIP} = 0.66 \times \theta_{PIP}$$

The values are then fed into each joint's PD controller, and all fingers are moved according to the set points from the data glove and constraints resulting from simulated dynamical forces and all finger-object collisions.

6.2 Electrotactile Simulation

Nowadays, in order to provide patients with artificial tactile sensations, researchers have utilized transcutaneous electrical nerve stimulation technology to realize sensory substitution. In our lab, researchers have proposed a low-power portable multichannel electrical stimulator, SJTU-Stimulator, for producing cutaneous electrotactile stimulation. Several tactile-feeling experiments were conducted to provide four subjects different kinds of pressure feeling and vibration feeling by modifying the electrotactile pulse width and frequency. The subjects reported they could distinguish among three pressure levels at an accuracy of $87.1\% \pm 2.1\%$ and among three vibration levels at $82.9\% \pm 5.0\%$. It is suitable to be used as a simple force feedback device.

7 Experiments

The simulator provides a low-cost solution to fast setup for training with force feedback and visual feedback. Without any expensive tactile sensors or any other software dependencies, the contact force can be directly obtained from the physics engine. The remainder of this section describes the force measured by virtual sensors, the force feedback and a training paradigm.

7.1 Virtual Sensor

To validate the effectiveness of the virtual sensor, both static load and dynamic load are applied to a simple sphere (see Fig. 5). According to Newton's law, the normal force equals to load force plus sphere's gravity. In order to clearly compare similarity, the component force caused by load is extracted as

$$F_C = F_N - mg$$

where F_N is the normal force, which is also the virtual sensor reading. Note that because of the moving average smoothing, there is a phase lag between load forces and sensor readings.

7.2 Force Feedback

When people grasp an object, they can determine how much force to be exerted by adjusting the finger joint angle. If a finger is already blocked by the object, further flexing will result in higher contact force. In this experiment, we demonstrate the relations between the joint angle and the contact force. The desired angles were measured by data glove as inputs of virtual hand. Note that in Fig. 6(a), if the joint has reached its preset highest force, it will not be able to eliminate the gap between the desired joint angle and the actual one; and this joint goes backward under the action of external force.

7.3 Training Paradigm

This training system in Fig. 7 provides force feedback besides visual feedback. Seeing the virtual hand or objects within the virtual environment would make

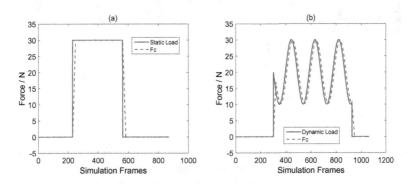

Fig. 5. The sensor reading under (a) static load and (b) dynamic load.

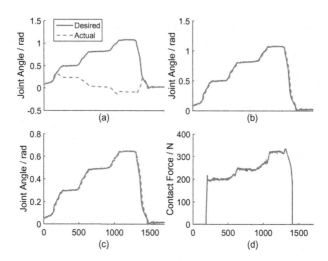

Fig. 6. The relations between angles of (a) MCP joint, (b) PIP joint, (c) DIP joint and (d) contact force on the fingertip over time.

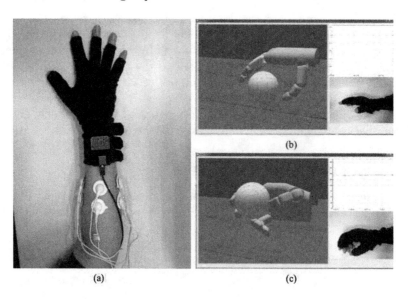

Fig. 7. The training paradigm using a data glove, an electrotactile stimulator and a virtual reality application.

patients concentrate on training process, while feeling objects by touching them would greatly enhance interaction and sense of reality. When the subject has grasped an object, he can feel electrical stimulation of a certain level; and the more firmly he grasps the object, the stronger current he can feel. With a data glove and an electrotactile stimulator integrated in the system, patients can send and receive information at the same time. The bidirectional real-time interfaces lead to a user-in-the-loop training paradigm.

8   Conclusion

A virtual reality application has been developed to specially train functional use of impaired hand, leveraging the features provided by delta3d, especially its graphics and physics components. Users can interact with or grasp objects within virtual environment using real-time interfaces like a data glove and an electrotactile stimulator. The force information obtained from the physics engine is realistic and accurate enough during the virtual grasping, and thus can be used to reinforce users' immersive experience. Future directions in this research will be more formal testing on real patients to determine if the extra force information improves the rehabilitation outcomes. And to enhance the immersive experience to a higher level, a head-mounted display (HMD) like the Oculus Rift will replace the current computer display.

Acknowledgments. This work is supported by the National Basic Research Program (973 Program) of China (Grant No. 2011CB013305), the National Natural Science Foundation of China (Grant No. 51375296).

References

1. Lambrecht, J.M., Pulliam, C.L., Kirsch, R.F.: Virtual reality environment for simulating tasks with a myoelectric prosthesis: an assessment and training tool. J. Prosthet. Orthot. **23**, 89 (2011)
2. Kuiken, T.A., Li, G., Lock, B.A., et al.: Targeted muscle reinnervation for real-time myoelectric control of multifunction artificial arms. Jama **301**, 619–628 (2009)
3. Hauschild, M., Davoodi, R., Loeb, G.E.: A virtual reality environment for designing and fitting neural prosthetic limbs. IEEE Trans. Neural Syst. Rehabil. Eng. **15**, 9–15 (2007)
4. Holden, M.K.: Virtual environments for motor rehabilitation: review. Cyberpsychol. Behav. **8**, 187–211 (2005)
5. Acosta, A.M., Dewald, H.A., Dewald, J.P.A.: Pilot study to test effectiveness of video game on reaching performance in stroke. J. Rehabil. Res. Dev. **48**, 431 (2011)
6. Lange, B., Chang, C.Y., Suma, E., et al.: Development and evaluation of low cost game-based balance rehabilitation tool using the Microsoft Kinect sensor. In: 2011 Annual International Conference of the IEEE Engineering in Medicine and Biology Society, EMBC, pp. 1831–1834. IEEE Press, Boston (2011)
7. Lange, B., Koenig, S., McConnell, E., et al.: Interactive game-based rehabilitation using the Microsoft Kinect. In: Virtual Reality Short Papers and Posters (VRW), pp. 171–172. IEEE (2012)
8. Popescu, V.G., Burdea, G.C., Bouzit, M., et al.: A virtual-reality-based telerehabilitation system with force feedback. IEEE Trans. Inf. Technol. Biomed. **4**, 45–51 (2000)
9. Jack, D., Boian, R., Merians, A.S., et al.: Virtual reality-enhanced stroke rehabilitation. IEEE Trans. Neural Syst. Rehabil. Eng. **9**, 308–318 (2001)

10. Sveistrup, H.: Motor rehabilitation using virtual reality. J. Neuroeng. Rehabil. **1**, 1 (2004)
11. delta3d. http://www.delta3d.org
12. OpenSceneGraph Manual. http://www.openscenegraph.org/
13. ODE Manual. http://ode-wiki.org/wiki/index.php?title=Manual
14. Smith, R.: Constraints in rigid body dynamics. Game Program. Gems **4**, 241–251 (2005)
15. Drumwright, E., Hsu, J., Koenig, N., Shell, D.: Extending open dynamics engine for robotics simulation. In: Ando, N., Balakirsky, S., Hemker, T., Reggiani, M., von Stryk, O. (eds.) SIMPAR 2010. LNCS, vol. 6472, pp. 38–50. Springer, Heidelberg (2010)
16. Hsu, J.M., Peters, S.C.: Extending open dynamics engine for the DARPA virtual robotics challenge. In: Brugali, D., Broenink, J.F., Kroeger, T., MacDonald, B.A. (eds.) SIMPAR 2014. LNCS, vol. 8810, pp. 37–48. Springer, Heidelberg (2014)
17. 5DT Data Glove 14 Ultra Manual. http://www.5dt.com/products/pdataglove5u.html

A Novel Tactile Sensing Array with Elastic Steel Frame Structure for Prosthetic Hand

Chunxin Cu, Weiting Liu[✉], Xiaodong Ruan, and Xin Fu

The State Key Laboratory of Fluid Power and Mechatronic Systems, Zhejiang University,
Zheda Rd. 38, Hangzhou, China
{cxgu,liuwt,xdruan,xfu}@zju.edu.cn

Abstract. A novel tactile sensing array based on elastic steel frame structure for prosthetic hand is proposed. The sensing units are mounted on the curving surface of a 3D printed sensor holder with five along the axial direction and four in the circumferential direction. Each sensing unit consists of a steel frame, a silicon gauge and flexible printed circuit board (FPCB). Based on the piezoresistive effect, the sensing unit can measure the normal force and the shear force along the longitudinal direction. The finite element analysis is conducted to evaluate the mechanical performance of the sensing unit. Meanwhile, the solution of detecting the resultant force is proposed.

Keywords: Tactile sensing array · Elastic steel frame · Silicon gauge · 3D print · FEA

1 Introduction

With the great demands for smart control, tactile sensor is indispensable for prosthetic hand. In common life, the modality of grasping is frequently utilized for prosthetic hand, which requires the ability of detecting normal force and shear force at the same time from the tactile sensor. Although having been widely investigated, the current tactile sensors are still far away from effective application in commercial prosthetic hand due to their complex structure, rigorous fabrication and poor stability.

So far, different tactile sensors have been developed based on several principles such as piezoresistive, capacitive, piezoelectric, optical methods [1–4] and etc. Among them, silicon based sensor fabricated by MEMS technology is favored for its high sensitivity, high stability and good compatibility with current electronic technology. In MEMS, the structures like cantilevers and beams are widely used to transmit force. For example, Beccai et al. have designed and fabricated a silicon three-axial force microsensor with a mesa located at the center of a cross-shaped beam [5]; Noda et al. have developed a tactile sensor composed of standing piezoresistive cantilevers and bridges to detect both normal and shear forces [6]. However, the silicon based sensing elements are usually fragile especially under an unexpected overload [7], which demands careful packaging design and decreases the reliability of the sensory system.

In our previous work [8], a tactile sensor with a silicon gauge fixed on a steel sheet was developed and tested with the result of good linearity, sensitivity and large force

© Springer International Publishing Switzerland 2016
N. Kubota et al. (Eds.): ICIRA 2016, Part II, LNAI 9835, pp. 471–478, 2016.
DOI: 10.1007/978-3-319-43518-3_45

range. Besides, based on the hybrid method of the traditional machining process and the MEMS technology, the prototype of the previous sensor takes advantages of easy fabrication and low expense which encourages further optimization and application of the structure.

In this study, we design a tactile sensing array mounted on a curving surface. Continuing the hybrid idea, we combine the silicon gauge with the elastic steel beam and optimize the structure by adding a cuboid mesa on the steel beam which does good to shear force detection. Moreover, simulation work is conducted to evaluate the mechanical performance of the sensor unit.

2   Design and Fabrication Method

The proposed design of the tactile sensing system is shown in Fig. 1. The system consists of a sensor holder, a tactile sensing array and a flexible printed circuit board (FPCB).

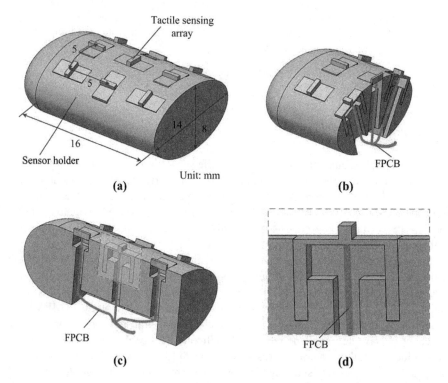

Fig. 1. Design of sensor structure. (a) Schematic of the whole sensor; (b) Cross section view of the sensor in a plane perpendicular to the axis direction of the holder; (c) Cross section view of the sensor in the vertical symmetry plane; (d) Local view of the sensing unit being fixed into the grooves of the sensor holder.

The holder is ellipse shaped in the cross section with the sizes of major axis and minor axis as 14 mm and 8 mm respectively, which shares the similar shape and dimensions of the distal phalanx of an adult's finger. On the curving surface of the holder, a 3 × 3 groove array intended for mounting the sensing array is located with both circumferential and axis spacing as 5 mm. In each groove, a channel is reserved extending to the back side of the holder to keep a path for the FPCB. Considering the relatively complex structure with the curving surface, grooves and channels, the holder is printed directly from the computer-aided design (CAD) file by the ProJet 3610 series professional 3D printer (3D System Inc.). The material of the structure is VisiJet Crystal with true plastic look, good mechanical properties and low density as 1.02 g/cm^3, which provides advantage of lightweight for tactile sensor integrated on the prosthetic hand. As shown in Fig. 1(a), there are five sensing units fixed along the axial direction and four along the circumferential direction, which is designed for detecting shear forces in both directions. The principle of shear force detecting will be discussed detailedly in the below context.

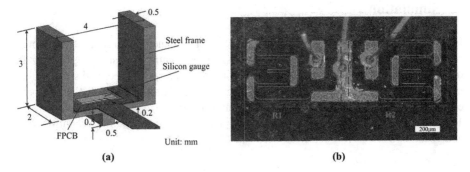

Fig. 2. Design of sensing unit. (a) Schematic of sensing unit structure; (b) Microscope photograph of the silicon gauge.

Plotted in Fig. 2(a), each unit of the sensing array contains an inverted U shaped steel frame which could be divided into a beam and two supporting braces. The beam has the dimensions as 3 mm, 2 mm and 0.2 mm respectively in each direction while the supporting braces have the thickness of 0.5 mm with the height as 3 mm. On the beam, a 0.5 mm thick cuboid mesa is designed to transmit force especially the shear force. According to its modest complexity and dimensions, the whole structure is available from mature machining process using the 65 M spring steel due to its high elasticity and well machining properties. On the reverse side of the beam, a silicon gauge is fixed in the center acting as the sensing element. The silicon gauge is fabricated by MEMS processing technology and has been applied in the previous work, which shows good linearity, sensitivity and large force range. As shown in Fig. 2(b), the silicon gauge is 1.5 mm long and 0.5 mm wide which brings no difficulty to be manipulated by manual work. The whole silicon gauge can be divided into two independent parts marked as R_1 and R_2 with the initial resistance as 3.4 kΩ for each. The gauge factor G for both R_1 and R_2 is 100, which means the relationship between the resistance values change and the strain values is shown as below:

$$\begin{cases} \dfrac{\Delta R_1}{R_1}/\varepsilon_1 = G \\ \dfrac{\Delta R_2}{R_2}/\varepsilon_2 = G \end{cases} \tag{1}$$

The silicon gauge aligns along the longitudinal direction of the steel beam, which complies with the largest stretch strain inside the beam under a force load. To make sure reliable adhesion and avoid stress relaxation, the mature technology of glass sintering is adopted to mount the silicon gauge well on the reverse surface of the steel sheet, which guarantees the same deformation between the silicon gauge and the steel beam.

3   Simulation

In this study, the sensor is just evaluated on its performance under static load, which corresponds to the stable manipulation process for prosthetic hand such as grasping an object for a while. Due to the good elastic property of the spring steel and small deformation, the static analysis of the sensor structure is based on the hypothesis of linear elasticity. By means of the finite element analysis (FEA) tool (ANSYS Workbench 14.5), the mechanical behaviors of the sensor under various loading conditions are analyzed.

3.1   Under Various Force Conditions

As all sensing units have the same structure, a single unit is analyzed under the conditions that an external normal force or shear force is applied on the mesa respectively. As shown in Fig. 3(a), the sensing unit is simplified into a two-dimensional model while the bottom surfaces of two supporting braces are constrained in all degrees of freedom. The size of the finite elements meshed is 0.05 mm which provides enough computational precision for the model. The Young's modulus of the spring steel is 197 GPa with the Poisson's ratio as 0.3 used for FEA.

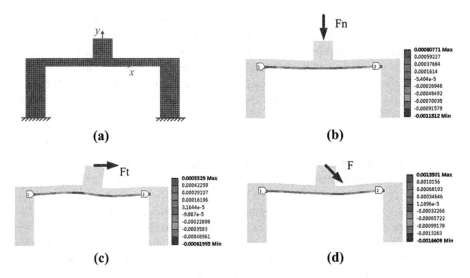

Fig. 3. Simulation results. (a) The meshed FEA model of the sensing unit; (b) X-directional normal strain distribution under a 10 N normal force; (c) X-directional normal strain distribution under a 10 N shear force; (d) X-directional normal strain distribution under a combined force with 10 N in both normal and shear directions.

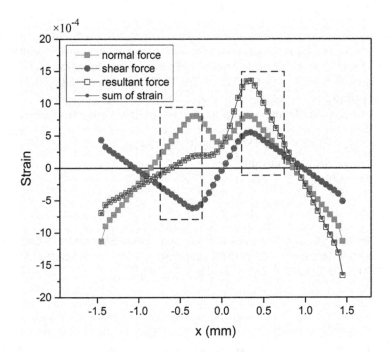

Fig. 4. Simulation results for x-directional normal strain distribution along the x-axis under various loading conditions with values set as 10 N for both normal and shear force and 10 N in two directions for the resultant force.

Due to the piezoresistive effect, the silicon gauge's resistance values linearly depend on the change of the normal strain in the longitudinal direction marked as x-direction in Fig. 3. As the force sensing range for human hand perception is within 10 N [9], the force value is set as 10 N in both normal and shear directions. Then the results under the normal, shear and resultant force loading are shown in Fig. 3(b), (c) and (d) respectively. As the silicon gauge is on the reverse side of the steel beam, only the normal strain distribution on the reverse surface is concerned. Meanwhile, the silicon gauge actually contains two independent sensing resistors as R_1 and R_2 which are located symmetrically about the axis of steel frame. Hence, R_1 and R_2 share the same value change under a single normal force and the reversal value change under a single shear force. Besides, the good elasticity of the steel beam makes it reasonable that a resultant force could be split into a single normal force and a single shear force.

In Fig. 4, the x-directional normal strain distribution along the reverse side is plotted under different force loading conditions as 10 N for both the single normal force and shear force and 10 N in both directions for the resultant force. The curve about the sum of strain is realized by adding the values from the single normal and shear force directly so that it is compared with the curve under the resultant force to further ensure the good elasticity of the sensing structure. The two blue dot rectangles are plotted to indicate the locations of the two sensing resistors R_1 and R_2 and which values belong to them.

3.2   Inverse Solution for Resultant Force

From the analysis results mentioned above, we can infer that it is possible to work out the values of a resultant force loaded on the sensing unit as long as the measurement values show good linearity in the full range. The x-directional normal strain is analyzed in the force range from 0–10 N for both normal loading and shear loading. As shown in Fig. 5, the strain values are obtained by calculating the average of strain values of those nodes disturbed in the silicon resistor's location range. It indicates that the sensing unit has good linearity in measuring both normal force and shear force with the sensitivity as $5.5 \times 10^{-4}\,\mathrm{N}^{-1}$ and $4.1 \times 10^{-4}\,\mathrm{N}^{-1}$ respectively. The relationships between the strain values and forces are shown below:

$$\begin{cases} \varepsilon_n = k_n \cdot F_n \\ \varepsilon_t = k_t \cdot F_t \end{cases} \tag{2}$$

The subscripts "n" and "t" represent for normal force and shear force respectively. Meantime, considering the symmetrical properties of the output values, the variant values of R1 and R2 could be represented as followed:

$$\begin{cases} \varepsilon_1 = \varepsilon_n + \varepsilon_t \\ \varepsilon_2 = \varepsilon_n - \varepsilon_t \end{cases} \tag{3}$$

From the Eqs. (1), (2) and (3), we can determine the normal force and shear force to form the resultant force:

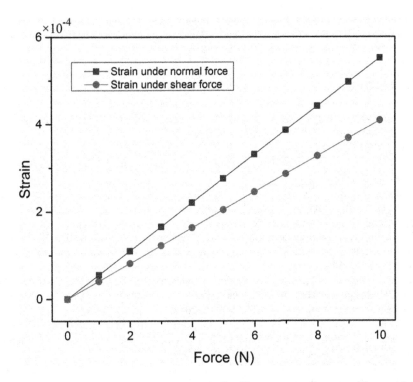

Fig. 5. Strain values change of sensing resistors in the silicon gauge under normal force and shear force.

$$\begin{cases} F_n = \dfrac{1}{2Gk_n}\left(\dfrac{\Delta R_1}{R_1} + \dfrac{\Delta R_2}{R_2}\right) \\ F_t = \dfrac{1}{2Gk_t}\left(\dfrac{\Delta R_1}{R_1} - \dfrac{\Delta R_2}{R_2}\right) \end{cases} \tag{4}$$

4 Conclusions

A tactile sensing array with novel structure composed of the silicon gauge and steel frame has been designed and the fabrication method has been proposed. The finite element analysis of the mechanical structure has been implemented, which shows good linearity and force detecting range in both normal and shear directions. In addition, the inverse solution to solve the resultant force has been obtained, which indicates the ability of detecting grip force and lift force for the sensing array when the prosthetic hand is grasping an object. However, a prototype needs to be realized and tested in a real environment in the future work.

Acknowledgements. This work is supported in part by the National Basic Research Program (973) of China (2011CB013303) and the Science Fund for Creative Research Groups of National Natural Science Foundation of China (51221004).

References

1. Büscher, G.H., Kõiva, R., Schürmann, C., Haschke, R., Ritter, H.J.: Flexible and stretchable fabric-based tactile sensor. Robot Auton. Syst. **63**, 244–252 (2015)
2. Lee, H.-K., Chung, J., Chang, S.-I., Yoon, E.: Normal and shear force measurement using a flexible polymer tactile sensor with embedded multiple capacitors. J. Microelectromech. Syst. **17**, 934–942 (2008)
3. Seminara, L., Capurro, M., Cirillo, P., Cannata, G., Valle, M.: Electromechanical characterization of piezoelectric PVDF polymer films for tactile sensors in robotics applications. Sens. Actuators, A **169**, 49–58 (2011)
4. Xie, H., Jiang, A., Wurdemann, H.A., Liu, H., Seneviratne, L.D., Althoefer, K.: Magnetic resonance-compatible tactile force sensor using fiber optics and vision sensor. IEEE Sens. J. **14**, 829–838 (2014)
5. Beccai, L., Roccella, S., Arena, A., Valvo, F., Valdastri, P., Menciassi, A., Carrozza, M.C., Dario, P.: Design and fabrication of a hybrid silicon three-axial force sensor for biomechanical applications. Sens. Actuators, A **120**, 370–382 (2005)
6. Noda, K., Hashimoto, Y., Tanaka, Y., Shimoyama, I.: MEMS on robot applications. In: International Conference on Solid-State Sensors, Actuators and Microsystems Conference, TRANSDUCERS 2009, pp. 2176–2181. IEEE (2009)
7. Yousef, H., Boukallel, M., Althoefer, K.: Tactile sensing for dexterous in-hand manipulation in robotics—a review. Sens. Actuators, A **167**, 171–187 (2011)
8. Gu, C., Liu, W., Fu, X.: A novel silicon based tactile sensor on elastic steel sheet for prosthetic hand. In: Zhang, X., Liu, H., Chen, Z., Wang, N. (eds.) ICIRA 2014, Part II. LNCS, vol. 8918, pp. 475–483. Springer, Heidelberg (2014)
9. Jones, L.A., Lederman, S.J.: Human Hand Function. Oxford University Press, London (2006)

Visualization of Remote Taskspace for Hand/Arm Robot Teleoperation

Futoshi Kobayashi[✉], Yoshiyuki Kakizaki, Hiroyuki Nakamoto, and Fumio Kojima

Department of Systems Science, Kobe University,
1-1 Rokkodai-cho, Nada-ku, Kobe 657-8501, Japan
futoshi.kobayashi@port.kobe-u.ac.jp
http://www.kojimalab.com

Abstract. The multi-fingered robot hand has much attention in various fields. Many robot hands have been proposed so far. A teleoperation system for the hand/arm robot allows intuitive manipulation of the hand/arm robot. In order that the hand/arm robot works various tasks well, it is important to present a remote taskspace visually. This paper deals with a visualization system of a remote taskspace for the hand/arm robot teleoperation. Here, the system uses the depth sensor placed in the remote taskspace and a video see-through head-mounted display that an operator is equipped with. For showing the effectiveness of the developed visualization system, some experiments are implemented.

1 Introduction

Recently, various robot hands have been developed in order to work various tasks so far [1–3]. We also have developed the universal robot hand I [4] and II [5] and constructed the hand/arm robot with the universal robot hand and an industrial manipulator. However, the hand/arm robot cannot carry out any tasks autonomously because the hand/arm robot does not have enough motion and sensing ability.

A teleoperation system for the hand/arm robot allows intuitive manipulation of the robot's hand and arm [6,7]. In the teleoperation system, the hand/arm robot can be operated by imitating the human motion. In order that the hand/arm robot works various tasks well, it is important to capture hand and arm motion of a human operator. Various types of the human motion capture (e.g., magnetic, optical, Kinect, inertial) have been developed so far. However, these methods have physical limitations. The magnetic motion capture uses only in a magnetic field. In the optical motion capture, the operator should be taken some images by multiple cameras. The Kinect is a well-known motion sensing input device by Microsoft, but this device cannot detect all motions of the operator because the occlusion occurs. Nowadays, an inertial measurement unit (IMU) is used for watching the human activities [8]. We have developed the motion capture system with IMU and the optical sensors for the hand/arm robot [9,10].

© Springer International Publishing Switzerland 2016
N. Kubota et al. (Eds.): ICIRA 2016, Part II, LNAI 9835, pp. 479–487, 2016.
DOI: 10.1007/978-3-319-43518-3_46

Here, the operator has to recognize the remote taskspace for controlling the hand/arm robot smoothly. In general, the visual information which the system captures by a camera in the remote taskspace is used in order to recognize the remote taskspace. However, it is difficult for the operator to recognize the remote taskspace because of the lack of the 3D information and the time delay due to the image pixels of the image.

This paper deals with a visualization system of a remote taskspace for the hand/arm robot teleoperation without presenting the visual image of the remote taskspace directly. Here, the system uses the depth sensor placed in the remote taskspace and a video see-through head-mounted display that an operator is equipped with. In the remote task space, the system measures the environment by the depth sensor and extracts the differences between the measured taskspace and the preserved taskspace. In the operator space, the system adds the extracted differences in the remote taskspace to the video image from the video see-through head-mounted display and presents the generated image to the operator. For showing the effectiveness of the developed visualization system, some experiments are implemented.

2 Hand/Arm Robot with Universal Robot Hand

The hand/arm robot consists of the developed universal robot hand II and an industrial manipulator as shown in Fig. 1. The universal robot hand has 5 fingers like a human hand. The height from the bottom of the palm to the top of the middle finger is 290 mm, the length from the thumb to the little finger is 416 mm when the hand is opened. The PIP joint and the DIP joint synchronize like a

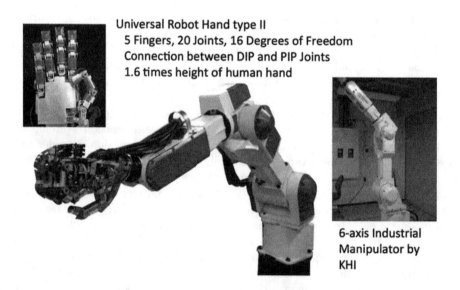

Universal Robot Hand type II
5 Fingers, 20 Joints, 16 Degrees of Freedom
Connection between DIP and PIP Joints
1.6 times height of human hand

6-axis Industrial
Manipulator by
KHI

Fig. 1. Hand/arm robot with universal robot hand II

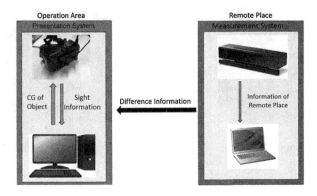

Fig. 2. Taskspace visualization system

human finger. The thumb has 4 DOF (IP joint, MP joint, CM1 joint and CM2 joint), the other fingers have 3 DOF (DIP-PIP joint, MP1 joint and MP2 joint) and the robot hand has 16 DOF.

This robot hand has the tactile sensors on the finger pads, and the multi-axis force/torque sensor in the fingertips. The tactile sensors have 3 layer structures (the electrode pattern seat, the pressure sensitive rubber, and the urethane gel), and this sensor can measure the pressure distribution by contact. The multi-axis force/torque sensors are made by BL AUTOTEC, LTD. and can measure the force and torque applied to the fingertips.

The industrial manipulator is manufactured by Kawasaki Heavy Industries, Ltd. This industrial manipulate has 6 D.O.Fs. The control system for the hand/arm robot consists on the hand controller, the arm controller, and the hand/arm coordinator. The hand and arm controller control the universal robot hand II and the industrial manipulator according to orders from the hand/arm coordinator, respectively. The hand/arm coordinator instructs the hand and arm controller according to facing situation.

3   Visualization of Remote Taskspace

A teleoperation system for the hand/arm robot allows intuitive manipulation of the robot's hand and arm. In order that the hand/arm robot works various tasks well, it is important to present a remote taskspace visually. The visualization system of a remote taskspace is shown in Fig. 2. Here, an operator space is similar to a remote taskspace as shown in Figs. 3 and 4. However, there are differences in two spaces from various factors as shown in these figures. Therefore, in the remote taskspace, the robot measures the taskspace by a depth sensor and gets the differences in two space. Then, in the operator space, the system presents the differences virtually through a video see-through head-mounted display.

Fig. 3. Remote taskspace   **Fig. 4.** Operator space

Fig. 5. Kinect depth sensor

3.1   Taskspace Measurement by Depth Sensor

The hand/arm robot exists in the remote taskspace. As I mentioned before, the robot measures the taskspace by a depth sensor Kinect (developed by Microsoft) as shown in Fig. 5. The Kinect can measure the depth from the Kinect to objects in the taskspace.

Fig. 6. Oculus Rift and Ovrvision

After measuring the depth by the depth sensor Kinect, the robot compares the reconstructed taskspace from the depth values and the preserved taskspace and acquires the differences in two spaces.

3.2 Taskspace Visualization by Head Mount Display

The human operator works in the operator space which is similar to the remote taskspace. The human operator is equipped with the video see-through head-mounted display. Here, the well-known Oculus Rift and the Ovrvision are used as the video see-through head-mounted display as shown in Fig. 6. The Oculus Rift is a virtual reality headset developed and manufactured by Oculus VR. This device has a high-resolution (960×1080 per eye) OLED display and high refresh rate. The Ovrvision is a stereo camera for the Oculus Rift. The resolution and the frame rate of the camera in this device are 640×480 and 60, respectively. The angles of view are 90° horizontally and 75° vertically. By using these device, the operator can watch the operator space as shown in Fig. 7.

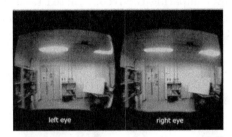

Fig. 7. Video image in operator space

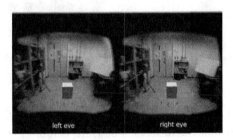

Fig. 8. HMD image in operator space

As I mentioned before, the differences between the operator space and the remote task space exists from various factors. When the operator controls the hand/arm robot in the remote taskspace, it is important for the operator to recognize the differences in two spaces visually. Therefore, the taskspace visualization system presents the differences in two spaces by the video see-through head-mounted display. Here, the system adds the reconstructed objects from the differences in two space to the video through the Ovrvision by the AR (augmented reality) technology as shown in Fig. 8. Then, the system presents the AR images by using the Oculus Rift.

4   Experiments

Some experiments are implemented in order to show the effectiveness of the developed visualization system for the hand/arm robot teleoperation. The first experiment confirms which the depth sensor can measure the taskspace. The second experiment confirms which the operator can recognize the remote taskspace virtually.

4.1   Experiment for Taskspace Measurement

The first experiment which the depth sensor can measure the taskspace. In this experiment, the system acquires the depth image by the depth sensor Kinect and extracts the objects by comparing the reconstructed taskspace from the depth image and the preserved taskspace. After extracting the objects, the size and the distance between the depth sensor Kinect and the object are calculated as shown in Figs. 9 and 10. Here, 4 objects with various size and distance are used in this experiment.

Fig. 9. Object size in remote taskspace

Fig. 10. Object distance in remote taskspace

The experimental results are presented in Table 1. The first line in this table represents real distance of the objects. The third column in this table represents the real width and height of the objects.

From this table, the system can calculate the object width with small error. Then, the error of the object width and the height become bigger as the distance becomes further. Otherwise, the distance of the object can be measured by the depth sensor Kinect. Consequently, the measurement system can reconstructed taskspace with various differences.

4.2   Experiment for Taskspace Visualization

The second experiment confirms which the operator can recognize the remote taskspace virtually. In this experiment, the human operator is equipped with the video see-through head-mounted display. In the remote taskspace, an object

Table 1. Experimental results for taskspace measurement

Object	Real value		Object position				
			1.0 m	1.5 m	2.0 m	2.5 m	3.0 m
#1	Width	0.490	0.486	0.474	0.469	0.453	0.411
	Height	0.645	0.615	0.630	0.656	0.695	0.739
	Distance	-	1.018	1.501	1.999	2.500	3.010
#2	Width	0.260	0.264	0.255	0.254	0.251	0.244
	Height	0.255	0.255	0.234	0.275	0.317	0.311
	Distance	-	1.016	1.508	2.017	2.529	3.013
#3	Width	0.390	0.376	0.390	0.383	0.379	0.369
	Height	0.490	0.500	0.521	0.523	0.575	0.627
	Distance	-	1.099	1.507	2.032	2.539	3.045
#4	Width	0.530	0.537	0.543	0.537	0.520	0.507
	Height	0.390	0.373	0.405	0.469	0.444	0.479
	Distance	-	1.019	1.521	2.024	2.519	3.034

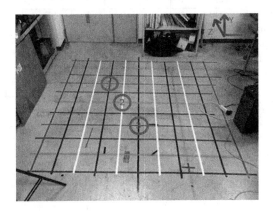

Fig. 11. Object location in experiment

is put in the 3 locations on the floor as shown in Fig. 11. After generating the AR image, the operator estimates the location of the object. The appearance of this experiment is shown in Fig. 12. Moreover, the AR image presented to the operator is shown in Fig. 13. The twelve subjects implement this experiment.

The experimental results are presented in Table 2. These table represents the estimated gap of the object position. In these table, the values of the cells show the number of the subjects with 3 object positions, respectively.

From these tables, the operator can estimate the position of the object with small gaps by the visualization system. Consequently, the operator may manipulate an object with the developed visualization system.

Fig. 12. Experimental environment

Fig. 13. Experiment results

Table 2. Estimated gap of object position

(a) Position #1

		Estimated Gap on axis x [mm]						
		-300	-200	-100	0	100	200	300
	-300				1			
	-200				1	1	1	
Estimated	-100							
Gap on	0		1	1		3		
axis y [mm]	100						3	
	200							
	300							

(b) Position #2

		Estimated Gap on axis x [mm]						
		-300	-200	-100	0	100	200	300
	-300				1			
	-200							
Estimated	-100							
Gap on	0				3	5	2	
axis y [mm]	100							
	200					1		
	300							

(c) Position #3

		Estimated Gap on axis x [mm]						
		-300	-200	-100	0	100	200	300
	-300							
	-200							
Estimated	-100							
Gap on	0		1		4			
axis y [mm]	100				3			
	200				1	2	1	
	300							

5 Summary

This paper proposed a visualization system of a remote taskspace for the hand/arm robot teleoperation. Here, the system uses the depth sensor placed in the remote taskspace for measuring the environment and extracting the differences between the measured taskspace and the preserved taskspace. In the operator space, a video see-through head-mounted display are used for presenting the operator space with the differences in the remote taskspace. Some experimental results shows the effectiveness of the developed visualization system.

References

1. Hollerbach, J.M., Jacobsen, S.C.: Anthropomorphic robots and human interactions. In: Proceedings of the 1st International Symposium on Humanoid Robots, pp. 83–91 (1996)
2. Mouri, T., Kawasaki, H., Yoshikawa, K., Takai, J., Ito, S.: Anthropomorphic robot hand: gifu hand III. In: Proceedings of the International Conference on Control, Automation and Systems, pp. 1288–1293 (2002)
3. Iwata, H., Sugano, S.: Design of anthropomorphic dexterous hand with passive joints and sensitive soft skins. In: Proceedings of the 2009 IEEE International Symposium on System Integration, pp. 129–134 (2009)
4. Nakamoto, H., Kobayashi, F., Imamura, N., Shirasawa, H.: Universal robot hand equipped with tactile and joint torque sensors (development and experiments on stiffness control and object recognition). In: Proceedings of the 10th World Multiconference on Systemics, Cybernetics and Informatics, vol. 2, pp. 347–352 (2006)
5. Fukui, W., Kobayashi, F., Kojima, F., Nakamoto, H., Maeda, T., Imamura, N., Sasabe, K., Shirasawa, H.: Development of multi-fingered universal robot hand with torque limiter mechanism. In: Proceedings of the 35th Annual Conference of the IEEE Industrial Electronics Society, pp. 2225–2230 (2009)
6. Diftler, M.A., Culbert, C.J., Ambrose, R.O., Platt Jr., R., Bluethmann, W.J.: Evolution of the NASA/DARPA robonaut control system. In: Proceedings of the IEEE International Conference on Robotics and Automation, pp. 2543–2548 (2003)
7. Mouri, T., Kawasaki, H.: A novel anthropomorphic robot hand and its master slave system. In: Hackel, M. (ed.) Humanoid Robots. Human-like Machines. InTech, Vukovar (2007)
8. Miller, N., Jenkins, O.C., Kallmann, M., Mataric, M.J.: Motion capture from inertial sensing for untethered humanoid teleoperation. In: Proceedings of the IEEE-RAS International Conference on Humanoid Robotics (2004)
9. Kobayashi, F., Hasegawa, K., Nakamoto, H., Kojima, F.: Motion capture with inertial measurement units for hand/arm robot teleoperation. Int. J. Appl. Electromagn. Mech. **45**, 931–937 (2014)
10. Kobayashi, F., Kitabayashi, K., Shimizu, K., Nakamoto, H., Kojima, F.: Human motion caption with vision and inertial sensors for hand/arm robot teleoperation. In: Proceedings of the International Symposium on Applied Electromagnetics and Mechanics (2015)

Author Index

Printed in the United States
By Bookmasters